D1753304

3D MODELING WITH ACIS

Also available from Saxe-Coburg Publications

Object Oriented Methods and Finite Element Analysis
R.I. Mackie

Computational Modelling of Masonry, Brickwork and Blockwork Structures
Edited by: J.W. Bull

Strength of Materials: An Undergraduate Text
G.M. Seed

Innovative Computational Methods for Structural Mechanics
Edited by: M. Papadrakakis and B.H.V. Topping

Derivational Analogy Based Structural Design
B. Kumar and B. Raphael

Computational Mechanics for the Twenty-First Century
Edited by: B.H.V. Topping

3D MODELING WITH ACIS

Jonathan Corney and Theodore Lim

SAXE-COBURG
PUBLICATIONS

© J. Corney and T. Lim

published 2001 by
Saxe-Coburg Publications
Dun Eaglais, Station Brae, Kippen
Stirling, FK8 3DY, UK

Saxe-Coburg Publications is an imprint of Civil-Comp Ltd

ISBN 1-874672-14-8

British Library Cataloguing in Publication Data
A catalogue record for this book is available from the British Library

Cover credits

The cover of this book shows images generated by users of CAD systems that incorporate the ACIS® 3D Geometric Modeler.

The front cover pictures are of a corrugated packaging machine, created by Pearson Packaging Systems, Spokane WA, USA using Autodesk Inventor™.

The back cover shows a temperature probe created by Paul Isabelle of I-Design, Quebec, Canada using CADKEY®.

Disclaimer

Many of the designations used by manufacturers and sellers to distinguish their products are claimed as trademarks. Where these designations appear in this book, and the authors were aware of the trademark claim, the designations have been printed with initial capital letters or all capital letters.

The authors and publisher have taken care in the preparation of this book, but make no expressed or implied warranty of any kind and assume no responsibility for errors or omissions. No liability is assumed for incidental or consequential damages in connection with or arising out of the use of the information or programs contained in this book.

Printed in Great Britain by Bell & Bain Ltd, Glasgow

To Heather, Zak and Jed: J. Corney

To my parents: T. Lim

Contents

Preface to 1st Edition   vii

Preface to 2nd Edition   xv

1 Geometric Modeling   1
   1.1   Introduction 1
   1.2   Boundary Representations 3
   1.3   Why COEDGEs? 7
   1.4   Representation of Geometry 8
   1.5   Interpolated Curves 9
   1.6   Sharing Geometry 10
   1.7   Identifying the Inside! 11
   1.8   Supporting Classes 12
   1.9   Creating Shapes 12
   1.10 Nonmanifold and Multi-dimensional Models 14
   1.11 Cellular Topology 14
   1.12 Moving and Scaling Shapes 15
   1.13 Model Management 17
   1.14 Displaying Models 17
   1.15 Pros and Cons of B-Reps 18
   1.16 Exercises 19

2 C++   21
   2.1   Object Oriented Programming 21
   2.2   Types, Pointers and References 23
       2.2.1   Types 23
       2.2.2   Pointer: Type Modifier 24
       2.2.3   Cast: Type Modifier 25
       2.2.4   Reference: Type Modifier 26
       2.2.5   Const: Type Modifier 26

2.3	Enumeration		28
2.4	C++ Functions		29
	2.4.1	Function Prototypes	29
	2.4.2	Default Arguments	30
	2.4.3	Reference Arguments	30
	2.4.4	Pointer Reference Arguments	31
	2.4.5	NULL Reference Argument	32
2.5	Class Terminology		33
	2.5.1	Access Specifiers for Class Members	33
	2.5.2	Inheritance and Virtual Functions	34
	2.5.3	Creating Objects with Constructors	35
2.6	Control		36
	2.6.1	While Loops	36
	2.6.2	For Loops	37
2.7	ACIS Classes		37
	2.7.1	Rollback	38
	2.7.2	Save, Restore and Identity Functions	39
	2.7.3	ACIS Attributes	40
	2.7.4	Deleting ACIS ENTITYs	41
	2.7.5	ACIS's Non-ENTITY Classes	42
2.8	Component Architecture		42
	2.8.1	Using components	45
2.9	Programming Interfaces		47
2.10	Exercises		48

3 Scheme   49

3.1	Scheme Basics	49
3.2	Expressions	50
3.3	External Representations	52
3.4	Defining Variables	53
3.5	Defining Functions	53
3.6	Conditional Statements	54
3.7	Recursion and Lists	55
3.8	For-each	57
3.9	Defining Local Variables	57
3.10	Set!	59
3.11	Lambda	59
3.12	Begin	60
3.13	Do	60
3.14	Features of Scheme Programming	61
3.15	Exercises	62

4 Basics · · · 63
- 4.1 Making Primitives with API Functions 64
- 4.2 Saving and Restoring Models 72
- 4.3 ENTITY_LISTs 76
  - 4.3.1 Casting ENTITYs 76
  - 4.3.2 Constructors and Destructors 77
  - 4.3.3 Tombstones 78
  - 4.3.4 ENTITY_LIST Arguments 78
  - 4.3.5 Save and Restore using the Scheme AIDE 79
- 4.4 Adding and Subtracting Shapes 80
  - 4.4.1 Mathematical Classes 81
  - 4.4.2 Set Operations using the Scheme AIDE 83
- 4.5 Calculating Mass Properties 85
- 4.6 Extruding Shapes 88
- 4.7 Blending 91
- 4.8 Blending in the Scheme AIDE 93
- 4.9 Sectioning 94
- 4.10 Slicing with the Scheme AIDE 96
- 4.11 Coordinate Systems 97
- 4.12 Part Manager 101
- 4.13 Exercises 101

5 Using the Direct Interface · · · 103
- 5.1 Counting Faces 103
- 5.2 Accessing FACE Details 105
- 5.3 Counting EDGEs 108
- 5.4 Getting Vertex Coordinates 110
- 5.5 Creating Primitives from ENTITYs 113
- 5.6 Parametric Surface Inquiries 117
- 5.7 An EDGE Marching Algorithm 119
- 5.8 Sense and Sharing of FACE Geometry 122
- 5.9 Exercises 125

6 ENTITY Intersections and Booleans · · · 127
- 6.1 Comparisons of Points 127
- 6.2 Point Intersection and Containment 130
- 6.3 Lowercase Geometry Intersections 130
- 6.4 EDGE–FACE Intersection 133
- 6.5 FACE–FACE Intersection 135
- 6.6 Overview of the Boolean Algorithm 138
- 6.7 Generating an Intersection Graph 143
- 6.8 Imprinting and Division of FACEs 146
- 6.9 Stitching and Separation 147
- 6.10 Quick Intersection Tests 150
- 6.11 Nonregularized Booleans 152
- 6.12 Selective Booleans 154

		6.13 Exercises	159

7 Rendering, Ray Firing and Faceting — 161
- 7.1 Ray Firing . . . 162
- 7.2 Faceting ENTITYs . . . 164
- 7.3 Faceted Hidden Line . . . 169
- 7.4 Precise Hidden Line . . . 172
- 7.5 Rendering in C++ . . . 173
- 7.6 Rendering using Scheme Extensions . . . 175
  - 7.6.1 Setting the Background . . . 176
  - 7.6.2 Setting the Lights . . . 176
  - 7.6.3 Setting the Materials . . . 177
  - 7.6.4 Setting the Facet Refinement . . . 178
- 7.7 The Advanced Rendering Component . . . 179
- 7.8 Spin! . . . 181
- 7.9 Creating Postscript Images . . . 183
- 7.10 Picking and Interaction . . . 184
- 7.11 Exercises . . . 185

8 Spline Surfaces — 187
- 8.1 Some Spline Concepts . . . 187
- 8.2 Bézier Curves . . . 189
- 8.3 B-splines . . . 193
- 8.4 B-spline Blending Function . . . 198
- 8.5 Rational Splines . . . 205
- 8.6 Creating Spline Surfaces . . . 207
- 8.7 Interpolation . . . 210
- 8.8 Deformable Surfaces . . . 213
  - 8.8.1 Deformable modeling theory . . . 213
- 8.9 Skinning and Lofting . . . 215
- 8.10 Net Surfaces . . . 218
- 8.11 Exercises . . . 220

9 Bulletin Boards and Rollback — 221
- 9.1 Counting the BULLETINs . . . 222
- 9.2 Creating DELTA_STATEs . . . 225
- 9.3 Rolling back the modeler . . . 228
- 9.4 Part history . . . 230
- 9.5 Saving and restoring of history . . . 231

10 Laws and Graphs — 233
- 10.1 Law Expressions . . . 233
- 10.2 Creating Laws in C++ . . . 235
- 10.3 Planar Offsets . . . 241
  - 10.3.1 Laws Offsetting: A practical example . . . 242
- 10.4 Helical Offsets . . . 247

10.5	Basic Sweeping	249
10.6	Complex Sweeping	251
10.7	Creating EDGEs with Laws	257
10.8	Creating FACEs with Laws	261
10.9	Space Warping and Scaling	267
	10.9.1 A Tapering Law	269
	10.9.2 A Twisting Law	269
10.10	Representation and Analysis of Graphs	272
	10.10.1 VERTEX-EDGE Graphs	272
	10.10.2 FACE-EDGE Graphs	273
	10.10.3 CELL-Adjacency Graphs	273
10.11	Exercises	278

11 Model Modification — 281

- 11.1 Simple Tweaking Procedure — 282
- 11.2 Local Operations — 287
  - 11.2.1 Creation and Deletion of ENTITYs — 288
  - 11.2.2 Self-intersections — 290
  - 11.2.3 Multiple Solutions — 292
- 11.3 Offsetting and Shelling — 295
- 11.4 Blending — 297
  - 11.4.1 Blending Terminology — 297
- 11.5 Variable Radius Edge Blends — 300
- 11.6 Vertex Blends — 302
- 11.7 Entity-Entity Blends — 303
- 11.8 Healing — 305
  - 11.8.1 The Healing component — 306
- 11.9 Exercises — 311

12 Attributes — 313

- 12.1 Adding a Simple String Attribute — 314
- 12.2 Using Generic Attributes — 316
- 12.3 Adding your own Attributes — 319
- 12.4 Attribute Pitfalls — 327
- 12.5 Exercises — 328

13 Customizing ACIS — 329

- 13.1 Adding your own ENTITYs — 329
- 13.2 Creating your own API Function — 342
- 13.3 Creating your own Scheme extension — 346
- 13.4 Exercise — 347

14 Debugging and Error Checking — 349

- 14.1 Using Outcomes — 349
- 14.2 Defining an Error Catching Macro — 351
- 14.3 Using the Debug Functions — 353

A	**Getting and Compiling the Examples**	**357**
	A.1 Getting the Examples	357
	A.2 Compiling the Examples	357
	A.2.1 Visual C++ 6.0	357
	A.3 Viewing SAT Files with the ACIS AppWizard	359
	A.4 How to get the ACIS 3D Geometric Modeler	360
	A.4.1 Nonprofit Research and Educational Institutions	360
	A.4.2 Commercial Software Developers	360
B	**ACIS Data Structure Class Summary**	**361**
	B.1 ENTITY	363
	B.2 BODY	364
	B.3 LUMP	365
	B.4 SHELL	366
	B.5 FACE	367
	B.6 LOOP	368
	B.7 EDGE	369
	B.8 COEDGE	370
	B.9 VERTEX	371
	B.10 CURVE	372
	B.11 STRAIGHT	373
	B.12 ELLIPSE	374
	B.13 PCURVE	375
	B.14 INTCURVE	376
	B.15 SURFACE	377
	B.16 TORUS	378
	B.17 PLANE	379
	B.18 SPHERE	380
	B.19 SPLINE	381
	B.20 APOINT	382
C	**ACIS Hacking Tips**	**383**
	C.1 General Points	383
	C.2 Finding Example Code	384
	C.3 The ACIS-Alliance Mailing List	384
	C.4 support@spatial.com	385
D	**Bibliography**	**387**
	Index	**388**

Preface to 1st Edition

ACIS - the 3D Geometric Modeler - is now an integral part of over a hundred different products and looks increasingly like becoming an industry standard. This book provides an introduction to ACIS and its associated data structures, interfaces and components. Using numerous examples the book explains how programs which model and manipulate three-dimensional objects, or worlds, can be created using both the C++ and Scheme interfaces.

Three-dimensional modelers, like ACIS, are now central to so many applications that they often go almost unnoticed. The technology, in one form or another, is embedded in every computer program, from CAD/CAM to Virtual Reality, which has to represent, display or manipulate three-dimensional shapes. Despite the many potential applications for 3D geometry the number of commercially supported modelers has always been small.

Maybe this is not surprising since modeling is a difficult blend of highbrow mathematics and complex programming which, to make matters worse, is plagued with difficult practical problems like numerical accuracy. In other words it is so difficult that few people, or companies, feel up to the task of creating a modeler from scratch; most would rather get a ready-made one off the shelf.

The stock of "off the shelf" (or so called **kernel**) modelers was, for a long time, limited and very expensive. Many of the companies developing modelers also embedded them in their own bespoke CAD/CAM systems and were naturally reluctant to sell their technology to groups intent on creating rival commercial systems. However, in the late 1980s an **open** modeler, known as **ACIS**, was launched which was tailor made for incorporation into other people's software.

The ACIS modeler is delivered as a library of software which can be used in the development of any application requiring the representation of 3D geometry. Although Spatial Corp who market ACIS, supply copious documentation, this book fills the gap between the extreme detail of the information in the manuals and the empty head of the new user, who wants something slim, to skim.

About this Book

Many people learn far quicker from experimenting with examples of programs than by reading reference manuals. So this is essentially a pattern-book that allows the reader to take a *copy and modify* approach to developing ACIS software. Studying the examples will also give a feel for the sort of functionality ACIS offers and an idea of the complications involved in accessing it. This book is **not** a substitute for the manuals; instead it says "here's some code and this is what it does".

Each program, or code fragment, is supported by a small amount of discussion, or commentary, about some interesting aspect of geometric modeling, ACIS or program language detail that is touched on by the example.

The book covers each of ACIS's three distinct interfaces. These are referred to as:

- **Direct Interface:** The lowest level of access to the C++ classes of the ACIS data structure through their public member functions.

- **Application Procedural Interface:** These so-called **API** functions provide a large number of algorithms for manipulation and creation of ACIS's 3D data structures. They also provide support for the rolling back (i.e. undoing) of modeling operations and for both error and argument checking.

- **Scheme AIDE**[1]**:** Allows ACIS APIs to be driven through extensions to the **Scheme** language, a dialect of LISP.

The book starts with three foundation chapters covering geometric modeling, C++ and Scheme (all discussed in the context of ACIS), subsequent chapters detail how different aspects of the modeler's functionality can be exploited via the APIs and Scheme Extensions.

ACIS's API routines are divided into groups called components[2].

Intended Audience

The book has been written with four types of reader in mind:

- Novice users (i.e. advanced undergraduate or postgraduate students) who are new to geometric modeling or ACIS.

- Occasional users who know the principles but can't recall the syntax.

- Programmers (commercial or academic) wondering "Where do I start?"

- People who want to get an overview of what ACIS can do. A typical example might be someone developing a Computer Aided Learning package who is trying to get a feel for the capabilities of geometric modelers.

---

[1] ACIS Interface Driver Extension.

[2] The term *husk* used to describe a collection of API functions in earlier versions of ACIS has been superseded by the word *component*.

Preface

Structure of Examples

Merely to read large lumps of code will cure most forms of insomnia: therefore none of the programs in this book are more than a couple of pages long. Furthermore, all the examples produce some form of tangible result. Scheme AIDE programs have no difficulty doing this since images are generated by default. The C++ programs generally produce an output by saving the results of the modeling operations as *.sat files which can then be viewed using the test harness or Scheme AIDE.

All the example programs (and solutions to some of the exercises) are available from the web site detailed in Appendix A.

Scope

The total number of ACIS classes and functions is vast and for the most part their details are best left to the reference manuals. This book provides examples which use around 20% of all the Direct, API and Scheme AIDE interfaces to the ACIS modeler. Supporting utilities like the mathematical classes, ENTITY lists, bulletin boards and attributes are also covered. For the most part the functions, or classes, not mentioned represent natural extensions of material covered in the book.

The book covers version 6.0 of ACIS which is supported across a range of platforms from PCs running Windows or LINUX through to UNIX workstations and MacOS on Apple Computers. Because new versions of ACIS are frequently released it is possible that some file or function names will be different on the reader's system from those stated in this book. To accommodate these small changes all the examples will be ported to each major new release and archived at the ftp site detailed in Appendix A.

> Readers of this book should ensure that the examples they copy from the ftp site match the major version of ACIS installed on their system.

How to Read this Book

The titles of the various chapters give a good indication of the type of subjects and functions found in each. In principle each chapter can be read more or less independently. However, if you are new to ACIS or boundary representation modeling then the introductory chapters (1-3) should be read first in order to get a firm grip on the two key concepts (i.e. the ACIS class hierarchy and the boundary representation data structure).

Chapter 1: Geometric Modeling An informal overview of what boundary representation is, described in terms of the ACIS entities.

Chapter 2: C++ The chapter covers the ways in which C++ has been used in the design of the ACIS Kernel, API functions and husks. It assumes some prior, but not expert, knowledge of C++. This chapter ends with a description of how the mechanism of inheritance has been used to provide a uniform "management" interface for all ACIS classes in an open and extendable way.

Chapter 3: Scheme Gives enough details to get a novice started, an explanation of Scheme terminology and syntax, with examples of how different constructs can be used in the context of the Scheme AIDE's command language.

Chapter 4: Basics Explains how to create, manipulate, modify, analyze and save models using ACIS API functions and their equivalent Scheme extensions.

Chapter 5: Using the Direct Interface Shows how the data structure of ACIS models can be traversed and accessed via Direct Interface calls and Scheme extensions.

Chapter 6: ENTITY Intersections and Booleans Examines two subjects core to any modeling system.

Chapter 7: Rendering, Ray Firing and Faceting Starting with firing single rays, the chapter covers progressively more complex and refined image generation techniques, ending with some examples of the Scheme AIDE's rendering and viewing commands.

Chapter 8: Spline Surfaces A look at the spline curves and surfaces supported by ACIS including those generated by lofts, or as net surfaces.

Chapter 9: Bulletin Boards and Rollback Gives examples of how "undo" functions can be created and changes in models recorded.

Chapter 10: Laws and Graphs Applying parametric equations to define and modify models. The Chapter ends by briefly reviewing ACIS's graph representation and analysis utilities.

Chapter 11: Model Modification Examples of local operations, blending and healing.

Chapter 12: Attributes Demonstrates several different ways in which applications data can be incorporated in models as user defined "attributes".

Chapter 13: Customizing ACIS Shows how users can create their own Entities, API functions and Scheme Extensions.

Chapter 14: Debugging and Error Checking Covers some of the facilities provided to support the trapping and tracing of errors.

Appendix A Details of how the examples can be obtained and compiled under Windows.

Appendix B Tables for fast look-up of class member functions.

Appendix C Miscellaneous tips: the ACIS-Alliance ListServer, searching the Test Harness for code fragments and how to obtain a copy of ACIS.

Appendix D A bibliography: with some suggestions for further reading.

Preface

Typographic Conventions

Central to the way in which ACIS represents shapes are a number of C++ classes with names such as ENTITY, EDGE and VERTEX. In the plural these are referred to as ENTITYs, EDGEs and VERTEXs. Although not correct English, this convention ensures that there can be no confusion regarding the name of an ACIS C++ class. The occasional reference to "entities" in other parts of the text simply implies the normal English use of the word.

The important exception to the uppercase naming convention occurs when the Scheme AIDE's extensions are discussed. In that context ACIS ENTITYs are referred to by lowercase names (i.e. face, entity, edge, etc.)

In the example programs and code fragments, keywords and type names are shown in **bold**, comments in *italics* and variables in the regular type face. Programs available from the ftp site are bracketed between pairs of horizontal lines. For example, the following code can be found in the file **rbdblock.scm**.

rbdblock.scm

```
(load 'setview.scm) ; From scm/examples
(define pickdemo
  (lambda ()
    (begin
      (front)       ; Create a Front view
      (print "Pick first corner")
      (define p1 (pick:position (read-event)))
      (rbd:line #t p1) ; Display the blocks diagonal
      (iso)         ; Isometric view
      (print "Pick second corner")
      (define p2 (pick:position (read-event)))
      (rbd:clear)
      (solid:block p1 p2))))
```

Code fragments (rather than complete programs) appear in the same fonts but without the horizontal lines. When a program variable is referred to in the text it appears in this font. For example, in the following code fragment:

```
// Variable declarations
  double x_coordinate, z_coordinate;
  double y_coordinate = 5.669;
  ENTITY_LIST el;
```

the first line declares x_coordinate and z_coordinate to be of type **double**.

Elsewhere in the text key words (such as class, function or file names) are highlighted in **bold** only the first time they are mentioned. Other words or phrases are occasionally highlighted or *italicized* for emphasis.

Text output directly from a program or file is displayed as:

```
      1 body record,       32 bytes
      4 attribute records,    184 bytes
      1 lump record,       32 bytes
      1 shell record,      40 bytes
      1 face record,       44 bytes
      1 surface record,    168 bytes
   Total storage 500 bytes
   "solid body"
```

Equations are displayed as:

$$y = mx + b$$

Generally the text uses US spelling throughout. However, because ACIS has been developed in both Britain and the USA there are occasional inconsistencies, for example the word *modeler* in the text and the function **api_start_modeller**. If a function or file name occurs at the start of a sentence it is left lowercase to avoid any confusion.

Acknowledgments

This book began with some remarks made to Charles Lang, so thanks must go first to him for his unflagging support and enthusiasm. However, without the active support of Dick Sowar at Spatial Corp and the considerable patience of Professor John Simmons, this book would never have been written so we are also indebted to them.

Of those who contributed examples and time to the project Erwin Argyle can probably see more familiar lines of code than most! Many examples here originated from his "tip of the week" e-mails to the ACIS-Alliance or material he generated for training courses. His perceptive comments pervade the text and have impacted on many aspects of the book. We must also mention Ian Braid who provided many useful comments about both the grammar and the coding!

The Edinburgh ACIS community all read drafts. Gordon Little found several very subtle errors which were always revealed with the phrase "Is it true to say...?" Doug Clark provided some especially detailed comments and helped keep some the most dyslexic sentences out of view. Bob Tuttle both reviewed early drafts and helped port some of the examples between different releases of ACIS. The chapter on spline geometry was influenced by several conversations with Frank Mill, Nick Sormaz and Ghassan Farsi. Jonathan Salmon contributed several examples and provided a fund of Scheme knowledge. Edinburgh University student Ser-Chong Chia contributed the programs (faceren.scm and spin.scm). Mike Sommerville, Jing Jing Fang and Ariffin Razak will all recognize lines of codes that they originally developed for their research work at Heriot-Watt University. Graham Seed provided detailed and thoughtful reviews of several chapters but most notably the one on C++. The tips on using the API_BEGIN/END macros originated from a posting on ACIS-Alliance by Steve Abrams and Rich Brandt. Quite late in the book's development Ralph Martin

Preface

provided a plethora of helpful comments, which he appeared to generate almost over night!

Many other people at Spatial Corp have contributed in both big and small ways, in no particular order: Joe Biegelsen, Steve Heye, Matt Atherton, Athena West, Steve Balderrama, Ray Gorman and Duane Heimann all answered questions during my time in Boulder. Spatial Corp, via George Greenwood, also supplied several illustrations found in various chapters of the book. The **hump.scm** program was developed from an example found in the WinScheme Editor's help system. Figure 4.1 is copied from an illustration in Bowyer & Woodwark's "Introduction to Computing with Geometry".

Since this book was set using the LaTeX2e system we must offer sincere thanks to Eberhard Mattes for his development and free distribution of the emTeX program and the many other package writers who have contributed to the CTAN sites around the world.

Despite our best efforts we have undoubtedly introduced errors to the material contributed or corrected by the people mentioned here, so these are mine alone. A list of known errors can be found on the book's homepage page[3]; please e-mail any errors not contained on this list.

---

[3] http://www.hw.ac.uk/mecWWW/research/jrc/research.htm

Preface to 2nd Edition

The production of this second edition was motivated by the need to update the original material and the desire to explore some of the new functionality that has become available to ACIS users over the last three years.

All the examples in the first edition were recompiled after modification and run under ACIS 6.3. Not surprisingly much of the original code differs only in the header file names and the inclusion of component initialization/termination calls. However in other areas (such as the Facetted Hidden Line APIs) the changes to both text and code are much larger.

While the updating of the existing material has been a necessary chore the creation of two new chapters describing the Laws system and the functionality available in the various components for editing models has been a pleasure.

The laws functionality first appeared in ACIS 3.0 and enables mathematical expressions to be used directly in the production and manipulation of models. Chapter 10 introduces some of the basic concepts and demonstrates the application of this new technology in various operations such as sweeping and space warping.

The material in both new chapters is structured so that, hopefully, the reader can follow the clear development of a 3D technology. We feel this makes for an easier read than a presentation that is divided up along marketing lines. Thus even though the Laws functionality is spread over several different components (e.g. Kernel, Warping and Advanced Surfacing) it is more coherent to present it in the context of other functions that do similar things.

This philosophy is also evident in the chapter on model modification. Here the common theme of a SURFACE change followed by re-intersection with surrounding FACEs is used to link material that is spread across the Blending, Local Operations, Shelling and Healing components.

Many people have contributed to the production of this book. Charles Lang has once again been dogged in his support of the project. Various Heriot-Watt students (past and present) will recognize bits of their projects: the BODY offsetting example was originally produced by Raymond Sung and similarly Keith Edmond will recognize the Deformable Surface code.

Much of the material on Local Operations, Blending, Shelling and Healing was gathered during a visit to Cambridge in December 1998. We are indebted to various members of Three-Space, D-Cubed and Spatial Corp(Ian Braid, David Plowman and John McKernan) for the time they gave up to answer many naïve questions about their components. During this visit the "early lofting" illustration was also acquired (see page 216) from Malcolm Sabin who had found a battered copy of Shelley's original book in a dusty corner of Cambridge University.

Special thanks should also go to Ralph Martin at Cardiff University, who (in the space of ten days) generated more comments than anyone else who saw the draft. At Spatial Corp Linda Mark and Ophelia Larson spent many hours pursuing images for the book's cover and both Jim Gordon and Jakob Berchtold took time out to improve the technical content. We must also thank Barry Topping and the staff of Saxe-Coburg Publications for their support and Jelle Muylle for the great cover he created.

Many of the "bugs" spotted in the text and code of the first version have been fixed and so we are indebted to the people who reported them. As ever, despite our best efforts, there may still be errors. So please take the time to e-mail us (J.R.Corney@hw.ac.uk, T.Lim@hw.ac.uk) with details of any mistakes you come across. We will maintain a web page that summarizes the known bugs in the text at: www.hw.ac.uk/mecWWW/research/jrc/research.htm.

Geometric Modeling

1.1 Introduction

The desire to create symbolic models of the physical world has motivated generations of mathematicians. However, it is only comparatively recently that the knowledge of geometry accrued over the years has been used in automated systems known as **Geometric Modelers**. This *mechanization of geometry* has been achieved by encoding the mathematics of shape in computer programs which allow the details of complex operations to be largely hidden from the user. The resulting systems allow users to, say:

1. make a torus and sphere with given dimensions,
2. overlap them by so much,
3. unite them to form a single object,

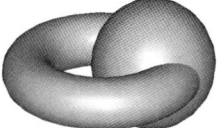

without having to know the details of toroidal surface equations or how to go about displaying them. In a very limited form this sort of programming is already commonplace, buried deep behind the graphical user interfaces that allow rectangular areas (such as pulldown menus and scroll bars) to be displayed and manipulated. In these "flatlands" of angular 2D polygons, every Window Manager must be able to represent points and lines together in structures called rectangles. Because such systems also have algorithms for intersection, point containment and rendering they could be called 2D Rectangular Geometric Modelers.

Thus what constitutes a Geometric Modeler varies with the application. At the very least a computer system for describing shape should be able to display pictures of the objects it is representing, and a large number of the modelers used in applications like virtual reality, games and animation do this very well with astonishingly simple representations (typically meshes of planar polygons, often just triangles).

Modelers based on these simple representations work very well in many applications where accuracy is not of paramount importance. But because the images they generate are based on approximations to the actual shape of an object, they can only ever produce inexact answers to questions about the physical properties of a shape.

In contrast to many of these systems, **ACIS** is an *exact* geometric modeler which represents a shape in terms of a network of interrelated curves and surfaces. Perhaps the first, and most fundamental, thing to grasp about ACIS is that it defines shapes by modeling their **boundary**. In the jargon of Geometric Modeling this approach to representing shape is called **boundary representation**, or B-rep, modeling because the program is creating and recording the equations of the curves and surfaces that lie on the boundary between the **inside** (i.e. solid) and the **outside** (i.e. air) of a solid, 3D, object[1].

Table 1.1   Types of Geometric Modeler.

Type	**Representation**
Wire frame	Vertices and edges only, no surfaces or faces
Set-theoretic	Surfaces (i.e. half-spaces) only, no vertices or edges, implicit boundary
Boundary representation	Vertices, edges, faces, explicit boundary
Polyhedral model	Only planar geometry, sometimes only triangular faces
Octree	Thousands of cubes arranged in a hierarchy
Surface	Spline curves and surfaces used throughout, no notion of volume
R-rep	Set of points defined by a function

Boundary representation modeling is not the only way of defining the shape of an object to a computer. In academe research continues into other methods of representation (summarized in Table 1.1) each of which has its own specific advantage over the others in terms of elegance, or suitability, for certain applications.

However, in the commercial world of exact geometric modelers, B-reps are used *almost* exclusively and this chapter ends with some speculation as to why this is. Prior to this, the reasoning behind ACIS's data structure is explored and some important pieces of terminology explained.

---

[1] Sadly there are very few statements one can make about geometry that are absolutely true in all cases. Even a simple notion like a *boundary* starts to blur when one thinks of infinitely thin shapes (such as 2D sheets or 1D wires). Such objects can be created easily with a boundary modeler but they have no *inside*; a point can only be *on* or *outside* their boundary (in 3D).

1.2 Boundary Representations

The B-rep data structure must hold sufficient information to allow precise questions to be asked about the shape of an object. ACIS does this with a data structure capable of describing models of arbitrary complexity. Although initially the large number of entities and even larger number of interconnections and cross-references might seem daunting, most people find it takes surprisingly little effort to remember the entire structure once they understand the logic behind the design. Consequently, before any ACIS code is written the programmer should establish a qualitative understanding of the way shapes are described in terms of their boundary.

How does one represent the boundary of an object? There is an easy answer, accompanied by an obvious problem. First, notice that for many man-made shapes it is often easy to identify patches of familiar surfaces such as planes, cylinders or spheres. For instance, the picture in the margin shows an object whose boundary is composed of one cylindrical and six planar surfaces.

One BODY

Where these surfaces meet, or intersect, curves are formed. Similarly where curves meet, points are formed. These three geometric entities, **points**, **curves** and **surfaces**, are the basic constituents needed to define an object's shape.

So the easy answer is to represent these surfaces (and curves) using the equations of planes and cylinders (or lines and arcs), and the obvious problem is that these algebraic expressions can define only **unbounded** geometry.

With the exception of individual points, circles and spheres, classical analytical geometry only represents curves and surfaces that extend off to infinity. The solution, adopted by B-reps, is to *explicitly* define the boundaries of curves and surfaces:

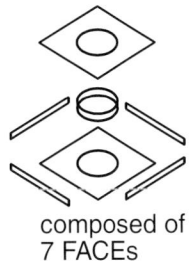

composed of
7 FACEs

- The boundaries of a curve are defined by pairs of points lying on the curve.

- The boundaries of a surface are defined by collections of curves lying on the surface.

This allows finite **segments** of curve and **portions** of surface to be defined. This solution brings with it the need to *organize* the relationships between these different geometric elements; we have to record:

- Which points **bound** which curves.

- Which curves **bound** which surfaces.

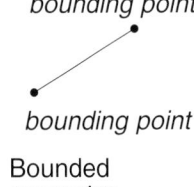

bounding point

bounding point

Bounded geometry

This sort of information, about what is associated with what, is known as **topology** and can be visualized as a web of interconnections between nodes representing individual points, curves and surfaces. Such a network of relationships can also be used to *share* bounding entities and so prevent the duplication of data. For instance, when representing a 3D solid, points will often represent the boundary of more than one curve and likewise curves will frequently bound more than one surface. One only has to think of a cube (where each corner lies at the end of three edges and each edge lies between two faces) to realize that the sharing of boundary data between entities is the norm, not the exception. In other words the points, curves and surfaces must

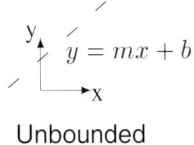

$y = mx + b$

Unbounded geometry

be organized in such a way that their associations (what bounds what), or adjacencies, are recorded. ACIS does this by giving the bounded bits of an object's surface different names from the unbounded ones.

- A connected portion of a **surface** is called a **FACE**.[2]
- A connected segment of a **curve** is called an **EDGE**.
- A **point** at the boundary (i.e. end) of an EDGE is a **VERTEX**.

So FACEs, EDGEs and VERTEXs are composite entities which refer to collections of other things. They are known generically as **topological** entities because they define how things interconnect. Generally the number of cross-references is large because all topological entities are defined in terms of other bits of topology or geometry. For example:

- The shape of a FACE is defined by a surface whose boundary is represented by a collection of EDGEs associated with it.
- The shape of an EDGE is defined by a curve whose boundary is represented by a pair of VERTEXs associated with it.
- The location of a VERTEX is defined by a point.

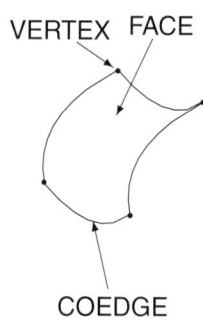

These relationships form a network of interconnections. The exact design of this network (or graph) was the subject of much debate during the early days of solid modeling. Although many differences remain between researchers, one consensus that emerged early on was that it would be a "good thing" to hold the information about how entities are connected separately from the entities themselves. In other words topology should be divorced from geometry. ACIS makes this distinction and the resulting structure enables one to write programs in terms of, say, FACEs in general rather than specific planar, cylindrical or spherical faces.

Although theoretically only three topological entities (i.e. FACE, EDGE and VERTEX) are needed, in practice a number of others are incorporated into the ACIS data structure either to increase the speed of modeling operations or to provide high-level handles (i.e. means of reference).

For example, most EDGEs on a 3D solid bound (i.e. lie between) two FACEs. However, there are many modeling operations which require the boundary of an individual FACE to be traversed (i.e. every EDGE visited). To do this quickly each EDGE is associated with a number of **COEDGE** entities. These are conceptual entities which lie on one unique FACE (whereas an EDGE proper can be shared by a number of FACEs). By chaining COEDGEs together, structures called **LOOPs** are formed which represent a high-level handle on parts of a FACE's boundary. The need for COEDGEs is discussed further in Section 1.3.

Analogous to the way COEDGEs are grouped together to form LOOPs, FACEs are grouped together to form SHELLs. Typically most objects have only one SHELL, but a 3D solid containing one or more voids will have two or more SHELLs.

---

[2] The use of uppercase letters for the names of the topological entities reflects the naming convention used for ACIS classes.

1.2 Boundary Representations

At a higher level still, collections of SHELLs are amalgamated to define the *separable* pieces of an object, known as LUMPs. Lastly LUMPs can be collected into a group known as a BODY. Table 1.2 gives an informal summary of the ACIS entities used to represent a shape by describing what they comprise and what they physically represent.

There are many ways of viewing this data structure: for instance, it can be thought of as being a tree, or hierarchy, with BODY at its root. A BODY can have a collection of LUMPs, each of which is comprised of one or more SHELLs formed from groups of FACEs. Likewise, FACEs are composed of LOOPs, formed from circuits of COEDGEs. Here, however, the hierarchy becomes less clearly defined. Up to this point, although every ENTITY might have many descendants, each has only one parent (i.e. each COEDGE belongs to only one LOOP which in turn belongs to only one FACE and so on).

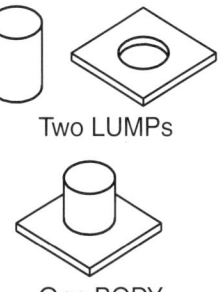

Two LUMPs

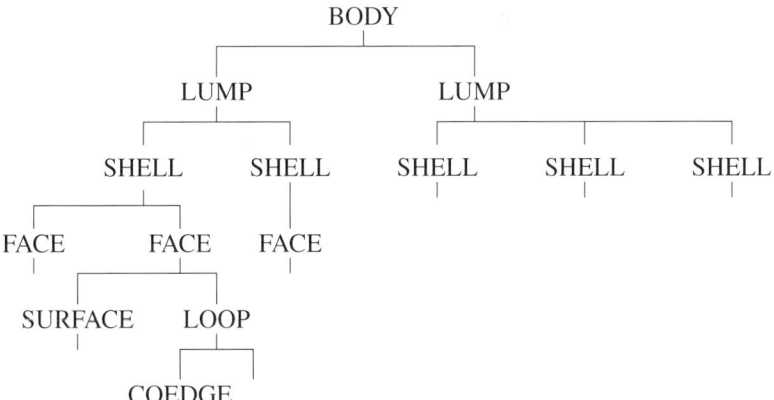

Figure 1.1: ACIS Representational Hierarchy.

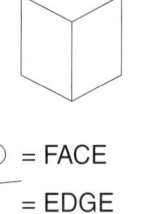

One BODY

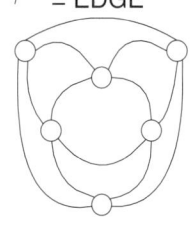

A cube's face-edge graph

Below COEDGE, however, this neat deconstruction breaks down into a network of interconnections which can be viewed as holding the model together. Each EDGE can be referred to by a number of COEDGEs, and similarly VERTEXs frequently have no unique owner but are shared (i.e. referenced) by a number of different COEDGEs and EDGEs. One way in which EDGE-FACE relationships can be visualized is as a form of graph, or network, where each node represents a FACE and every arc an EDGE in the model (see margin). Because these graphs define which FACE adjoins which, they are sometimes referred to as FACE-adjacency graphs.

So far the ACIS B-rep data structure has been described in general terms. However when used to model a physically realizable 3D solid there are some implicit restrictions on its content. In mathematics such shapes are said to be **manifold** and have data structures which conform to a number of rules, such as:

- Every EDGE must lie between two FACEs.
- FACEs and EDGEs do not self intersect.
- Every ENTITY in the model is bounded.

Table 1.2 Representational ACIS ENTITYs.

Entity	Comprises	Represents physically
BODY	Collection of LUMPs or WIREs	Highest level ENTITY in an ACIS model; can be a 1D, 2D or 3D shape with a complete or incomplete boundary
LUMP	Collection of SHELLs	A region of a BODY disjoint from any other LUMP
SHELL	Collection of FACEs and/or WIREs	Outer, or inner, boundary surface of a LUMP
FACE	One SURFACE and zero or more LOOPs	A portion of an individual surface
LOOP	Circuit (or list) of COEDGEs	Connected portion of a FACE's boundary which may be open or closed
COEDGE	EDGE and (on spline surfaces) a PCURVE	Records the occurrence of an EDGE in a FACE boundary
VERTEX	APOINT	Boundary (i.e. end point) of an EDGE
EDGE	Collection of COEDGEs, two VERTEXs and a CURVE	Holds the model together with adjacency information. Formed by the intersection of two, or more, surfaces
WIRE	Collection of EDGEs	A contiguous collection of EDGEs not attached to a FACE or enclosing any volume
SURFACE	Geometric definition	The shape of a FACE
CURVE	Geometric definition	The shape of an EDGE
APOINT	(x,y,z) position	The location of a VERTEX
STRAIGHT	Analytic curve (straight line)	Refinement of CURVE
ELLIPSE	Analytic curve	Refinement of CURVE
INTCURVE	Interpolated curve	Refinement of CURVE which has no explicit equation
PCURVE	Spline curve in (u,v) space	Parametric curve on a spline surface defining a COEDGE's shape
SPHERE	Analytic surface	Refinement of SURFACE
PLANE	Analytic surface	Refinement of SURFACE
CONE	Analytic surface	Refinement of SURFACE
TORUS	Analytic surface	Refinement of SURFACE
SPLINE	Spline surface	Refinement of SURFACE

It is important to note that these requirements are not inherent in the data structure. Consequently ACIS can be used to create and manipulate nonmanifold[3] shapes which can exist only as idealizations within the memory of the computer (i.e. they are **not** physically realizable). Section 1.10 gives a brief overview of how these types of shape are represented by the B-rep data structure.

1.3 Why COEDGEs?

Why bother with COEDGEs? The ACIS data structure appears to be holding three objects for every edge in the model. Surely it would be more efficient to have only one EDGE record?

The reason for the three edge objects has its roots in some of the mathematical theorems which underlie the boundary representation and provide conditions for ensuring the validity and integrity of objects modeled in this way. Ensuring the **integrity** of a model's topology and geometry is central to any representation scheme.

One aspect of topological rigor is the enforcement of **Möbius's Rule**. This states that edges shared by two FACEs are bidirectional entities. Implicit in this rule is the idea that edges not only bound surfaces but also have a topological function of linking adjacent FACEs.

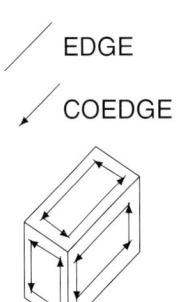

LOOPs of COEDGEs

All commercial, and academic, B-rep modelers obey Möbius's rule and the differences in their data structure can be measured by the amount of redundant information stored and the difficulties encountered in FACE traversal.

Several early modelers (in the 1970s) adopted or modified a so-called **winged-edge** data structure in which each edge record contained directional information about its preceding and neighboring edges which lay on the two FACEs adjacent to it. This was a very concise representation but made FACE traversal awkward since the edge direction constantly had to be checked.

In 1982 a **split-edge** data structure was proposed which created an edge record for each FACE adjacent to an edge. Each edge record was oriented in opposite directions and each had a reference to its partner in order to record which FACE was adjacent to which. This design doubled the number of edge records but made FACE traversal straightforward.

However, the directionality and adjacency requirements for an edge in a manifold solid are different from those in wire-frame and nonmanifold bodies. In order to be able to model all types of object, ACIS adopts a derivative of these two previous schemes (not unlike the Half or Hybrid Edge Representations proposed elsewhere in the late 1980s). The main effect of the COEDGE/EDGE scheme is to separate the orientation role of an edge (which is carried out by the COEDGEs) from the topological role of recording adjacency information (done by the EDGE) objects.

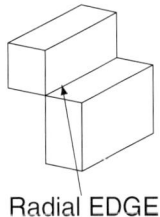

Radial EDGE

This distinction also supports the modeling of nonmanifold objects that have more than two FACEs sharing an EDGE. Consider, for instance, the two blocks united along a single EDGE (shown in the margin). In this case one EDGE is part

---

[3]The name manifold arose from the observation that such objects can be constructed from a 2D sheet (of paper) cut and folded in the correct manner ... created by *many folds*.

of the boundary of four different FACEs. This situation creates a so-called "radial EDGE", in which an EDGE is shared by a total of four COEDGEs, each of which lies on a different FACE.

1.4 Representation of Geometry

So far we haven't said much about exactly how the geometry of surfaces and curves is represented. There are many possible forms that, say, the equation of a line can take. Probably the *explicit* form, $y = mx + c$, is the best known and least useful because of its numerous *special cases* (imagine, for example, what happens to the gradient m when the line is vertical).

Explicit:
$y = f(x)$
Implicit:
$f(x, y) = 0$
Parametric:
$x = f(t), y = f(t)$

A better proposition for geometric modeling is the **implicit** form of geometric equations which effectively define **half-spaces** (i.e. in the case of a line an infinite boundary that cuts 2D space in two). Given the coordinates of a point on the boundary an implicit equation evaluates to zero, and on either side returns more, or less, than zero. Although much more stable than the explicit forms, implicit equations cannot be easily derived for all the surfaces used by ACIS. Parametric representations, on the other hand, can be created for all surfaces incorporated in the modeler.

Parametric curves are defined by equations given in terms of a single independent variable often known as **t**. Similarly, every point on a parametric surface can be referred to by a two-parameter (normally **u**,**v**) mathematical function. In other words every surface on a model has an embedded (u,v) coordinate system.

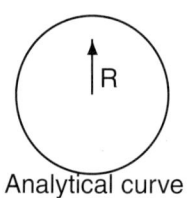
Analytical curve

ACIS uses analytic, implicit and parametric representations of geometry. The analytical curves and surfaces have both a low level implicit representation and a higher level parametric one,[4] while others, such as splines, have only a parametric representation within ACIS.

When dealing with surfaces, or curves, which have both implicit and parametric representations, ACIS's modeling algorithms can exploit the more specific form to increase efficiency. For example, if the intersection between a plane and cone surface is required, a specific routine, **int_plane_cone**, can be used to ensure a fast result. However, the same calculation can be carried out for any pair of surfaces through the general parametric interface when no specific procedures are available.

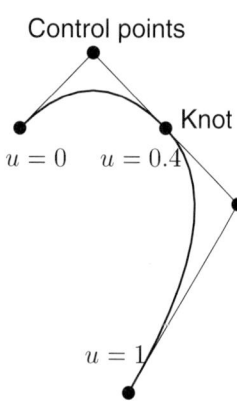
Composite curve formed by two quadratic segments, "knotted" together.

This approach means that ACIS isn't limited to, say, analytical geometries, but can be easily extended to incorporate any surface for which a **parameterization** can be created.

In detail, the parametric interface is used to support general smooth curves and surfaces. Such curves and surfaces, termed splines,[5] are constructed by specifying numbers of **control points** which determine their shape. There are many forms of spline surface. Open any computational geometry book and you can very rapidly find out more than you care to know about: Hermite, Bézier and Overhauser splines.

---

[4]They are said to have a *parameterization* which maps from their implicit to parametric forms. This is distinctly different from the PCURVE's *parameterization* which maps from a 2D parametric surface to a 1D parametric curve.

[5]ACIS uses the word spline in a slightly more general manner than described here; see the manual for full details.

The ACIS Kernel uses a form known as **nonuniform B-splines** which can be either rational (i.e. NURBS) or non-rational (i.e. NUBS).

Perhaps the easiest mental model of a NURBS curve is to imagine it as constructed from segments of curves, which join *smoothly* together. The point where one segment starts, or ends, is known as a **knot**. The locations of a spline's knots are defined in terms of the curve's **u** parameter. The nonuniform part of the name refers to the spacing of the knots. A nonuniform spacing implies that one can have multiple internal knot values and unequal spacing between them. This provides greater flexibility in controlling a curve's shape.

Rational splines allow the pull of individual control points to be varied by a weighting factor. This allows conic curves (i.e. circles and ellipses) and quadric surfaces (cones, cylinders, spheres etc.) to be represented exactly. NURBS have two compelling advantages:

Local Control: Unlike some other forms of spline, the control points associated with a NURBS curve (or surface) only influence the shape in their immediate neighborhood.

Efficiency : NURBS have a fixed degree, no matter how many control points are used to define them, and so can be computed quickly.

The question that should pop into your head about now is "Hey, if these rational splines can model *any* conic exactly, why not simply use them all the time and have only one representation?" The answer is you could (and some people have) ... but generally you wouldn't want to! Spline surfaces take a lot of calculation and when they intersect even the fastest computers tend to pause for thought. Chapter 8 looks at some of the ways NURBS can be created and used in ACIS.

> The terminology of free-form, or complex, surfaces comes from some of their earliest engineering applications (see page 216). The draftsmen who specified the shapes of the early airplanes and ships used long flexible strips of metal, or wood, called *splines*. These splines were distorted by means of weights applied at specific distances along them. The mechanical properties of the metal meant that the metal deformed with second-order continuity which is also modelled by mathematical splines.

About splines

1.5 Interpolated Curves

Spline geometry frequently appears in ACIS models under the guise of an ENTITY known as an INTCURVE (i.e. Interpolated Curve). These usually define the shape of EDGEs formed by the intersection of two surfaces. Typically an INTCURVE object holds a spline approximation to its true shape and pointers to the two (intersecting) SURFACEs on which it lies. The approximate spline can be used for low precision operations like display, whereas more demanding operations, such as Boolean operations (see Chapter 6), use the two parent surfaces to generate more precise answers. This approach allows ACIS to hold an exact representation of a shape without the high cost of explicitly calculating the geometry of each intersection curve to a high accuracy. Some people find it helpful to envisage INTCURVEs as a sort of virtual geometry that is only evaluated on demand.

An exception to this general situation occurs when the INTCURVE is formed not by surface intersection but by an exact spline curve. Such curves can arise from, say, user input and, since they make no reference to any coincident pair of surfaces, are regarded as an exact definition of the geometry.

1.6 Sharing Geometry

The shape of a model's FACEs and EDGEs are defined by the geometry of the curves and surfaces they lie on. ACIS allows one curve or surface to support several EDGEs or FACEs. To allow geometry to be shared by different ENTITYs and yet maintain a clear separation between the representation of geometry and topology, ACIS defines two classes for each type of supported geometry.

Table 1.3 ACIS's geometry classes.

Name	Description
curve	Defines functions common to all curves such as position and parameter based inquiries (i.e. the parent class)
ellipse	Defines a circle (or ellipse) in any plane
straight	Defines an infinite straight line
intcurve	Defines a curve formed by the intersection of two surfaces or an exact spline
pcurve	Defines a curve on a spline, or blend, surface in terms of (u,v) parameters
surface	Defines functions common to all surfaces such as position and parameter based inquiries (i.e. the parent class)
plane	Defines an infinite planar surface
cone	Defines conical surfaces and cylinders
torus	Defines a toroidal surface
sphere	Defines a spherical surface
spline	Defines procedural surfaces such as exact splines and blends

1.7 Identifying the Inside!

The two types of class are distinguished by capitalization of their names:

Uppercase: Geometry classes (such as CONE, SPLINE and PCURVE) keep a record (known as a **use_count**) of how many other ENTITYs refer to the surface, or curve, and supply member functions for memory management and save/restore utilities.

Lowercase: Geometry classes (such as cone, spline and pcurve) define the constructors and methods required to display and manipulate the shape by various modeling operations.

Two FACEs, one SURFACE

Two EDGEs, one CURVE

In other words, the equations which represent a type of curve, or surface, reside in the lowercase class, while the uppercase class supplies all the administrative functions.

For example, the two flat FACEs split by the slot (shown in the margin) will both refer to the same instance of the **PLANE** class, which will have a **use_count** of two. The **PLANE** object itself, however, will only hold one instance of a **plane** class.

Table 1.3 lists the lowercase geometry classes currently supported by the ACIS Kernel.

1.7 Identifying the Inside!

Although a solution to the problem of adding boundaries to infinite geometries has been described there is a second difficulty with the analytical equation of surfaces not mentioned earlier: they have no notion of side. Given a FACE, as it has been described up to this point, one would be unable to tell which side represents fresh air and which is solid material. To enable this sort of distinction, each FACE in an ACIS model is oriented and can supply an outwardly directed unit vector (known as a FACE normal) for any point lying on it.

Surface Normal vector

It is no surprise to learn that plane FACEs have an outward direction that is the same at any point within their boundary, whereas other surfaces, like spheres, have normals that vary from point to point. In all cases the ACIS convention is that the FACE normal is directed away from the material.

One direct consequence of the way in which ACIS shares geometry between ENTITYs (Section 1.6) is that a FACE's normal vector is not necessarily the same as its SURFACE's normal vector. One need for independent FACE and SURFACE normals is illustrated (in the margin) by a cross-section of a BODY containing two FACEs (F1 and F2) which share the same underlying SURFACE (S1) but have opposite orientations.

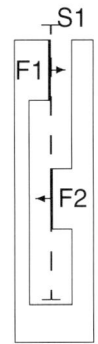

Concepts of FACE orientation are important in the modeling of nonmanifold objects. In general FACEs can be classified as either **single** or **double** sided. Double sided FACEs are further categorized as **internal** or **external**. This distinction allows ACIS to represent *conceptual* shapes with dangling (external) flaps or thin (internal) divisions and cellular models.

1.8 Supporting Classes

In addition to the topological and geometric entities needed to create a Boundary Representation, ACIS incorporates several other classes, summarized in Table 1.4, which provide support for specific functions. The following table reviews the applications for which these objects provide support.

Table 1.4   Supporting ACIS classes.

Entity	Comprises	Supports
TRANSFORM	A general affine transform	Specifies an amount of scaling, rotation and translation
box	An axis-aligned cuboid	Intersection calculations
ATTRIB	General operations to support lists of attributes attached to ENTITYs	Allows a list of application specific data to be attached to an ENTITY

1.9 Creating Shapes

The complexity of B-rep data structures makes them almost impossible for humans to construct without some form of computer assistance. In ACIS this help comes in a number of different forms:

Primitive instancing: allows basic shapes (e.g. block, cylinder, cone, sphere, etc.) to be created (see Table 4.1).

Sweeping: allows 2D profiles to be swept, or spun, to form 3D solids (see Sections 4.6 and 10.5).

Booleans: allow set operations (i.e. addition, subtraction and intersection) between model ENTITYs (see below and Chapter 6).

Skinning and Lofting: generates a FACE that interpolates a number of cross-sections (see section 8.9).

Stitching: joins pairs of BODYs along coincident EDGEs or VERTEXs.

Local operations: modify geometry and occasionally topology (see Chapter 11).

Blending: creates transition surfaces between ENTITYs (see Section 4.7 and Chapter 11).

Shelling: creates a thin-walled solid from a BODY by offsetting the FACEs (see Chapter 11).

1.9 Creating Shapes

While all these operations are useful, and sometimes indispensable, perhaps the most important are the **Boolean**, or set, operations. Because these particular operations strongly influence several details of the ACIS data structure, it is useful to have an appreciation of their underlying algorithm. There are three basic Boolean operations:

Unite: unites two BODYs to form a third. $A \cup B = B \cup A$

Subtraction: subtracts one BODY from another to form a third. $A - B \neq B - A$

Intersection: takes two BODYs and constructs a third representing the material common to both. $A \cap B = B \cap A$

Notice that while Union and Intersection are commutative, this is not true of Subtraction.[6]

The algorithm for carrying out set operations goes something like this :

1. Each boundary segment (i.e. the faces and edges) of one object (call it the tool) is compared with each boundary segment of the other (call it the blank).

2. Where segments are found to cross, they are split.

3. The parts of newly split boundary segments are marked as being inside or outside the boundary of the other BODY.

    - When doing a **union** the split parts marked **outside** are kept, while the internal parts are deleted.
    - When doing an **intersection** operation only **inside** bits of the boundary are retained.
    - When doing a **subtraction** the same rules as for union are used but applied with the *sense* or *orientation* of the tool's boundary segments reversed, thus flipping the outside/inside classifications.

4. After all the boundary crossings have been calculated the kept boundary segments are reconnected and the rest deleted.

Chapter 6 (Booleans and Intersections) explains why the actual mechanics of carrying out one of these set operations are more easily said than done. Here, however, you need only to appreciate the role of WIREs and boxes in the process:

- The first step (or stage) is called **imprinting** and can be envisioned as a line drawn round the perimeter of one object which has been pushed into the other. This line is represented in ACIS as an **intersection graph** and is formed from the curves generated by curve/surface and surface/surface intersection operations. These curves are embedded in a number of EDGEs which together form a WIRE body.

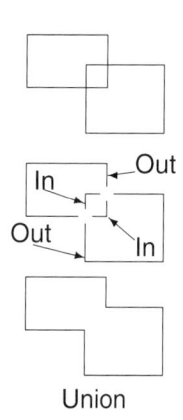

---

[6]Although set theory suggests that the order of the arguments given to union and intersection operations is unimportant this is not true in practice. Both procedures modify one of the input BODYs to form the result (known as the blank) and remove the other. Furthermore the resulting shape is calculated in the space of the surviving BODY (i.e. the blank) whose inverse transform is applied to the other BODY (i.e. the tool). Since tool and blank BODYs are determined by the order of the arguments given to the Boolean routines they are not commutative as implemented.

- The process requires the calculation of every intersection between every FACE on both objects. Rather than do this when it is not needed ACIS associates a **bounding box** with every ENTITY so that a series of rapid overlap tests can be performed, thus eliminating quickly FACEs and EDGEs which do not touch.

Booleans are always carried out between *pairs* of BODYs. A Boolean between three shapes could be done by operating on the first two and then combining the third with the result.

1.10 Nonmanifold and Multi-dimensional Models

Models of complete 3D solids represent a special class of shape, a small subset of the range of geometries ACIS can represent. Not only can 1D, 2D or 3D objects (with complete or incomplete boundaries) be created but they can coexist in the same model. In practice this means that models can have:

- Incomplete boundaries (e.g. a cube with only five FACEs or a STRAIGHT EDGE with only one VERTEX).
- Idealized geometry (e.g. a 3D solid could contain an internal slit modeled as a double-sided planar FACE).

Although such nonmanifold objects don't exist in the real world, they do arise quite naturally during the intermediate steps of the Boolean process (see Chapter 6) and perhaps more importantly they model the way many people (particularly engineers) often envisage or idealize shapes.

For example a composite material, such as a reinforced concrete beam, could be represented as a 3D block with several internal 2D FACEs defining the boundary between cement and metal. A similar situation arises in finite element modeling, where loads are often idealized to POINTs on a surface.

Table 1.5 illustrates some of the different types of nonmanifold and mixed dimension models which can be represented by the ACIS B-rep data structure. The Boolean operations described earlier in this chapter can be used in conjunction with any form of ACIS BODY.

1.11 Cellular Topology

The data-structure of nonmanifold BODY's can be difficult to navigate; EDGEs can be adjacent to any number of FACEs, whilst FACEs themselves can be internal or external and single or double sided.

So rather than deal with the raw topology many applications (such as selective Booleans, page 154) find it easier to organize nonmanifold topology into structures of 3D and 2D CELLs.

Figure 1.2 illustrates the difference between 2 and 3D CELLs. Figure 1.2(a) shows the nonmanifold union of two orthogonal blocks. The resulting BODY has a number of internal FACEs which partition its volume into five distinct 3D CELLs (labeled

1.12 Moving and Scaling Shapes

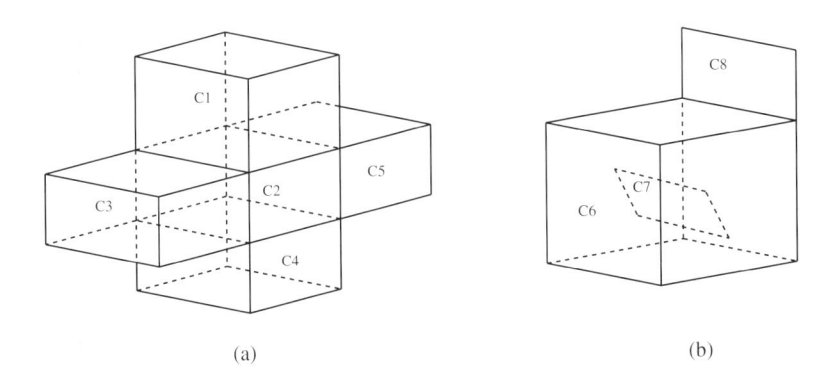

Figure 1.2: Examples of 3D and 2D Cellular Topology.

C1-5). Not surprisingly 2D CELLs do not enclose volumes and are typically formed by dangling (e.g C8), or detached (e.g. C7), sets of FACEs, Figure 1.2(b).

ACIS creates cellular models by attaching attributes (see Chapter 12) to ENTITYs in the data structure. Because attribute data can itself have attributes attached to it, individual CELLs can have application data associated with them. For example the steel reinforcement in a concrete beam might be modelled by a 3D CELL with different material properties (e.g. density) to those surrounding it.

By default CELLs are simply organized into linear lists attached to a BODY's LUMPs. However the adjacency relationships between CELLs can be explicitly represented by a graph structure in which each vertex (in the graph) is a CELL and each edge (in the graph) represents a number of shared FACEs or EDGEs, see page 154.

1.12 Moving and Scaling Shapes

If anything keeps writers of solid modelers awake at nights it is probably numerical accuracy. Although the question "Is the point (56.7899567, 45.6734577, 123.99998) the same as the point (56.7899567, 45.6734578, 123.99998)" is not hard for a computer to answer, it does little to tell you if the two points were *meant* to be the same!

Typically these numbers arise not from the dimensions input by the user but from the actions of Booleans, transformations or other modeling operations.

Every time a component is rotated a number of degrees, each coordinate is multiplied by a long floating point number which, inevitably, produces a small rounding error. These small errors have a cumulative effect which can be lessened by making sure transforms are applied only when needed and not simply for cosmetic reasons.

ACIS does this by associating a **TRANSFORM** object with each BODY. When a BODY is displayed, coordinates are read from the data structure and then transformed before being projected onto the screen. Any further rotations or translations of the model can be applied to the TRANSFORM rather than the model. In this way the number of additions and multiplications, and so rounding errors, are kept to a minimum.

Table 1.5 Types of shape representable by ACIS.

Shape	Type	Notes
	Manifold	All ENTITYs are bounded, so the shape represents a physically realizable solid
	Incomplete or semi-bounded	Two of the FACEs have a missing EDGE
	Nonmanifold	One EDGE has four COEDGEs, shared between two SHELLs
	Multi-dimensional	One FACE incorporates two LOOPs one of which comprises a single EDGE, degenerated to a VERTEX (the same situation is found at the apex of a cone). A WIRE formed by a STRAIGHT EDGE is attached to the FACE at this point
	Multi-dimensional	One EDGE has four COEDGEs, two of which lie on a double sided, external FACE (i.e. a flap)
	WIRE BODY	The BODY is composed of two WIRE LUMPs, each consisting of a single EDGE whose shape is defined by the INTCURVE resulting from the intersection of a sphere and a torus (see page 137)
	2D (sheet) BODY	The BODY is composed of a LUMP with a SHELL consisting of one double sided FACE whose shape is defined by a B-spline surface. The boundary of the FACE is defined by a LOOP of EDGEs whose shapes are specified by PCURVEs

1.13 Model Management

The elements of an ACIS model are C++ objects, each derived from a *base class* known as ENTITY. The descendants of ENTITY are easily recognized by their uppercase names. Because all the topological (e.g. FACE, LOOP, LUMP) and geometric (e.g. SURFACE, CURVE, PLANE) entities mentioned in this chapter are children of ENTITY, they inherit common behaviors which allow global operations to be implemented easily. In this way every ENTITY in an ACIS model responds to the same commands for:

- creation and deletion,

- save and restore to file,

- backup (typically for "undo" and "redo" purposes) via objects known as **BULLETIN**s which hold copies of every altered ENTITY and pointers to those which have been *created* or *deleted*,

- attachment of user defined data via an ENTITY class known as **ATTRIB** (i.e. attribute).

The ENTITY class hierarchy and its common behaviors are described further in Chapter 2.

1.14 Displaying Models

This chapter has given a brief tour of B-rep modeling and said very little about particular uses of the technology. However, 3D graphics are used in almost every application of geometric modeling and consequently they deserve special treatment. Because ACIS provides API level support for faceting (i.e. the Faceter and Interactive Hidden Line (IHL) components), the description here concentrates on this particular method of image generation. Several other approaches to 3D graphics are possible. For example the Precise Hidden Line or Advanced Rendering components described in Chapter 7.

Somehow it seems almost absurd that the first thing many display algorithms do is generate an approximate representation from the exact geometry which has been so carefully constructed. However, unless you are plotting out engineering drawings, the images generated from inexact descriptions will often be so good that the loss of detail is not noticed.

When displaying, or rendering models, ACIS programmers can use the exact geometry of an object to generate a mesh of planar facets. Approximating the real shape of an object to one covered in small triangular faces takes a bit of computation, but buys you several advantages:

1. Every facet which faces away (surface normals are attached to each polygon) from a given viewpoint can be removed, since they will be invisible. This process can dramatically reduce the number of facets in the scene to be displayed.

2. The visibility of the remaining polygons does not need to be explicitly calculated. If the facets are shaded in order of depth, from background to foreground, the closer ones will simply cover up any invisible bits of the ones further away.

3. Silhouette (or horizon) lines don't have to be explicitly calculated since they will naturally be formed by the edges of facets.

4. Surface color calculation only needs to be done at the corners of each triangle rather than across an entire face.

5. Many PCs now have built-in 3D graphics cards which essentially process triangles in hardware, so faceting is needed to take advantage of such facilities.

Although the price paid for these speed advantages is undoubtedly quality, two techniques help reduce the angular appearance of the shapes being displayed:

1. The color of adjacent triangles can be interpolated to remove any sharp changes in tone (i.e. Phong or Gouraud Shading).

2. The size and shape of the facets generated can vary across a model's surface, allowing areas of high curvature, or those close to the "horizon" (given a viewpoint), to be covered by more facets than other parts.

The process of creating facets and rendering is covered in Chapter 7.

1.15 Pros and Cons of B-Reps

Before plunging into the details of C++ and ACIS's functions and classes, it is appropriate to end this chapter with a last glance at Table 1.1. Although there are several other methods of defining the shape of solid objects it is beyond question that *boundary modelers*, such as ACIS, are *currently* dominating the CAD/CAM market. There are several long answers as to why this is, but here is a very short one. WYSIWYG: with a B-rep model "What you see is what you get" or "If you can see it on the screen, it exists in the data structure" (well, almost). In other words, every EDGE displayed on the screen is explicitly held in the model's data structure, making the editing and interrogation of objects straightforward. This may appear to be a trivial point but it makes the attachment of attributes, such as machining tolerances, to FACEs very easy.

There is, of course, another school of thought which says that the problem with B-reps is WYSIAYG: "What you see is all you get", and that holding and maintaining the consistency of so much information is an overly complex way of doing things. Constructive Solid Geometry (CSG) modelers[7], for instance, have a data structure that looks positively frugal next to a B-rep one because they typically hold only a set of equations defining the surfaces (i.e. half-spaces) of an object.

---

[7] Also known as set-theoretic modelers.

However, there is a difference, often forgotten in academic circles, between things that are hard and things that are impossible. It is undoubtedly hard, for reasons that will become apparent later, to write a robust B-rep modeler, but it is not impossible, as the existence of ACIS and other B-rep modelers proves. Many of the other modeling representations are simpler to implement so surely they must be better? No, not if someone else has already done the hard bit. I'm not bothered that it is hard to design and manufacture a working internal combustion engine, because I'm not going to do it. I am simply going to buy one and use it.

1.16 Exercises

1. Sketch out a Boundary Representation for a coffee mug of your choosing.

2. Explain the difference between ACIS's PLANE and plane classes (hint, see section 1.6).

3. Euler's formula for a polyhedron (without through holes or hole loops) states that :

$$V - E + F = 2$$

   where:

   V = number of VERTEXs
   E = number of EDGEs
   F = number of FACEs

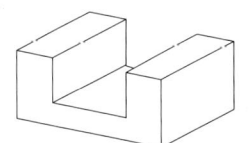

   Verify that this relationship:

   (a) does hold for a block with a slot through it;
   (b) does **not** hold (as stated) for BODYs with through holes. Try, for instance, to sum V, E and F for a square piece of flat plate containing a circular through hole.

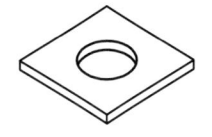

4. Why would one **not** expect the coordinates of a BODY's VERTEXs to change after a translation transform has been applied (hint, see section 1.12).

C++

2

Solid modelers are large and complex lumps of software that pose real problems irrespective of which programming language is used. However, because ACIS was written from the "ground up", its creators were free to exploit a modern language called **C++**, which offered ways of organizing the data and functions needed by a solid modeler in a new, and more concise, way. Now, with 20-20 hindsight, it is clear that the choice of this object-oriented descendant of the **C** language has provided a coherency and expressive power which modelers written in older languages find hard to emulate. Rather than explaining how to program in C++ (a subject that could fill this and several other books), this chapter describes the different ways in which the functionality offered by the language is exploited in the design of the ACIS modeler. We start with some of the more abstract ideas about how an object oriented language, like C++, can be used to support geometric programming and end with the details of how ACIS has used the semantics of the C++ language to implement these ideas.

2.1 Object Oriented Programming

Every high-level programming language has some way of grouping **data** together under a common reference, or handle (e.g. FORTRAN has records, C has structures). Object oriented programming languages extend this idea by making it easy to group **functions** and **data** together in collections known as **classes**. This doesn't sound like much but it allows the programmer to control exactly which functions are used to change specific bits of data and encourages consistency in the way information is created, manipulated and deleted. The application of this technology to any system that has to maintain and represent geometric entities is easy to envisage. A rectangular object could ensure that its "corner point" data was only changed via code that also automatically updated the position of the *other* vertices. Although the same result could be achieved, with a lot more effort, in any of the older languages mentioned earlier, C++ provides built-in facilities for doing this, making the job of coding such a rectangle very easy.

Each **object** is an instance of a **class**.

Classes provide a generic definition of an object's data and functions. ACIS defines classes for all the topological and geometric entities in its data structures. Some of these classes are very complex, others transparently simple.

For example, one of the most basic classes in ACIS is called **position**. A *position object* holds three numbers representing its x, y and z components. These three numbers are referred to as **member data** because they are *members* of the position class. The position class also defines a number of functions that can be, sensibly, associated with individual, or pairs of, positions, such as transformation, translation and equality tests. Not surprisingly these procedures are known as **member functions**.

Several different classes can define member functions, or operators, with the same name. A prime example of this is the **+** operator which is defined by both the ACIS **vector** and **box** classes. The operator is said to be **overloaded** and its behavior is determined by the class of its arguments rather than its name.

The grouping of functions and data together is only part of the C++ story; equally important is the way in which groups of classes can themselves be bundled together and organized into **hierarchies**. This allows member functions to be shared across a range of related classes through a mechanism known as **inheritance**. Using the inheritance mechanism ACIS's vector class makes some of its member functions available to the **unit_vector** class which lies below it in the hierarchy. This facility allows every **unit_vector** object to use member functions of the more general vector class (say, for calculating cross products) without having to redefine them.

These mechanisms provide two aspects of C++ expressive power which can be summarized as:

Expression: Objects can directly express abstract mathematical entities.

Abstraction: Similarities among objects can be exploited by programmers and lead to more concise code.

Solid modelers are well placed to exploit both these features since geometry has several natural hierarchies.

The base class of ACIS's hierarchy is called **ENTITY**. All the geometrical and topological classes needed to define a boundary representation model (e.g. EDGE, FACE, LOOP, etc.) are derived from ENTITY and referred to collectively as ENTITYs. The ACIS ENTITY hierarchy is wide but shallow, never being more than four classes deep in any one place[1].

The tops of hierarchies are often pretty empty places, because it can be very difficult to define anything that is generally true about lots of things. This, however, is not the case with ACIS's class hierarchy, for while there are very few mathematical things one can *usefully* say about all ENTITYs, there are many administrative functions that are universally applicable. Consequently it is in the ENTITY base class that two of ACIS's most important architectural ideas can be found:

Uniform management of user extensions: Any user of ACIS is free to derive their own classes from the ENTITY base class. Because operations that support

---

[1] While this is probably not intentional it definitely helps rapid understanding of the structure. The documentation of deep hierarchies of classes raises several difficulties not least because books are linear and trees are not. It often takes several passes through a manual to grasp the relationship between classes in deep hierarchies.

general utilities, such as save and restore, are defined in the base class they are automatically inherited by any derived classes.

Association of user data with existing ENTITYs: In addition to supporting common functions, the base class also contains a pointer to the start of a linked list of classes called ATTRIB (i.e. attribute). Users are free to define their own derivations from the ATTRIB base classes (creating what is simply a special form of ENTITY). Members of the ATTRIB class are unique because every class derived from ENTITY can support a direct reference to one.

These topics are covered in more detail in Chapter 9 which deals with ACIS's generic rollback mechanism known as the Bulletin Board and in Chapter 12 on the Attribute Mechanism.

2.2 Types, Pointers and References

Before we can talk about ACIS objects in any detail we need to review a few key bits of C++ syntax.

2.2.1 Types

C++ is a *strongly typed* language. The compiler checks both the argument list and the return types of every function during the compilation process. So it is critical to ACIS programming that types are correctly defined.

Although there are many possible types defined by the C++ language, the ones seen most often in ACIS code are:

Objects: of classes such as BODY, FACE, EDGE, surface, curve, transf, etc.

Numbers: of type **double** and **int**.

Functions: such as **api_make_sphere** (which have a type determined by the value they return).

A **variable** can be declared anywhere in a C++ program by simply prefixing its name with a **type** declaration. Variables can also be initialized at the time of their declaration. For example, consider the following code fragment:

```
// Variable declarations
  double x_coordinate, z_coordinate;
  double y_coordinate = 5.669;
  ENTITY_LIST el;
```

The first line declares x_coordinate and z_coordinate to be variables of type **double**. The second line does the same for y_coordinate but initializes it with a value of 5.669. The last line declares the variable el to be an **ENTITY_LIST** object.

If a variable is an object then the functions and data associated with it can be accessed using the direct member access operator ".". For instance the ENTITY_LIST class holds a list of pointers to ENTITYs and the length of the list is returned by its member function **count()**. So the command:

```
// Assumes el is an ENTITY_LIST
int length = el.count();
```

> The symbol // starts a comment which is terminated by the end of the line.

calls a member function of the ENTITY_LIST class, which returns an integer number that is assigned to the variable length. Although objects can be used directly it is more common in ACIS to manipulate them via pointers or references (known generically as type modifiers).

2.2.2 Pointer: Type Modifier

A pointer is the address (in memory) of an object and in ACIS these have two prominent roles:

1. Pointers are used to define links between different topological and geometric objects.
2. Pointers allow data and objects to be referenced through public access functions.

A **pointer type** is defined by adding an **asterisk** (*) to the **type name** of the thing it refers to. Thus the statement:

```
// Declares ref is a pointer to a double precision, floating point number
double *ref;
```

> A pointer is the **address** of a variable.

declares the variable ref to be a pointer to a number of type **double**. Pointers to things are generated by the *address of* operator, &. So the following lines:

```
// Assigns an address to a pointer
double *ref;
double num = 3.14;
ref = &num;
```

cause ref to become a pointer to a **double** with the value 3.14.

The preceding examples have created pointers for numbers, but pointers in C++ can be used to reference entire objects using exactly the same syntax. Thus a pointer to an ACIS BODY object is declared as:

2.2 Types, Pointers and References

```
// Declares that b1 is a pointer to a BODY object
BODY *b1;
```

Given a pointer to an object, its public functions and data can be accessed using the indirect member access operator, −>. For example, the BODY class has a member function called **lump()** which returns a pointer to the first object in a list of lumps that make up a given BODY. The −> operator allows this to be called via a pointer to the BODY in the following way:

```
// Assuming b1 is a pointer to a BODY object
LUMP *component = b1−>lump();
```

Here the variable **component** is declared to be a pointer to a LUMP object which is returned by a BODY object's member function **lump()**.

Generally all references within the ACIS data structure are made by pointers, with values only being returned by objects representing primitive geometry, such as points.

Pointers can also be de-referenced using the * operator. For example, the previous code fragment could, rather obtusely, be written as:

```
// Assuming b1 is a pointer to a BODY object
LUMP *component = (*b1).lump();
```

In this context ***b1** can be read as *the object pointed to by* **b1**.

2.2.3 Cast: Type Modifier

The type of a variable or a pointer can be changed at any time with a **cast**. For example, although it is a dangerous thing to do, C++ allows you to reinterpret the bit pattern of a floating point number as an integer in the following way:

```
// Declares that n1 is a double
double n1;
// Casts n1 to an integer
int num = (int) n1;
```

Casting is more commonly done with pointers. In ACIS cast operations frequently occur when a pointer has been typed as **ENTITY** and the programmer wishes to denote a more specific type. Consider the following:

```
// Assume a pointer to an ENTITY returned from, say, a ray firing routine
ENTITY *hit_entity;
// Change it to a FACE pointer, assuming the type of ENTITY has been
// determined elsewhere
```

```
FACE *f1 = (FACE*)hit_entity;
```

Generally casting is considered to be a *bad thing* and in this example is acceptable only because the FACE class is derived from the ENTITY base class and we are sure hit_entity does point to a FACE.

2.2.4 Reference: Type Modifier

Variables can also be manipulated via a **reference** to them. For example:

> A reference is an **alias** for an object.

```
// Declare a variable of type integer
// with an initial value of 45
int k = 45;

// Declares a reference, called r, to the variable k
int &r = k;
```

A reference is a new name for an existing thing. Although pointers and references both provide access to an object, they are different. Unlike a pointer, a reference acts like another name for the object and so can only refer to *one* object. All operations applied to a reference act on the thing to which it refers.

```
// Declare a variable of type integer
int k = 45;
int &r = k;
// Add one to the value of r
r++;
// k now equals 46

// Similarly the next line initializes nv with the address of k
int *nv = &r;
```

2.2.5 Const: Type Modifier

The * and & operators used to create pointers and references are known as *type modifiers*. A third type modifier frequently seen in the ACIS code is called **const**.

The **const** type modifier allows programmers to declare either that a value cannot be modified or that a member function makes no change to member data of its class.

Const applied to variables

The most straightforward use of the **const** keyword is to declare certain variables immutable. For example, the value of π could be declared in the following way:

2.2 Types, Pointers and References

// Declares a double called pi that cannot be changed
const double pi = 3.1415926535898;

Pointers can be declared **const** in two different ways such that:

1. The value of the object referred to by the pointer cannot be changed.

2. The pointer cannot be changed to point to a different object.

The first use can be illustrated by assuming that **pi** has been declared as above. A programmer could then make the following statement:

// Declares a pointer to a constant double
const double *pie = π

meaning that **pie** is a pointer to a **double** that is constant. In this way any erroneous modification of π can be prevented.

In a similar way the second use protects against the wrongful modification of pointers by making them constant:

// Declares a constant pointer to a double
double* const pie = π

which declares **pie** to be of type constant pointer to a double, meaning the value of **pie** can change but not the address. Lastly if the pointer and the thing pointed to, are both constant the declaration:

// Declares a constant pointer to a constant double
const double* const pie = π

can be made to ensure it.

Complex **const** declarations are often more easily read from right to left.

Const applied to functions

The following function is a member of the COEDGE class:

// A constant (i.e. read only) function
public: EDGE* COEDGE::edge() const;

Here a member function called **edge**, which returns a pointer to an EDGE object, has been declared to be of type **const**. This means that the function cannot change any part of the class it is a member of, but can only return a pointer to an item of member data.

const declarations are synonymous with access functions.

Const in argument and return specifications

By making function arguments a **const** reference, the efficiency of call by reference is obtained (no copying of the argument is needed) with the safety of call by value (the argument cannot be modified). The construct is typically seen in the constructor functions of ACIS ENTITYs classes. For example:

// PLANE constructor's reference argument
public: PLANE::PLANE(plane const &);

Here the argument of the PLANE class's constructor is a **const** reference to an instance of the **plane** class (the lowercase name indicates it is not derived from ENTITY).

An example of a **const** return value can also be found in the PLANE class:

// PLANE class inquiry function which returns a constant reference
public: position const& PLANE::root_point() **const**;

This statement declares one of the PLANE class's access functions called **root_point** which has no arguments but returns a reference of type **const** to a **position** object. In other words the returned reference to position cannot be changed by any subsequent operations.

2.3 Enumeration

C++ provides a way to define mnemonic names for sets of integer identifiers. The following declaration appears in one of the ACIS header files:

```
// Declare an enumerated type called bl_continuity with values 0-3
// This enumerated type encodes the continuity with which a blend face
// meets adjacent BODY FACEs.
enum bl_continuity { unset_continuity,
            position_continuous,
            slope_continuous,
            curvature_continuous
            };
```

The key word **enum** creates a new type, **bl_continuity**, and defines four unalterable variables of that type. This allows one to say:

// Assigns the value 2 to the variable type_of_cont
bl_continuity type_of_cont = slope_continuous;

which declares the variable type_of_cont to be of type **bl_continuity** with initial value **slope_continuous**. This is simply syntactic sugar; type_of_cont is really an *integer*.

2.4 C++ Functions

When you are not accessing the objects of an ACIS data structure through their member functions you'll probably be using the API library of C++ functions. C++ functions have the general declaration syntax:

    Returned-Type Function-name (Arguments);

2.4.1 Function Prototypes

ACIS, like most libraries, is too large to keep in a single file, so in common with other languages C++ provides mechanisms for allowing functions defined in one file to be called in another file. Unlike, say, FORTRAN, C++ checks that a function's argument and returned types are compatible with the call to it. In order to do this the **name**, **arguments** and **returned-type** of every function must be declared before it is called. This declaration is known as a **function prototype**. For example the prototype for the **api_make_sphere** function is found in the *header file* **cstrapi.hxx** and reads:

```
// function prototype for api_make_sphere
   outcome api_make_sphere( double, BODY*& );
```

To ensure consistency between the function definitions and calls in the users' program the API prototypes are recorded in header files which can then be incorporated into a program using the **#include** preprocessor directive. A programmer wishing to create a sphere would put the following line (specified in the function's documentation) at the top of the file containing the call to **api_make_sphere**:

```
// Include API function prototypes
   #include "constrct/kernapi/api/cstrapi.hxx"
```

Examination of **cstrapi.hxx** reveals that in addition api_make_sphere the header file also declares many other API function prototypes related to the creation (or construction) of ENTITYs.

2.4.2 Default Arguments

C++ provides a mechanism for specifying default arguments in the function prototype statement. For example:

> // Function declaration with default arguments
> **curve** *__trans_curve__(**transf const** & = *(**transf** *)NULL, **logical** = FALSE);

declares that a function named **trans_curve** returns a pointer to a **curve** object. Its first argument is a reference to a **const transf** object (i.e. so its value cannot be changed) and the default value (indicated by the = sign) is a pointer to a NULL reference argument (see Section 2.4.5). The second argument is a logical one whose default value is FALSE. These default values would be used if the function was called with no, or a number of missing, arguments.

2.4.3 Reference Arguments

A pointer given as an argument allows a function to access, and in some cases modify, the object it refers to. In contrast, a reference pointer given as an argument actually allows the pointer itself to be changed!

Consider the function **api_edge** which creates a copy of an existing edge given as an argument. The example given in the manual goes something like this :

> // Creates a new edge which is a copy of the specified edge
> **EDGE*** InputEdge;
> **EDGE*** OutputEdge;
>
> **api_edge**(InputEdge, OutputEdge);

Also notice that the arguments specified in the API manual read:

> Args Given: **EDGE*** InputEdge;
> Args Returned: **EDGE***& OutputEdge;

Why the difference? The declaration **EDGE***& OutputEdge; should be read from right to left: OutputEdge is a reference to a pointer to an object of type EDGE. So when the **api_edge** function is called the value of the second pointer is changed to refer to the new EDGE.

The second use of a reference return type seen in the ACIS system enables functions to be made the target of an assignment.

For example, the top level ENTITY class has a innocent-looking Inquiry Function which is documented as follows:

2.4 C++ Functions

```
// Returns the pointer to the bulletin board entry
// for this ENTITY
BULLETIN*& rollback();
```

Unlike other Inquiry functions, which are typed **const** (see Section 2.2.5), **rollback** can be used to set, as well as return, the pointer to the bulletin board entry for this ENTITY! In other words both the following constructs are equally valid.

```
// Extract an ENTITY's last BULLETIN
ENTITY *some_type_of_entity;
BULLETIN *bull = some_type_of_entity->rollback();

// But you can also set an ENTITY's BULLETIN pointer
BULLETIN *a_bulletin;
some_type_of_entity->rollback() = a_bulletin;
```

2.4.4 Pointer Reference Arguments

There is a subtle difficulty with pointer reference arguments. Consider the API function **api_copy_entity** which, as the name suggests, makes a copy of an ENTITY. The example in the manual goes:

```
// Creates a new ENTITY which is a copy of the specified ENTITY
ENTITY* entity;
ENTITY* new_entity;
api_copy_entity (entity, new_entity);
```

with the argument specified as:

```
Args Given:     ENTITY* entity;
Args Returned:  ENTITY*& new_entity;
```

Given this specification it is surprising, at first glance, that a call with two BODY pointers can create any problems. Assume the following:

```
// Copy a BODY object
BODY *old_body;
BODY *new_body;
api_copy_entity(old_body, new_body);
```

The first pointer (old_body) will be automatically cast to an ENTITY*; the second, however, will give rise to the following warning message:

```
'api_copy_entity' : cannot convert parameter 2 from
'class BODY *' to 'class ENTITY *& '
```

If, in an attempt to kill the warning, the programmer inserts a cast, so that the program reads:

BODY *old_body;
BODY *new_body;
api_copy_entity(old_body,(**ENTITY***)new_body);

a different warning message appears:

```
A reference that is not to 'const' cannot be bound
to a non-lvalue
```

One solution is to force a copy of the pointer's reference (instead of the pointer) to be created:

BODY *old_body;
BODY *new_body;
api_copy_entity(old_body,(**ENTITY** *&)new_body);

This cast creates a copy of new_body's reference as type **ENTITY*&** and then passes it as a reference argument. Any changes to **api_copy_entity**'s last argument now alter the pointer new_body.

2.4.5 NULL Reference Argument

Several **API** functions require the caller to pass a reference argument as NULL, usually when some sort of default behavior is required. However simply supplying the argument NULL will results in the following type of error message:

```
'api_cover_circuits' : cannot convert parameter 3
from 'const int' to 'const class surface &'
```

The following code fragment demonstrates how this error is avoided by casting the NULL argument to be a surface pointer. The example is interesting not only because of the NULL Reference Argument used in the API call, but also for the array of ENTITY_LIST objects created as the function's second argument.

```
// Array of pointers to ENTITY_LISTs
```
ENTITY_LIST* ent_list_array[2];
ENTITY_LIST cutting_faces;

ENTITY_LIST *circuits = **new ENTITY_LIST**;
ent_list_array[0]= circuits;
circuits−> **add**(slicer_e);

api_cover_circuits(1, ent_list_array, *(**surface***)NULL, cutting_faces);

FACE *slicer_f = (**FACE***)cutting_faces[0];

The third argument of **api_cover_circuits** is a **cast** of the form ***(surface*)NULL**. This initializer causes a NULL reference to be passed as the argument. It works by first casting NULL to be a pointer to a surface and then de-referencing it, forcing the argument to become a NULL surface.

2.5 Class Terminology

In order to understand ACIS data structures, some of the C++ terminology already mentioned needs to be expanded on. Recall that the procedures associated with an object are known as **member functions** and the information held by (or in) it is termed **member data**. Functions and data are declared to be members of a class using the scope operator **::**. In the following example the scope operator is used to signify that the function **next** is a member of the **BULLETIN** class.

```
// Returns the pointer to the next bulletin on the bulletin board
public: BULLETIN* BULLETIN::next () const;
```

The declaration is prefixed by the access specifier **public:** described in the next section.

2.5.1 Access Specifiers for Class Members

Classes can control the availability of their members by prefixing their declarations with one of the four *access specifiers*. These are:

public: A public member is accessible from every part of the program. The classes derived from ENTITY (i.e. with uppercase names) have public member functions for accessing and setting private data.

In contrast the Kernel Geometry Classes (which have lowercase names) have both public data and functions. For example the **sphere** class includes the following declarations:

```
// The center of the sphere
public position centre;
// Deletes a sphere
public: sphere::~sphere();
```

private: A private member can only be accessed by the member functions and **friends** of a class. The data members of all the ENTITY sub-classes are declared private.

protected: A protected member appears to be public to the functions of any **derived** subclass but private to anything else. The ENTITY class has several protected functions, for example:

```
// ENTITY Destructor:
protected: virtual ENTITY::~ENTITY ();
```

Declaring the destructor protected ensures that it can only be called indirectly through, say, the **lose** function which makes correct provision for rollback (i.e. undo) operations.

friend: A friend is not a member of a class but has access to the non-public members of it. Functions, member functions of other classes or even entire classes can all be declared as friends. In the ENTITY class, for instance, the declaration:

```
// Bulletin board system function that calls the roll_notify method
friend: void roll_once (BULLETIN_BOARD*);
```

ensures that the **roll_once** function can access the **roll_notify** method.

This public/private mechanism is used to keep the inner workings of ACIS secret from applications programs (and programmers)! In ACIS nearly all the member data held by classes is declared private and can only be accessed through public member functions.

2.5.2 Inheritance and Virtual Functions

Virtual member functions are used as place holders for functions whose default definition (if one exists) is redefined lower in the class hierarchy.

Virtual functions that do have definitions can serve as default definitions for classes derived from them. The ENTITY class does this in several cases where it provides a default definition of a function's behavior which is overridden when a derived class declares its own version of the virtual function.

For example, the ENTITY class defines a function called **identity** in the following way:

```
// level default value = 0 (i.e. current level)
public: virtual int ENTITY::identity (int = 0) const;
```

The statement's syntax tells one that the function **identity**, which is a **public virtual** member of the class ENTITY, takes an integer argument and returns an integer. Because it is declared **const** we know it is an access function which does not change any of the class's internal data. Further down the hierarchy in the **SURFACE** class header file we find the same **identity** function redefined:

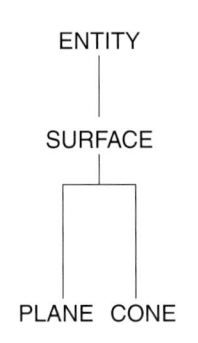

ENTITY
|
SURFACE
|
PLANE CONE

Fragment of the ACIS class hierarchy

2.5 Class Terminology

```
// level default value = 0 (i.e. current level)
public: virtual int SURFACE::identity (int = 0) const;
```

Not all **virtual** functions have default behaviors. For example, the member function for finding an outward facing vector normal to a surface is defined in ACIS's **surface** class as:

```
virtual unit_vector eval_normal(par_pos const &);
```

Without knowing the type of surface no code can be incorporated in the surface class. This type of construct allows one to write programs that deal only with abstract surfaces and leaves the choice of exactly which routine to use (for evaluating the normal) to the computer at run time.

2.5.3 Creating Objects with Constructors

Objects can usually be created in several different ways using a special member function called a constructor. For example the *vector* class we mentioned earlier has a constructor that can be used in three different ways:

```
// Vector object called v1 with no coordinate values
vector v1;

// Vector object called v2 with the stated values
vector v2(56.7,89.7,34.5);

// Create an array of doubles
double coords[3];

// Assign values to the array
coords[0] = 56.7;
coords[1] = 89.7;
coords[2] = 34.5;

// Vector object called v3 based on array values
vector v3(coords);
```

However, constructors are used more commonly to return a pointer to a newly created object with the **new** operator:

```
// Reserve space for a uninitialized BODY object
BODY *b1 = new BODY;
// Create a WIRE BODY
WIRE *w1;
BODY *b2 = new BODY(w1);
```

One of the important differences about these different ways of constructing an object is that things created with **new** are not deleted by the default scoping rules. They exist until they are explicitly deleted by an operator, like **delete** or a function such as **api_del_entity**.

Generally C++ objects are deleted using a member function known as a **destructor**, prefixed with a ~. Although all ACIS classes declare destructors, those derived from ENTITY provide other methods for their removal (see Section 2.7.4).

2.6 Control

C++ has the following control structures:

- **if-then-else**
- **while**
- **do-while**
- **for**
- **switch**

The following sections illustrate how the **while** and **for** loops can be used to traverse ACIS data structures using the public member functions of various classes.

2.6.1 While Loops

While loops continue an operation until a test expression equates to logical *false*. For example, if going through a list, the test expression might return *true* until the end (represented by a NULL pointer) is reached.

```
// Count the number of FACEs in a SHELL of a BODY called block
int count = 0;
FACE *ff = block->lump()->shell()->face_list();
while(ff != NULL)
{
    count++;
    ff = ff->next_in_list();
}
```

A SHELL's FACEs are held in a linear list.

However, a similar construction (using a do-while loop) can be used to move round circuits of, say, COEDGEs which have no explicit end.

2.7 ACIS Classes

```
// Count the number of COEDGEs in a LOOP called lp
int count = 0;
COEDGE *first = lp->start());
COEDGE *ce = lp->start();  // lp is a pointer to a LOOP
 do {
    count++;
    ce = ce->next();
 } while (ce != first);
```

A LOOP's COEDGEs are held in a circular list.

See Chapter 5 for a discussion of the member functions used in these code fragments.

2.6.2 For Loops

The **for** loop combines initialization, termination condition and incrementing into one construction with the general form shown below. Most people will be familiar with the numerical applications of this sort of control structure. For example:

```
// Calculate 10 factorial
int fact = 1;
for( int i = 1 ; i <= 10 ; i++ ){
     fact = fact*i;
}
```

In ACIS another common application is to control movement around the data structure. The following example counts the number of loops on a face.

```
// Count the LOOPs of COEDGEs bounding a FACE
int loopc = 0;
for(LOOP *lp = f->loop(); lp != NULL; lp = lp->next(), loopc++);
```

A FACE's LOOPs are held in a linear list.

2.7 ACIS Classes

ACIS defines a class for each topological and geometric entity and organizes them into a hierarchy with a base class known as ENTITY. A summary of the most commonly used ENTITY classes is given in Appendix B. The focus of this section is the ENTITY base class whose member functions support the following operations for every subclass of ENTITY:

1. Rollback: maintaining a "before and after" record of every change to an ENTITY. These change records are known as **BULLETINs**.

2. Save and Restore: providing support for the mapping of ENTITY pointers to, and from, indexed positions in lists of ENTITYs.

3. Identification: allowing every ENTITY's place in the hierarchy and type to be returned.

4. Addition of User Attributes: providing a pointer to the start of a list of user defined ATTRIB objects.

5. Creation and Removal: providing memory management functions, which access a private free list, allowing ENTITYs to be constantly created and deleted without fragmentation of the store.

The ENTITY class holds two important items of member data:

1. A pointer to the last **BULLETIN** object generated by the ENTITY.

2. A pointer to the start of a list of **ATTRIBUTE** objects.

Users are free to derive their own classes from the ENTITY base class and a number of macros (described in Section 13.1) are provided to help do this. The class hierarchy is illustrated in Figures 2.1 and 2.2.

2.7.1 Rollback

Every time a variable of a class derived from an ENTITY is created, changed or deleted, its **before** and **after** states are recorded in a **BULLETIN** Class. For example, a BULLETIN which records the creation of an ENTITY has its before pointer set to NULL and its after pointer set to the new ENTITY. Because a single modeling operation, such as unite, will cause many ENTITYs to be changed, they are recorded in an ordered list known as a **BULLETIN_BOARD**. Modeling operations can be undone, or **rolledback**, by traveling sequentially through the list, **reversing** the contents of each BULLETIN on the BULLETIN_BOARD. BULLETIN_BOARDs themselves are grouped together into lists know as **DELTA_STATEs**. Again DELTA_STATEs can be undone simply by reversing the contents of the list. Although ACIS provides these functions to support the rollback of ENTITYs, it doesn't happen automatically. A developer adding a new ENTITY class to the system must ensure the function **backup** is called whenever it is created, changed or deleted. This topic is covered in more detail in Chapter 9.

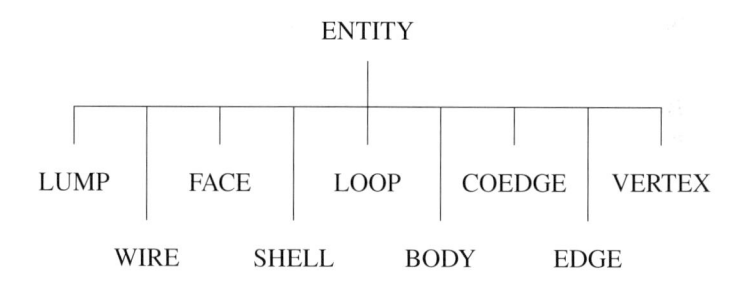

```
                          ENTITY
         ┌────────┬────────┬────────┬────────┐
       LUMP     FACE     LOOP    COEDGE   VERTEX
       WIRE    SHELL     BODY     EDGE
```

Figure 2.1: ACIS's topological class hierarchy.

2.7 ACIS Classes

2.7.2 Save, Restore and Identity Functions

An ACIS data structure is written to, and read from, disk using several of the ENTITY class's virtual function. Because these functions are defined by all classes derived from ENTITY any new, user defined, ENTITYs can be saved (and restored) automatically by standard APIs such as **api_save_entity_list** (see pages 72 and 333).

Essentially ACIS models are written to disk by a three step process:

1. A list is created of all the ENTITYs (including their dependent ENTITYs) to be saved.

2. A copy of this list is created in which every ENTITY pointer of the copied ENTITYs is replaced with an index number (i.e. position in the list) of the ENTITY to which it referred.

3. This *pointerless* list is then written to disk.

The program **entity_contents.cxx** demonstrates how the ENTITY function **copy_scan** supports step one of the save process by listing an ENTITY's dependent ENTITYs (e.g. a BODY ENTITY will hold, or be dependent on, a reference to a LUMP ENTITY which in turn holds pointers to other LUMP and SHELL ENTITYs).

entity_content.cxx

```cpp
// This program creates a cone and then uses an ENTITY class function
// (copy_scan) to create a list of all the ENTITYs in its data structure.
// The program ends after printing out a summary of the list's content.
#include <stdio.h>                              // Input/Output functions
#include "constrct/kernapi/api/cstrapi.hxx"    // Declares make frustum function
#include "kernel/kernapi/api/kernapi.hxx"      // Declares start/stop modeller functions
#include "kernel/kerndata/top/alltop.hxx"      // Declares topology Classes
#include "kernel/kerndata/lists/lists.hxx"     // Declares ENTITY_LIST class


void main(){
    api_start_modeller(0);         // Initialization functions
    api_initialize_constructors();

    BODY* cone;                    // Make a cone
    api_make_frustum(50,20,30,10,cone);

    ENTITY_LIST list;              // Create an ENTITY_LIST
    list.add(cone);                // Add cone's BODY pointer to it
    list.init();                   // Initialize the list iterator

    //Fill the list with a pointer to every ENTITY in the BODY
    for(int c= 0; c >= 0; c++){    // For as long as it takes !
        ENTITY* an_entity = list.next();  // Get the next ENTITY from the list
        if(an_entity != NULL){     // If there is still something left in the list
```

```
                    an_entity−>copy_scan(list);  // copy_scan adds to "list" copies of all
                                                  // pointers to other ENTITYs contained in "an_entity"
                } else  // else reached the end of the list
                    break;  // so break out of "for" loop
        }
        int number_of_entities = list.count();

        printf("List length =%d\n",number_of_entities);

        for(c = 0; c < number_of_entities; c++)
            printf("ENTITY Number %d is a %s \n",c,list[c]−>type_name());

        api_terminate_constructors();
        api_stop_modeller();
    }
```

The program **entity_contents.cxx** creates a BODY ENTITY called **cone** and then adds it an ENTITY_LIST called **list** (see page 76 for more information about the ENTITY_LIST class). A for-loop is then used to invoke the **copy_scan** method of every ENTITY in the list. Because copy_scan adds dependent ENTITYs to the end of **list** the process continues until every ENTITY in the **cone** BODY is recorded (note that ENTITY_LISTs do not keep duplicates so the for-loop will terminate). This process results in 27 ENTITYs being added to the list and, at the end of the program, this list is written to the screen.

When **entity_contents.cxx** is compiled and executed the last lines of its output reads:

```
ENTITY Number 20 is a ellipse
ENTITY Number 21 is a vertex
ENTTIY Number 22 is a ellipse
ENTITY Number 23 is a point
ENTITY Number 24 is a point
```

The program also illustrates the use of the ENTITY member function **type_name** which is used to generate the name of each class being printed out.

2.7.3   ACIS Attributes

The ENTITY class holds a pointer to an **ATTRIB** object which by default is NULL (i.e. no attributes are associated with the ENTITY). Because each ATTRIB object includes a **next()** function (which returns a pointer to another ATTRIB object) each ENTITY can support a list of ATTRIB objects. The use of the ENTITY class's ATTRIB pointer is supported via two member functions:

- **attrib()**: A public access member function which returns an ENTITY's ATTRIB pointer.

2.7 ACIS Classes

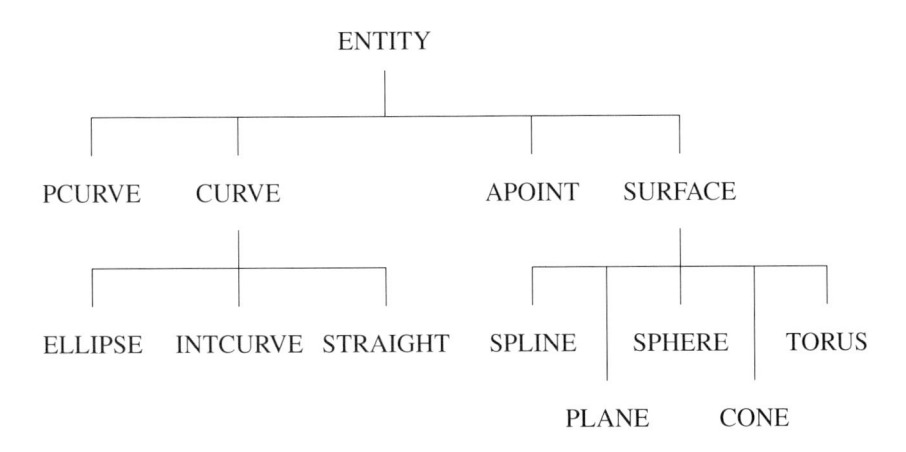

```
                        ENTITY
         ┌────────────────┼────────────────┐
    PCURVE  CURVE              APOINT   SURFACE
            ┌──┼──┐                 ┌────┼────┐
       ELLIPSE INTCURVE STRAIGHT  SPLINE SPHERE TORUS
                                         │
                                   PLANE   CONE
```

Figure 2.2: ACIS's geometric class hierarchy.

- **set_attrib(ATTRIB*)**: Changes the attribute pointer to point to a new attribute.

The creation and use of ATTRIB classes is described in Chapter 12.

2.7.4 Deleting ACIS ENTITYs

ACIS objects can be created in two different ways:

1. Direct invocation of the constructor.

2. Indirect invocation of the constructor via the **new** operator.

Directly created objects are regarded as local variables by C++ and are deleted in the normal way (i.e. by automatically invoking their destructor) when they move out of **scope**.

Objects created with **new**,[2] on the other hand, are unaffected by scoping and remain until they are explicitly deleted. There are four distinct ways in ACIS one can program the death of an ENTITY. They are:

delete: This C++ operator returns the memory grabbed by **new** to the pool of available memory. This is a one way trip; and once called you can't change your mind. It doesn't inform the current bulletin board and it doesn't follow any references to other ENTITYs (e.g. delete a FACE in this way will not delete its EDGEs). Consequently **delete** should *never* be applied to an ENTITY.

lose(): A member function defined by ENTITY. It is much more controlled than the brutal **delete** and does three things:

---
[2]To create objects using ACIS's internal memory management system use the **ACIS_NEW** macro.

1. Places a pointer to the ENTITY being removed on a BULLETIN_BOARD.
2. Calls the **lose()** method of any directly dependent ENTITYs (e.g. **lose** BODY will call **lose** TRANSFORM).
3. Uses **delete** to remove any non-entity data that is associated with the object (e.g. bounding boxes).

At the end of this process you haven't got the memory back because the ENTITY is retained to support any rollback (i.e. undo) operations involving the bulletin board.

api_del_entity: Carries out the same process as **lose** but this time follows the links to every ENTITY it can find. In this way a call to **api_del_entity** to remove, say, a COEDGE object will take the entire model with it (i.e. every EDGE, LOOP, FACE, VERTEX, etc. in the model will be removed by having their own **lose** methods invoked).

api_del_ent: Like **api_del_entity** except that it only goes *down* the data structure. So a call to remove a COEDGE will take the two VERTEX objects and the CURVE, but leave any LOOP it is part of.

Even if you are not using the rollback facilities (i.e. **api_logging** has been set to FALSE) deleted ENTITYs will still be posted on the current BULLETIN_BOARD. The only way to really free up the memory is to create a DELTA_STATE and call **api_delete_ds** (see Chapter 9 and Section C.1).

2.7.5 ACIS's Non-ENTITY Classes

By no means are all the classes defined within ACIS derived from the ENTITY class; there are many others.

Perhaps the most notable of these are the public Kernel Geometry Classes which define the parametric and position-based inquiry methods of different types of curve and surface. Unlike their uppercase namesakes these classes are not derived from the ENTITY base class and so do not have their creation or deletion recorded by BULLETINs. See Section 1.6.

2.8 Component Architecture

ACIS's software architecture groups functionally related APIs and classes into units known as **components**. This modular structure allows programmers to create applications which contain **only** the functions they require. For example a program for viewing ACIS models might require the software components for facetting and rendering models but not those for doing Booleans or offsetting.

There is a component associated with each top level directory in the ACIS directory tree (i.e. cstr, eulr, ga, etc.) The main components of ACIS version 6.0 are listed in Tables 2.1, 2.2 and 2.3. Because almost every component uses functions found in other components there is a network of dependency relationships between them.

2.8 Component Architecture

Table 2.1 Main ACIS Components (a-h).

Directory	Name	Description
abl	Advanced Blending	Optional blending operations
ag	AG Spline	Provides spline API
ar	Advanced Rendering	Optional rendering operations
base	Base	Provides functionality common to all components
blnd	Blending	APIs for standard blending operations
bool	Boolean	APIs for unite, intersect, and subtract operations
br	Basic Rendering	APIs for default rendering
catia	CATIA Translator	Optional component for reading CATIA files
clr	Clearance	APIs to determine the minimum distance between BODYs or FACEs
covr	Covering	APIs for covering wires and sheets
cstr	Constructors	APIs for basic topology construction and analysis
ct	Cellular Topology	APIs for cellular models
ds	Deformable Modeling	Optional component for minimum energy based modelling of splines
eulr	Euler Operations	APIs to expand, flatten, separate, and combine LUMPs
fct	Faceter	Generate faceted (polygonal) representation
ga	Generic Attributes	APIs for adding attributes to ENTITYs
gi	Graphic Interaction	APIs for GUIs
gl	OpenGL	APIs for rendering using OpenGL
heal	Healing	Optional component for fixing flaws in models that have been imported from other systems

Table 2.2 Main ACIS Components (i-r).

Directory	Name	Description
iges	IGES Translator	Optional component for reading and writing IGES files
ihl	Interactive Hidden Line	APIs for creating facetted hidden line views
intr	Intersectors	APIs for calculating ENTITY intersections
kern	Kernel	Provide definitions of basic classes (i.e. ENTITY, geometry, maths etc.)
law	Laws	Supports user defined mathematical functions
lop	Local Ops	Optional component for editing and tweaking
mesh	Meshing	Optional component for managing networks of polygonal elements
ofst	Offsetting	APIs for WIRE and FACE offsetting
part	Part Management	APIs for managing collections of ENTITYs
phl	Precise Hidden Line	Optional component for performing hidden line removal
pid	Persistent ID	Allows identifiers to persist across saves
proe	Pro/E Translator	Optional component for translating Pro/E files
rbase	Rendering Base	Supports a common interface to all renderers
rbi	Repair Body Intersections	Optional component for repairing self-intersections
rem	Remove Faces	Optional component for removing FACEs

For example the **boolean** component (bool) uses functions found in both the **euler** (eulr) and **constructor** (cstr) components which in turn use functions found in the **intersectors** (intr) component.

These relationships give rise to the network (i.e. graph) of dependency relationships shown in Figure 2.3[3].

---

[3]For the sake of clarity, the figure omits some of the smaller components.

2.8 Component Architecture

Table 2.3   Main ACIS Components (s-x).

Directory	Name	Description
sbool	Selective Booleans	APIs for selective Boolean operations
scm	Scheme Interpreter	Scheme Interpreter and extensions
shl	Shelling	Optional APIs for creating shelled (hollow) BODYs
skin	Advanced Surfacing	Optional APIs for fitting a surface through a set of curves
step	STEP Translator	Optional translator for STEP files
swp	Sweeping	APIs for sweeping profiles
trans	Translator Utility	Optional component that provides functions common to all translators
vda	VDA-FS Translator	Optional translator for VDA-FS files
vm	VisMan	APIs for the Visual display and manipulation of models
warp	Space Warping	APIs for warping (twisting, bending) ENTITYs
xgeom	Translation Geometry	Optional component providing geometric translation functions common to translators

2.8.1 Using components

Using the APIs, classes or functions defined within a particular component requires only three things:

1. Inclusion of the relevant component header files. For example in the program **entity_contents.cxx**:

    #include "constrct/kernapi/api/cstrapi.hxx"

    declares the API functions associated with the **construction** component (where **constrct** is a sub-directory of the top-level directory **cstr**). While:

    #include "kernel/kernapi/api/kernapi.hxx"

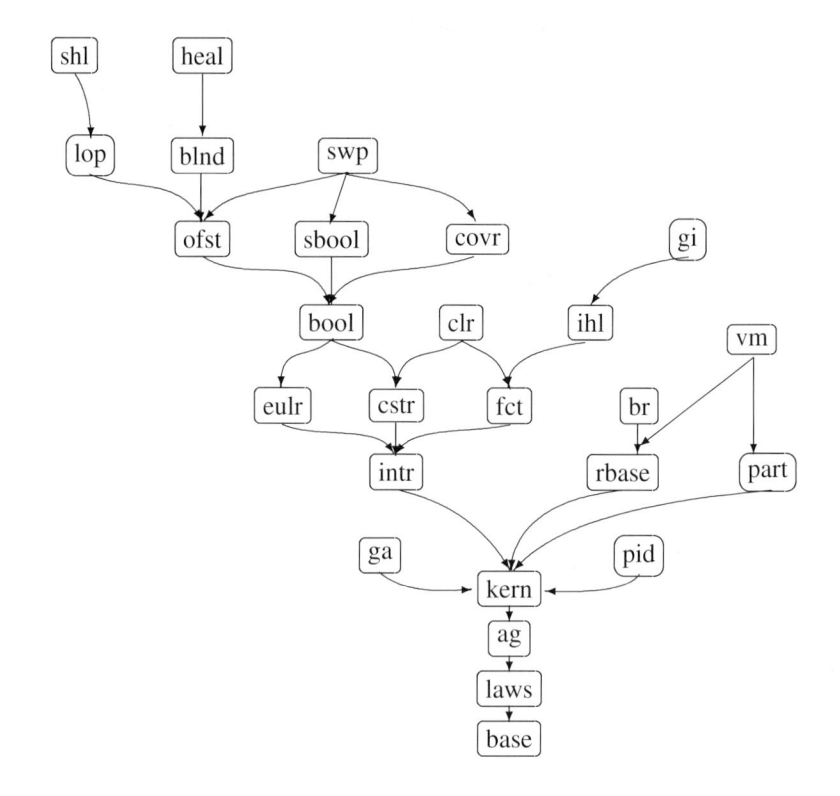

Figure 2.3: Dependencies of main Components.

declares the **kernel** components API (where **kernel** is a sub-directory of top level directory **kern**).

2. Initialization of each top level component. Each component will automatically initialize those components below it in the dependency hierarchy. Consequently in the program **entity_contents.cxx** only the **construction** component is explicitly initialized with a call to:

   api_initialize_constructors();

   The other components below it in Figure 2.3 (i.e. intr, laws, base, ag and kernel) do not need to be explicitly initialized.

3. Termination of each top level component at the end of the program. As with initialization each component will automatically terminate those components below it in the dependency hierarchy. Termination is done with an API call such as:

api_terminate_constructors();

One important point to note about components is that if you forget to initialize them your program will still compile and link without any warning! You can then execute your code and spend some time trying figure out why the program does not work as expected!

2.9 Programming Interfaces

There are two distinct levels of access to ACIS:

- The **direct interface**: which uses the public member functions of the classes directly.

- The **API** (Application Procedural Interface): which uses a library of high-level functions to create, manipulate and enquire about ACIS objects. The API functions also support error checking and logging (for rollback) in a uniform manner. For example every API function returns an **outcome** object which contains a success, or failure, flag, roll-back data and any error messages.

The following code fragments outline the differences between the two methods. For example, a model of a sphere can be created through the API interface in the following way:

```
// Create a ball by calling api_make_sphere
double radius = 50;
BODY* ball = NULL;
outcome check = api_make_sphere(radius, ball);
```

This creates an instance of a BODY object referred to by a pointer called ball. Access to the data structure inside this model can be obtained by the **direct interface** which uses pointers returned by public member functions. For example, the BODY class defines a member function for returning a pointer to the first of a list of LUMP objects in the following way:

```
// Returns a pointer to the beginning of the list of bounding lumps of a body
public: LUMP* BODY::lump () const;
```

This syntax declares **lump** to be a function which returns a pointer to a LUMP object. Notice that the function *itself* is of type **const**. Recall that this means it can make no change to any part of the data held in the LUMP class; only return a reference to some part of it. It could be used in the following way to get a reference to the first (and only) LUMP object in the ball:

```
// Get a LUMP object from BODY called ball
double radius = 50;
BODY* ball;
outcome check = api_make_sphere( radius , ball );

LUMP *flump = ball->lump();
```

Generally speaking creation, destruction and modification of ACIS ENTITYs is carried out with API functions. The direct interface is used mainly to "read" the model and provide arguments for the API functions.

2.10 Exercises

1. Learn C++.

2. Write a **position** class and compare it to the ACIS class of the same name.

3. Why are pointers to ENTITYs not set to NULL after **api_del_entity** has been called?

4. Describe the differences between the ACIS classes **PLANE** and **plane**.

Scheme   3

This chapter introduces the language in which ACIS's high-level, rapid prototyping interface is written, **Scheme**. Despite their many advantages C++ programs are not quickly written; header files have to be included, components initialized, complex syntax debugged and so on. In contrast the Scheme AIDE allows programs to be created interactively at great speed. So although it does not rival C++ as a software engineering tool, it does provide a vehicle for quickly trying out ideas before serious development gets under way. Scheme is a dialect of Lisp and if C++ is one of the richest programming languages then Scheme is refreshingly poor!

There are at least three things that will immediately endear Scheme to anyone struggling with C++:

- It does not have any pointers.

- It does not have any header files.

- It is interactive and can be programmed through an **interpreter** (rather like Basic or Prolog).

The Scheme interpreter is written in C++ and drives the ACIS modeler through **extensions** to the standard language. Each of these Scheme extensions is also written in C++ and typically supports high-level functionality such as rendering and part management.

3.1 Scheme Basics

Unlike C++, the things you absolutely have to know in order to write Scheme programs can be written on the back of a postage stamp.

- The program is driven by making interactive Scheme procedure calls.

- Every procedure call is enclosed, with its arguments, in parentheses.

- The things inside the parentheses are called *expressions*.
- Expressions have a basic **prefix** format in which the procedure name is followed by any number of arguments.

    (procedure_name <arguments>)

- Comments start with a ;

About starting the Scheme AIDE

1. Locate the Scheme AIDE executable (called something like **acis3dt**) below the **bin** subdirectory of the main ACIS directory.

2. Invoke it and specify a **load-path** (using "-p" startup option) in which the default startup file, called **acisinit.scm**, can be found. In Windows this can be done from the **Run** option on the **Start** menu with a command something like:

    C:/Acis/bin/NT/acis3dt.exe -p C:/Acis/scm/examples/

3. The acisinit.scm file can be used to load in useful utilities for view rotation/zoom and creation of default views. The init file provided in the scm/examples sub-directory loads programs that allow left mouse button rotate and right mouse button zoom in any display windows created.

If you are using a Windows PC the **3DScheme** environment from Schemers Inc. (www.schemers.com) cannot be too highly recommended. In addition to automatic indentation, bracket matching and highlighting of keywords it has a truly fabulous help system that provides many insights into ACIS functionality. Scheme programming is an almost futile exercise without it and the cost is very modest!

3.2 Expressions

Because Scheme is interactive the fastest way of trying it out is simply to type Scheme expressions into the Scheme AIDE's interpreter. In the following:

    acis>

represents the interpreter's prompt. Each line of input should be terminated with a return. The following examples evaluate arithmetic expressions:

3.2 Expressions

```
acis>(* 45 68)
3060
acis>(* 45 68 77)
235620
```

The *prefix notation* used by Scheme not only allows the number of arguments to be arbitrary but also makes the *nesting* of expressions straightforward.

```
acis>(* (+ 40 5) (* 4 17))
3060
```

Theoretically there is no limit to the depth to which Scheme expressions can be nested, if you can keep track of the closing parentheses!

Both the above are examples of expressions that apply built-in procedures for doing multiplication and addition. The ACIS extensions to Scheme mean that a large number of geometric modeling procedures also appear to be built-in and are available through this simple, clean interface. For example one can type:

```
acis>(solid:block (position 0 0 0) (position 20 20 20))
#[entity 1 0]
```

to create a cube using the procedures **position** and **solid:block** (the `#[entity 1 0]` is the default *name* given to the resulting body). Behind the scenes a much more verbose bit of C++ code, shown here in **Sls_scm.cxx**, is taking this Scheme command and using it to run **api_solid_block**.

Sls_scm.cxx

```cpp
//Scheme Extension for make solid:block
ScmObject  P_Solid_Block(ScmObject p1, ScmObject p2)
{
   ENTER_FUNCTION("P_Solid_Block");

   // get the block diagonal corners
     position pt1 = get_Scm_Position(p1);
     position pt2 = get_Scm_Position(p2);

   // make the block
     BODY *block = NULL;
     start_entity_creation();
     outcome result =  api_solid_block(pt1, pt2, block);
     end_entity_creation(block, result);
```

Note that only users wishing to extend the Scheme AIDE will need to write this sort of code.

```
        return make_Scm_Entity(block);
}

SCM_PROC(2, 2, "solid:block", P_Solid_Block);
```

The C++ macros used to create Scheme extensions are discussed in Section 13.3.

3.3 External Representations

Each object created by a Scheme expression has an **internal** (i.e. C++) and an **external** (i.e. textual) representation. Although superficial the external representation plays an important user interface role. For example, notice in the programs on page 50, that if the results of evaluating an expression were not required as an argument for another expression (i.e. it was not nested) the Scheme interpreter *automatically* printed out the result's external representation. In the case of arithmetical calculations this external representation is simply a number.

ACIS objects also have an external representation which is printed out when nothing better can be found to do with them. The generic format is:

```
#[type_of_object <arguments>]
```

For example, consider the command:

>; *Creates a position object*
>acis> (**position** 10 -20 56)
>#[position 10 -20 56]

is the external representation of a **position** object with coordinates $x = 10$, $y = -20$ and $z = 56$ is output as shown above.

Objects derived from the ACIS ENTITY class have a similar external representation, the only complication being that they are collected together into units known as **PART**s. A default PART (numbered 0) is created at start-up and manifests itself in the following way:

>; *Create a solid block*
>acis> (**solid:block** (**position** 0 0 0) (**position** 10 10 10))
>#[entity 1 0]

Here, the **solid:block** procedure creates an ACIS ENTITY, the external representation of which has two arguments: the ENTITY number (1) and the part number (0). The ENTITY number can be used in subsequent expressions, for example:

>; *Delete an entity*
>acis> (**entity:delete** (**entity** 1))

3.4 Defining Variables

The values calculated by procedures, like "+" and **solid:block**, are not going to be very useful unless they can be named, or later identified, in some way. Scheme allows both variables and procedures to be labeled, or named, using a built-in operator called **define**. The **define** command can associate a name (i.e. a symbol) with *any* Scheme expression, and is commonly used with the syntax:

>   (**define** <label> <expression>)

and applied in the following way:

>   acis>(**define** prod (* 45 68))
>   prod
>   acis>prod
>   3060

Similarly

>   acis>(**define** p1 (**position** 10 -20 56))
>   p1

returns a **position** object that can be referred to as p1. Notice that the external representation is no longer displayed.

3.5 Defining Functions

Functions can also be created using the **define** command.

>   ; *Define the procedure "square"*
>   acis>(**define** (square x) (* x x))
>   square
>   acis>(square 5)
>   25

The arguments of functions, such as x in the above example, have what is called **local scoping**. In other words they are defined only in the body of their own function.

However, the symbols (or variables) created with the **define** command have global scope. At this point it is worth noting that one can drive the Scheme AIDE quite a long way using only the **define** command. Up to a point (around thirty to fifty lines of code) this will work fine, but eventually all the usual programming problems of variable name clashes, readability and peer group pressure will probably force you to do something more complex.

Consequently the rest of this chapter covers the more sophisticated Scheme constructions for defining conditional statements, recursion, lists, local variables, lambda functions, begin and for-each statements.

3.6 Conditional Statements

Conditional, or decision, expressions can be created using the keywords **if** and **cond**.

The **if** expression is really just a special case of the more general **cond** (i.e. conditional) construction.

```
(cond  (<condition-1> <consequence-1>)
       (<condition-2> <consequence-2>)
       (<condition-3> <consequence-3>)
       (else <consequence-4>)
)
```

Each condition is a Scheme expression which is evaluated (in order of appearance) to give a **true** or **false** (i.e. #t or #f) result. If a condition evaluates to #t then its associated consequence, also a Scheme expression, is evaluated and its value returned. If there are no valid conditions and no **else** clause then the value returned by the **cond** expression is unspecified. Consequently an else clause should always be present.

```
; Procedure for printing out an edge's type
(define (tell-me-edge-type edge)
  (cond
    ((edge:circular? edge)   (print "Circular-edge"))
    ((edge:elliptical? edge)(print "Elliptical-edge"))
    ((edge:linear? edge)     (print "Linear-edge"))
    ((edge:spline? edge)     (print "Spline-edge")))
    (else (print "Is this an edge ?")))
```

Note the use of the **else** clause in the above example to provide a default result when none of the other conditions have proved **true**. Compared to C++ this program looks strangely un-typed! Where are the checks to ensure that the variable presented as an argument to **tell-me-edge-type** really is an **edge**? Although unseen, type checking is done on the C++ side of the Scheme extension and an error is flagged if the wrong data types are used.

There is also a restricted form of conditional called **if** which can be used when there are only two possible results from the expression. It has the general form:

```
(if <condition> <consequence> <alternative>)
```

For example:

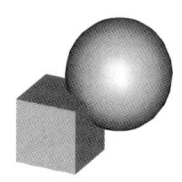

```
; Check if a cube and sphere overlap
(define cube (solid:block (position -30 -30 -30) (position 0 0 0)))
(define ball (solid:sphere (position 100 100 100) 25))
(define intersection (bool:intersect ball cube))
(if (solid? intersection)
  (print "They overlap")
    (print "They don't overlap"))
```

3.7 Recursion and Lists

There are also the usual operators available for forming logical compositions of expressions (i.e. **and**, **or**, **not**). For instance, in the previous example we could explicitly check, rather verbosely, if the thing whose type was being determined, is an edge at all!

```
; Procedure for printing out an edge's type
(define (tell-me-edge-type edge) (cond
    ((and (edge? edge) (edge:circular? edge))
                    (print "Circular-edge"))
    ((and (edge? edge) (edge:elliptical? edge))
                    (print "Elliptical-edge"))
    ((and (edge? edge) (edge:linear? edge))
                    (print "Linear-edge"))
    ((and (edge? edge) (edge:spline? edge))
                    (print "Spline-edge"))
    (else (print "Error: In tell-me-edge-type"))))
```

Although only a handful of Scheme keywords have been mentioned, they already allow small programs to be composed and either typed straight into the interpreter or loaded from a file using the `(load "filename")` command.

3.7 Recursion and Lists

The program **edgetypes.scm** introduces the Scheme **list** type and then uses a **recursive** function called **et** to call **tell-me-edge-type** (defined above) for each edge in the list.

edgetypes.scm

```
; Create a cube and a sphere then unite them to form a body called union
(define cube  (solid:block (position -30 -30 -30) (position 0 0 0)))
(define ball  (solid:sphere (position 5 5 5) 25))
(define union (bool:unite cube ball))

; Create a list of the edges in union, find its length, start the recursion
(define (et body)
    (define eelist (entity:edges body))
    (define list-length (length eelist))
    (work-through eelist list-length) )

; Definition of work-through
(define ( work-through alist index)
  (define edge (list-ref alist (- index 1)))
  (tell-me-edge-type edge)
  (if (<= 0 (- index 1))
    (work-through alist (- index 1))
    (print "No more edges")))

(et union)   ; Run the program
```

The program runs as follows:
```
(et union)
"Linear-edge"
"Linear-edge"
"Linear-edge"
"Linear-edge"
"Linear-edge"
"Linear-edge"
"Linear-edge"
"Linear-edge"
"Linear-edge"
"Linear-edge"
"Linear-edge"
"Linear-edge"
"Circular-edge"
"Circular-edge"
"Circular-edge"
"Circular-edge"
"No more edges"
```

A number of the Scheme AIDE's extensions return **lists**. For example in the **edgetypes.scm** program the expression:

(**define** eelist (**entity:edges** body))

uses the **entity:edges** extension to return a list, called eelist, of the edges associated with the body. Having created eelist its size can be determined using the **length** function and individual members accessed via a procedure called **list-ref**.

Scheme has two other important functions for the manipulation of lists known, for historical reasons, as **car** and **cdr**. Given a list as an argument, **car** will return the first thing in the list (i.e. the head of the list) and **cdr** the rest of the list (i.e. the tail).[1] Thus given a list of scheme objects, say edges called e1–e4 :

> Note: Lists are preceded by the special symbol ' and must end with the **empty list**, represented by (). The () does not have to be explicitly declared.

```
; Get the head and tail of a list
acis>(define e-list '(e1 e2 e3 e4 ()))
e-list
acis>(print (car e-list))
e1
acis>(print (cdr e-list))
(e2 e3 e4 ())
```

The **car** and **cdr** functions allow lists to be processed without first determining their length.

carcdr.scm

```
;Top level function to generate list of edges
(define (edge-types body)
   (define elist (entity:edges body))
   (work-through elist) )

;Recursively works through the list using car and cdr
(define (work-through alist)
   (define edge (car alist))      ; Get the head of the list
   (tell-me-edge-type edge)       ; Find out its type
   (if (null? (cdr alist))        ; If the list is empty
      (print "No more edges")     ; print message
      (work-through (cdr alist))))) ; Else make a recursive call with tail of list
```

> The empty list, (), is the only argument that will make the function **null?** return **true** (#t).

This produces the same result as before, only reversed.

---

[1] Internally lists are represented by nested expressions of Scheme data objects each known as a **pair**. Pairs have two parts, referred to as **car** and **cdr** respectively.

3.8 For-each

All the previous versions of the edge-checking program can be made to seem rather verbose by using the built-in operator **for-each** which has the following syntax (for single argument functions):

    (**for-each** `procedure list1`)

The function repeatedly calls the **procedure** with each element of the list being supplied as an argument in turn. It is possible to use the same command for n argument functions, if n lists of equal length are also supplied.

feach.scm

    ; Print out each edge type in a BODY
    (**define** (edge-types body)
      (**define** elist (**entity:edges** body))
      (**for-each** tell-me-edge-type elist))

In fact there is no real need for a procedure at all; one could simply type:

    ; Print out each edge type in union
    (**for-each** tell-me-edge-type (**entity:edges** union))

The built-in operator **map** provides a facility similar to **for-each** but returns the result (of applying a procedure to every member of a list) in the form of a list. The value returned by for-each is unspecified.

3.9 Defining Local Variables

The built-in function **let** allows the naming and evaluation of variables that can only be referenced within its **body** (i.e. between the (...)).

    (**let** ((variable-name-1)(expression-1)
        (variable-name-2)(expression-2)
        (variable-name-3)(expression-3))
        body-expressions)

When the **let** structure is evaluated each variable name is associated with the value of the corresponding expression. Using **let** we can clarify our running example by making the definition and scope of the variables explicit:

let creates variables whose scope is local to the let body.

letvar.scm (a)

```
; A rewrite of carcdr.scm to use let
(define (work-through alist)
 (let ((edge (car alist))      ; Variable edge = head of the list
       (tail (cdr alist)))     ; Variable tail = the rest of the list
  (tell-me-edge-type edge)     ; Find out its type
  (if (null? tail)             ; If the list is empty
   (print "No more edges")     ; print message
   (work-through tail))))      ; Else call function with tail of list.
```

One thing anyone using **let** should be aware of is that there is no guarantee that the expressions defining variables in a **let** statement will be evaluated *in the order written in the code*. Sometimes this does not matter; as for example in **letvar.scm** where the program's behavior will be the same regardless of whether edge or tail is evaluated first.

However when it does matter, variables should be created using the **let*** operator. Consider the program **count-faces**, shown below, which displays the number of faces on a body. Here the order in which the variables are defined is critical because you cannot count the number of things in a list before it is created!

letvar.scm (b)

```
;Count the number of faces on a given body
(define (count-faces body)
 ( let*(
       (face-list (entity:faces body)) ;1st variable : a list of faces
       (number (length face-list)))    ;2nd variable : length of list
  (display "Number of faces= ")        ;Let* body expressions
  (display number)
  (newline)))
```

There is a third form of **let**, known as **letrec**, which supports the creation of variables local to a recursion. In other words, they can be accessed from any depth of recursion.

3.10 Set!

The value of a variable can be updated using the **set!** operator.

```
; set! used to change a defined value
(define val 67)
;val = 67
(set! val 77)
;val = 77
```

The **set!** operator can also be used to extend lists. For example, the value of the built-in variable load-path is updated by saying:

```
; Update the load-path list using cons to append a string to it
(set! load-path (cons "C:/Spatial6.0/scm/examples"  load-path))
```

3.11 Lambda

In the same way that **let** makes the creation and use of *local* variables explicit, the keyword **lambda** can be used to create unnamed functions. The general form of a **lambda** expression is

```
( lambda (function-arguments) (function-body))
```

Notice that it is not necessary to specify a function name; as is done with **define**. Thus **lambda** can be used to define nameless (local) functions. Using **lambda** we can rewrite the edge-type function one last time.

letvar.scm

```
;Count the number of faces on a given body
(define c-faces
  (lambda (body)
    ( let*(
        (face-list (entity:faces body))   ;1st variable : a list of faces
        (number (length face-list)))      ;2nd variable : length of list
      (display "Number of faces= ")       ;Let* body expressions
      (display number)
      (newline))))
```

Many people find that mentally replacing the word **lambda** with **function** makes Scheme much easier to read.

3.12 Begin

In the same way that **let** says nothing about the order in which variable expressions are to be evaluated, it also says nothing about how the expressions in the **body** of the expression are to be evaluated. Sometimes this is okay; in the **tell-me-edge-type** program, for example, it mattered not a jot whether the linear edge test happened before or after the spline one. On the other hand it always matters when a program is printing something out. To force Scheme to carry out your instructions in the order you wrote them, use the **begin** keyword. The general form of a **begin** expression is:

```
(begin expressions or defines )
```

The expressions, or definitions, given as arguments to **begin** will be evaluated in the order in which they are written.

f_detail.scm

```
;Get a list of the faces
(define faces-no
  (lambda (body)
    ( let*(
        (face-list (entity:faces body))
        (number (length face-list)))
      (begin
      (display "Number of faces= ")
      (display number)
      (newline)
      number))))
```

The last expression in a **begin** statement is the value returned by the procedure. In the example **f_detail.scm** the procedure face-no evaluates to (i.e. returns) the number of faces. So one can say:

```
; Assign number of FACEs to fno
acis> (define fno (faces-no (solid:sphere (position 0 0 0) 10)))
```

To assign a value of 1 to the the variable fno and print it out.

3.13 Do

Do-expressions are a means of carrying out iterations in Scheme. They have the general syntax:

```
(do ((variable init-expression update-expression)
      ...)
     (test-expression exit-expression ...)
   continue-expressions ...)
```

The program **basepos.scm** shows how a **do**-expression can be used to increment the parameter value of a position on an edge.

basepos.scm

```
; Print out 10 different positions round the base of a cone
(define basepos
  (lambda ()
    (let*
      ((cone (solid:cone (position 0 0 0)
                         (position 0 0 30)
                         15 0))
       (edges (entity:edges cone))
       (base (car edges)))
      (do ((param 0 (+ param 0.1)))
          ((> param 1) 'finish)
        (display (curve:eval-pos (curve:from-edge base) param))
        (newline)))))
```

The program outputs the following points round the base of the cone:

```
acis> (basepos)
#[position 15 0 0]
#[position 12.1352549156242 -8.8167787843871 0]
#[position 4.63525491562421 -14.2658477444273 0]
#[position -4.63525491562421 -14.2658477444273 0]
#[position -12.1352549156242 -8.8167787843871 0]
#[position -15 0 0]
#[position -12.1352549156242 8.8167787843871 0]
#[position -4.63525491562421 14.2658477444273 0]
#[position 4.63525491562421 14.2658477444273 0]
#[position 12.1352549156243 8.81677878438711 0]
#[position 15 0 0]
finish
acis>
```

3.14 Features of Scheme Programming

1. Things need not happen in the order you have typed them unless you use **let*** and **begin**.

2. Too many parentheses can be just as bad as too few. For example:

```
acis> (newline)   ;OK
acis> ((newline)) ; Wrong
top-level: application of non-procedure:
```

3. Since errors in coding never result in a line-number, procedures should be very small to make debugging possible.

4. BODYs created in, say, a **let** expression are not deleted when they move out of scope. So, although Scheme might lose the reference to them, they still exist and have memory allocated to them.

3.15 Exercises

1. Learn Scheme!

2. Visit the Schemers Inc. website and download the free trial version of 3DScheme at www.schemers.com.

3. Using **tell-me-edge-type**, as a template write and test a procedure called **tell-me-face-type**.

Basics 4

This chapter contains examples which establish how the most fundamental modeling tasks are carried out. Specifically the programs demonstrate:

- The creation of "primitive" shapes (e.g. blocks, spheres, cones etc.).
- The saving and restoration of ENTITYs to disk.
- The creation of a new BODY through set operations (i.e. addition, subtraction and intersection) on existing ones.
- The analysis of a BODY's mass properties (i.e. volume and surface area).
- The movement of BODYs through the application of transformations.
- The modification of a BODY's shape by:
    - the extrusion (i.e. sweeping) of existing FACEs,
    - the rounding (i.e. blending) of existing EDGEs.
- The management of objects and coordinate systems.

The examples are coded in both C++ and Scheme and in addition to the basic functionality of the various routines used the discussion also touches on:

- Error checking facilities in the API.
- The use of ENTITY_LISTs.
- The ACIS "sat" file format.

As in previous chapters, each program has its keywords, functions and class names highlighted in **bold** and comments emphasized in *italic* lettering.

The layout of the C++ code in the examples has a structure, shown below, which is typical of all ACIS programs:

General layout of C++ code

```
#include   <System Header files>  // Declare system functions
#include   "ACIS Header files"    // Declare ACIS APIs and classes

void main()
{
   api_start_modeller(0)   // Create internal data-structures
   api_initialize_"components"
   //API and direct interface calls
   api_terminate_"components"
   api_stop_modeller()     // Remove internal data-structures
}
```

This general structure can be clearly seen in the first example program, **block.cxx**, which uses an API function from the "constructors" component to create a simple cuboid. Instructions for compiling block.cxx with Visual C++ (6.0) can be found in Appendix A (page 357).

About APIs
> The term API and the idea of encapsulating modeling functions inside a software wrapper that supports error checking can be traced back to at least the early 1980s when CAM-I, an international consortium of CAD/CAM companies, published a specification for solid modeling APIs. Attempts to specify a non-proprietary interface to modeling software continue to this day with the recent publication of a *representation*-independent API called Djinn (see "Djinn: A Geometric Interface for Solid Modelling", www.inge.com).

4.1 Making Primitives with API Functions

ACIS offers a number of different mechanisms, summarized in Table 4.1 on page 69, for creating the so-called modeling primitives. Seven primitive shapes are available: four with FACE geometries representing different surfaces of revolution (sphere, cone, torus and cylinder) and three polyhedra (i.e. cuboid, pyramid and prism).

4.1 Making Primitives with API Functions

The program **block.cxx** creates a block and writes its data structure to file using a debug utility.

block.cxx

```
// This program starts the modeller, creates a block,
// prints out its data-structure and stops the modeller
#include <stdio.h>
#include "constrct/kernapi/api/cstrapi.hxx"  // Declares construction APIs
#include "kernel/kernapi/api/kernapi.hxx"    // Declares start and stop APIs
#include "kernel/kerndata/top/body.hxx"      // Declares BODY class
#include "kernel/kerndata/data/debug.hxx"    // Declares debug routines

void main()
{
    api_start_modeller(0);

    api_initialize_constructors();

    BODY* block;

    api_make_cuboid(100,50,200, block);

    FILE* output = fopen("cube.dbg","w");

    debug_size(block, output);

    fclose(output);

    api_terminate_constructors();

    api_stop_modeller();
}
```

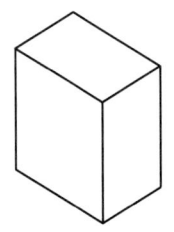

Creating a block is almost the solid modeling equivalent of writing "*Hello, World*" on the screen of your terminal. Compiling and running even this simple program establishes several useful facts (i.e. the machine you're using has a compiler on it, and you know how to edit a file). It also gives you a warm glow because you've created something that can be seen to work! Though having said that, the way in which this first program can be "seen to work" is rather low key. The behavior of the program is now described line by line:

Declare header files: The program starts by including a number of header files which contain various function prototypes and class definitions, the absence of which would cause the compiler to complain of unknown types. The ACIS documentation for each API function and class contains a list of the header files required to define it and its arguments.

Note that some compilers require only the header file name, not the entire path. For example, CodeWarrior on an Apple Macintosh (with correctly set search paths) simply requires:
#include "body.hxx"

Start modeler: This creates a number of supporting data structures and *must* be called before any other API functions.

Initialize components: This function initializes the "highest" component required by the program (see Figure 2.3). Lower components required by the library are initialized automatically.

Declare pointer: A variable called block is declared to be a pointer to a **BODY** object.

Call API function: The routine **api_make_cuboid** creates an ACIS data-structure representing a 100 by 50 by 200 block.

Open file: for writing.

Call debug utility function: The function **debug_size** causes a summary of the block's data structure, and memory taken by each part of it, to be written to file.

Close file:

Terminate components: Frees resources used by the components.

Stop modeler: This frees the memory used by its data structures.

The only way this program can be seen to work is through the file created by the non-API function **debug_size**(ENTITY*, FILE*). This is an example of one of ACIS's generic functions which takes an ENTITY argument. Because this function accepts an ENTITY argument it can be used to debug *any* object derived from this base class. The file generated by this function (called **cube.dbg**) should contain lines similar to the following:

cube.dbg

```
 1 body record,       32 bytes
 1 lump record,       32 bytes
 1 shell record,      40 bytes
 6 face records,     264 bytes
 6 loop records,     192 bytes
 6 surface records,  960 bytes
24 coedge records,  1056 bytes
12 edge records,     864 bytes
 8 vertex records,   192 bytes
12 curve records,   1344 bytes
 8 point records,    384 bytes
Total storage 5360 bytes
```

From this it is apparent that at least 85 objects have been created by the API function. Each of these has required memory to be allocated and pointers set. This is no mean task as we'll see later (Section 5.5) when a primitive is created "by hand" via the direct interface rather than the API.

4.1 Making Primitives with API Functions

Doing it Properly

The **block.cxx** program is actually very sloppy! What if something went wrong, say the computer didn't have sufficient memory to get the modeler started? All the API functions support this kind of error checking by returning an **outcome** object. The use of the outcome class is discussed more fully in Chapter 14, but at the simplest level a program can check for errors using the member function **ok** and an **if** statement. The code of **ball.cxx** illustrates this:

ball.cxx

```
// This program starts the modeler, creates a sphere and
// prints out its data structure before stopping the modeler.
// The result of each API call is checked for errors.
#include <stdio.h>                            // Declares print functions
#include <stdlib.h>                           // Declares exit function
#include "constrct/kernapi/api/cstrapi.hxx"   // Construction APIs
#include "kernel/kernapi/api/api.hxx"         // Declares outcome class
#include "kernel/kernapi/api/kernapi.hxx"     // API function prototypes
#include "kernel/kerndata/top/body.hxx"       // Topological Class BODY
#include "kernel/kerndata/data/debug.hxx"     // Debug routines

void main()
{
   outcome res = api_start_modeller(0);
   if(!res.ok()){
     printf("Error Starting Modeller\n");
     exit(1);
   }

   res = api_initialize_constructors(); // Initializes constructor library
   if(!res.ok()){
     printf("Error in api_initialize_constructors\n");
     exit(1);
   }

   BODY* ball;

   res = api_make_sphere(50,ball);
   if(!res.ok()){
     printf("Error Making Sphere\n");
     exit(1);
   }

   FILE* output = fopen("ball.dbg","w");

   debug_size(( ENTITY*)ball,output);
```

The function **exit(1)** causes the program to terminate with the argument 1 becoming its return value. Notice that **res** was only declared once.

```
        fclose(output);

        res = api_terminate_constructors();
        if(!res.ok()){
          printf("Error in api_terminate_constructors\n");
          exit(1);
        }

        res = api_stop_modeller();
        if(!res.ok()){
          printf("Error Stopping Modeller\n");
          exit(1);
        }
    }
```

Although direct use of the outcome class's member function **ok()** to return a Boolean (True/False) value is effective at catching errors, the **ball.cxx**'s code could hardly be described as succinct. However, Chapter 14 gives details of a macro that can be used to achieve the same results in a less intrusive way. Generally no error checking is shown in the programs described in this book. This is *not good practice* and is done only for the sake of brevity.

Primitive Creation Alternatives

Table 4.1 summarizes the APIs available to create simple shapes. The primitive creation APIs prefaced by "make", like **api_make_cuboid**, generate solids centered at the origin. In contrast the "solid" family of creation APIs allows construction of BODYs in any location. The routine **api_solid_cylinder_cone** is typical of these procedures in that it takes two **positions** as arguments in contrast to the explicit dimensions as required by, say, **api_make_frustrum**. In practice the positions supplied to a routine like **api_solid_block** usually arise from mouse picks in the system user interface.

The following program, **cone.cxx**, demonstrates the use of a "solid" construction function by creating a cone and printing out its surface area.

4.1 Making Primitives with API Functions

Table 4.1 Primitive creation functions.

Dimension Driven	Position Driven	Scheme extension
api_make_cuboid	api_solid_block	solid:block
api_make_frustum	api_solid_cylinder_cone	solid:cone solid:cylinder
api_make_prism	-	-
api_make_pyramid	-	-
api_make_sphere	api_solid_sphere	solid:sphere
api_make_torus	api_solid_torus	solid:torus

cone.cxx

```
// This program creates a cone and calculates its surface area
#include <stdio.h>                          // Declares print functions
#include "kernel/acis.hxx"                  // Declares system-wide information
#include "constrct/kernapi/api/cstrapi.hxx" // Construction APIs
#include "kernel/kernapi/api/kernapi.hxx"   // Declares kernel API
#include "kerndata/top/body.hxx"            // Topological Class BODY
#include "baseutil/vector/position.hxx"     // Position Class

void main()
{
    api_start_modeller(0);

    // Initializes the constructor library other dependent
    // components are automatically initialized
    api_initialize_constructors();

    BODY* hat;

    api_solid_cylinder_cone( position(40, 40, 40),   // Top position
                    position(100, 100, 100),          // Bottom position
                    10*M_PI, 20*M_PI,                 // Bottom major/minor radius
                    0,                                // top radius
                    NULL, hat);

    double area = 0;
    double accuracy = 0; // estimate of relative accuracy achieved
```

Area = 23247mm²

```
    // Calculate hat's surface area with a requested accuracy of 1%
    api_ent_area(( ENTITY*)hat, 0.01, area, accuracy);

    printf("The surface area of the cone is = %f\n", area);

    api_terminate_constructors();
    api_stop_modeller();
}
```

The **cone.cxx** program uses one of ACIS's Utility Classes called **position** to represent the two position vectors (in 3D Cartesian space) supplied as the API function's arguments.

Primitive Creation with Scheme Extensions

Undoubtedly the easiest way of creating a block is to use the AIDE's Scheme interpreter (as discussed in Chapter 3: see the box about starting the Scheme AIDE on page 50). To make a block (and see it), one need only type the following commands:

block.scm

; Make a viewing window and a block
; Note that (view:dl) can be used when (view:gl) is not available

acis>(**view:gl**) *;Make a viewing window.*

acis>(**solid:block** (**position** 0 0 0) (**position** 15 15 15))

acis>#[entity 1 1] *;Entity 1 in Part 0*

Default view of a cube

In the same way as the Scheme interpreter outputs results without prompting, the AIDE will, by default, display newly created objects in any **views** available. Unfortunately because the default view is one that looks directly along the z-axis the resulting picture of the newly created block is simply a square!

The default properties of the view can be easily modified with the **view:set** procedure, which takes up to four arguments known as **eye**, **target**, **up** and **view**.

eye: Specifies a position representing the location of the viewer's eye relative to the origin of the world coordinate frame.

target: Specifies a position representing the location of the center of the window relative to the origin of the world coordinate frame.

up: Specifies a **gvector** which is directed from the bottom to the top of the window.

4.1 Making Primitives with API Functions

view: An optional argument which specifies which **view** the new setting is to be applied to. If this is omitted the last view used (known as the *active view*) will be updated.

So by typing the command

    acis>(**view:set**
        (**position** 200 -400 200)
        (**position** 0 0 0)
        (**gvector** 0 0 1))

the parameters of the active view can be changed and the block seen in all three dimensions.

All this typing of coordinates is hard work, and if you don't care too much about the exact dimensions of the block it is much easier to specify some approximate dimensions using the mouse. The **rbdblock.scm** program uses Scheme AIDE functions for picking positions from a view window and displays a rubberband line during the course of the block's construction. In order to save typing, the program uses some views defined in the file **setview.scm**[1] called **front** and **back**.

A **gvector** represents a displacement vector in 3D Cartesian space. The name is chosen to distinguish it from Scheme's **vector** data structure which represents an ordered lists of objects.

rbdblock.scm

```
(load 'setview.scm) ; From scm/examples
(define pickdemo
  (lambda()
    (begin
      (front)       ; Create a Front view
      (print "Pick first corner")
      (define p1 (pick:position (read-event)))
      (rbd:line #t p1) ; Display the block's diagonal
      (iso)         ; Isometric view
      (print "Pick second corner")
      (define p2 (pick:position (read-event)))
      (rbd:clear)
      (solid:block p1 p2))))
```

Load and run the program by typing:
(**load** 'rbdblock.scm)
(**pickdemo**)

Although effective, this procedure only displays a rather plain, 2D rubberband during the block's construction. To see the rubberbanding of the entire block, in 3D, load the file **block.scm** from the **scm/examples** directory. This program allows the interactive creation of a block in a series of three mouse picks.

---

[1] If you are unable to load this file it could be that your load-path is incorrect; type
(set! load-path (cons ".;c:/Spatial6.0/scm/examples" load-path))
(load 'acisinit.scm) ...and try again (obviously the exact path names will vary).

```
(load 'setview.scm)    ; To define some views
(view:gl 0 0 500 500)  ; Create a large window in the corner of the screen
(iso)                  ; Make the view Isometric
(load 'block.scm)      ; Load the rubberbanding block procedure
(block)                ; Create the block with three mouse picks
(load 'rotsph.scm)     ; Load an interactive view changing procedure
(rotsph #t)            ; Switch it on
; Click and drag with your mouse button held down in the window
; to rotate your view of the block.
```

A user should now be able to create a block interactively and rotate the view. The facilities offered by **setview.scm** and **rotsph.scm** are so useful that most people add them to their **acisinit.scm** files so as never to be without them!

4.2 Saving and Restoring Models

As the use of ACIS grows so does the importance of its file format. From the very earliest days of CAD/CAM the benefits of having a common format for the exchange of data have been clear. If components modeled on one CAD system can be e-mailed to suppliers (using different CAD systems) for quotations, or analysis, without the need for manual re-input then everybody wins. Customers aren't locked into a single supplier and every software vendor who adopts the standard format has a ready market.

Data exchange was first made a reality by standards like IGES which although very verbose works pretty well when moving data between systems which have similar numerical precision. Attempts to provide IGES-like transfer of things other than 2D and 3D CAD data, such as FE meshes, resulted in the STEP standard. Standards take a long, long time to specify and even longer to be widely adopted. During the many years of STEP's gestation the ACIS file format has started to show signs of becoming a *de facto* standard for the exchange of 3D product data.

If the number of ACIS based applications continues to grow this could become the single most compelling reason for its use in third party products!

ACIS model files are commonly known as **sat** files[2] and are created using the function **api_save_entity_list**. The following program demonstrates how this is done for a single body. Later in this section both ENTITY_LISTs and the sat file format are examined in more detail.

---

[2] But note that there are actually two file formats associated with ACIS. The Standard ACIS Text (sat) and Standard ACIS Binary (sab).

4.2 Saving and Restoring Models

save.cxx

```cpp
// This program creates a prism and saves it to a file
#include <stdio.h>                         // Declares I/O functions
#include "constrct/kernapi/api/cstrapi.hxx"   // Declares construction API's
#include "kernel/kernapi/api/kernapi.hxx"     // Declares kernel api's
#include "kerndata/top/alltop.hxx"            // Declares BODY class
#include "kerndata/lists/lists.hxx"           // Declares ENTITY_LIST class
#include "kernel/kerndata/savres/fileinfo.hxx" // Declares fileinfo class

void save_ent( char*, ENTITY*); // Function prototype

void main()
{
    api_start_modeller(0);
    api_initialize_constructors();   // Initializes the constructor library

    BODY* pris;

    api_make_prism(100,150,200,7, pris);

    save_ent("save.sat",pris); // Call save routine defined below

    api_terminate_constructors();
    api_stop_modeller();
}

void save_ent( char *filename, ENTITY *ent)
{
    FileInfo info; // Create FileInfo Object
    info.set_product_id("HW-University"); // set info's data
    info.set_units(1.0); // Millimeters

    // Sets header info to be written to sat file
    api_set_file_info(FileId | FileUnits, info);

    FILE *fp = fopen(filename, "w");
    if (fp != NULL) {
        ENTITY_LIST *savelist = new ENTITY_LIST;
        savelist->add(ent);
        api_save_entity_list(fp,TRUE,*savelist); // TRUE for sat, FALSE for sab
        delete savelist;
    } else
        printf("Unable to open file!\n");

    fclose(fp);
}
```

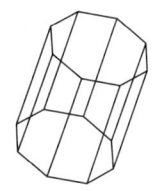

Note how the BODY pointer is automatically cast to an ENTITY pointer by C++. This happens because BODY is a derived class of ENTITY.

Common values for set_units:
-1.0 = unspecified
1.0 = mm
10.0 = cm
1000.0 = m
1000000.0 = km
25.4 = Inches
304.8 = Feet
914.4 = Yards
1609344.0 = Miles

There are five key steps in this program:

1. **Create an ENTITY:** in this case a BODY.
2. **Place the ENTITY in an ENTITY_LIST:** this allows the option of saving multiple ENTITYs.
3. **Creation of the FileInfo object:** to define the sat file's header information.
4. **Call to api_set_file_info:** to set the header information for inclusion in all save files subsequently generated.
5. **Call to api_save_entity_list:** to write an ASCII representation of the ENTITY to disk.

The **FileInfo** object holds all the header information recorded at the start of a SAT file (i.e. the ACIS version, date, tolerance, units and the name of the "system" doing the saving). The class has two public functions for setting the units and the originators name. Although defined these parameters will not take effect unless **api_set_file_info** is called (its first argument is a mask indicating which fields to set).

The function **api_save_entity_list** can also be used to write binary versions of ENTITYs either to disk files or to specific areas of memory by using a FILE pointer argument derived from an abstract base class known as **FileInterface**. In this way the **cutting** and **pasting** of models between different programs running on the same machine can be supported by the API interface.

Once compiled and run, the resulting program creates a file called **save.sat** in the current directory. The first few lines of the file should look something like this:

ACIS's file format can be extended to include user defined ENTITYs and attributes.

save.sat

```
600 0 1 0
13 HW-University 13 ACIS 6.0.1 NT 24 Fri Feb 23 09:41:15 2001
10 9.9999999999999995e-007 1e-010
body $-1 $1 $-1 $-1 #
lump $-1 $-1 $2 $0 #
shell $-1 $-1 $-1 $3 $-1 $1 #
face $-1 $4 $5 $2 $-1 $6 forward single #
face $-1 $7 $8 $2 $-1 $9 reversed single #
loop $-1 $-1 $10 $3 #
plane-surface $-1 0 0 50 0 0 1 1 0 0 forward_v I I I I #
face $-1 $11 $12 $2 $-1 $13 reversed single #
loop $-1 $-1 $14 $4 #
plane-surface $-1 0 0 -50 0 0 1 1 0 0 forward_v I I I I #
coedge $-1 $15 $16 $17 $18 forward $5 $-1 #
face $-1 $19 $20 $2 $-1 $21 reversed single #
loop $-1 $-1 $22 $7 #
```

4.2 Saving and Restoring Models

```
plane-surface $-1 84.261735139405005 -
140.88116512993818 0
 -0.72845669399800517 0.6850918514837967 0 0 0 1
 forward_v I I I I #
coedge $-1 $23 $24 $25 $26 forward $8 $-1 #
```

The format of the SAT files is described exhaustively in the "ACIS Save File Format" document[3], but the following points should enable a quick appreciation of some of what is going on.

- The first line records the version number of ACIS used to generate the file (in this case 6.1) and the number of BODYs saved in the file. To generate sat files in earler formats (i.e. ACIS 4.0) call **api_save_version** prior to **api_save_entity_list**.

- The second and third lines record information about how and when the file was created. The last two numbers give the values of **resabs** and **resnor** (which specify the modeler's numerical precision) in use when the file was created (see Section 6.1).

- On subsequent lines any number preceded by a $ sign represents a pointer value which has been replaced with a record number in the file. The record numbering starts at zero on the fourth line.

- The symbol $-1 indicates a NULL pointer.

- For example, the fourth line records a BODY object which has only a LUMP record located in the fifth line (i.e. record number 1). It has no TRANSFORM, WIRE or ATTRIB records.

The algorithm used for save and restore is supported by several virtual functions of the ENTITY class (one of which is demonstrated in Section 2.7.2). Every class derived from ENTITY (including user defined ones) must define a procedure for saving and restoring itself in sat file format. In this way inter-operability between developers is ensured and the "sat" file can become the *Esperanto* of the ACIS community by providing a common format for the exchange of data between different applications.

Restoring ENTITYs

The function for reading-in a sat file from disk is called **api_restore_entity_list** and its use is almost identical to the save entity routine:

> *// Restoring a BODY from file*
> **ENTITY_LIST** new_bits;
> **FILE*** save_file = **fopen**("save.sat", "r");
> **api_restore_entity_list**(save_file, **TRUE**, new_bits);
> **BODY** *bod = (**BODY***)new_bits[0]; *//First ENTITY in list*

---
[3] Available for download from www.spatial.com

Foreign (or unknown) attributes are retained by save and restore operations.

Occasionally if you're restoring from a sat file that has been generated by a different vendor's system, a warning about "unknown attributes" may be output. This means that some part of the model has attribute data attached to it which your system regards as meaningless. However, despite this the attribute data is retained by ACIS and saved with the model.

This ensures that the data will still be available should the file ever return to a system that requires the "unknown attributes".

A frequent mistake made by new ACIS programmmers is to forget that models restored from file may have transforms associated with them. A 'quick and dirty' fix which allows the complications of working with BODY transforms to be avoided is to call **api_change_body_trans**(BODY*,NULL). This will change the geometry of the new object according to the transformation and set its TRANSFORM to zero.

4.3 ENTITY_LISTs

An ENTITY_LIST maintains a set of ENTITYs which contains no duplications.

One of the arguments for **api_save_entity_list** is an object of class **ENTITY_LIST**. This class provides general functions for manipulating variable-length lists of ENTITYs. The contents of an ENTITY_LIST are hashed so that they can be located relatively quickly. Table 4.2 summarizes the member functions of the ENTITY_LIST class.

Although generally straightforward there are several behaviors, detailed in the following sections, of which programmers using ENTITY_LISTs should be aware.

4.3.1 Casting ENTITYs

Because the ENTITY_LIST contains only pointers to ENTITYs, when an item is retrieved from a list it must be tested to find out its type and cast to that type before any of its member functions can be used.

This is done using the virtual ENTITY class function **identity**, which takes an (optional) argument to specify the depth in the class hierarchy for which the type is required.

ENTITY (level 0)
↓
SURFACE (level 1)
↓
SPHERE (level 2)

For example the FACE class is at level 1, but the SPHERE class is at level 2, because it is derived from the more general (level 1) SURFACE class.[4] Using this method the type of an entity can be found in the following way:

// Test a list item's identity
if (list[i].**identity**(EDGE_LEVEL) == EDGE_TYPE)
    EDGE *e = (**EDGE***)list[i];

If no level is specified, the most specific type (i.e. lowest level) of the object is returned:

---

[4]Levels are specified as **const int** of the form *_LEVEL (where *== the class name) in the header files of the ENTITYs.

4.3 ENTITY_LISTs

```
// Test a list item's identity
if (list[i].identity() == EDGE_TYPE)
    EDGE *e = (EDGE*)list[i];
```

Table 4.2   ENTITY_LIST member functions.

Member function	Description	Notes
int **add**(ENTITY*)	Adds ENTITY to the list	No duplicates are added
int **count**()	Returns length of the list	Including deleted entries
int **iteration_count**()	Returns length of the list	Ignoring deleted entries
int **lookup**(ENTITY*)	Searches for ENTITY	Returns its index position in the list
int **remove**(ENTITY*)	Deletes an ENTITY from the list	Leaving a NULL pointer
ENTITY ***operator**[ ](int)	Returns a pointer to the ENTITY held in the position given	Returns LIST_ENTRY_DELETED if location is NULL
void **init**()	Prepares a list for traversal	Called once before **next**()
ENTITY ***next**()	Returns ENTITYs in list order	Ignores deleted items

4.3.2 Constructors and Destructors

In the example **save.cxx** (page 72) an ENTITY_LIST object is created using the C++ operator **new** which dynamically allocates memory for the object. Because objects created with **new** do not get deleted automatically by changes in scope, they must be removed explicitly using the **delete** operator. The following code fragment shows how the **save.cxx** program could be rewritten without the need for **new** or **delete** operators.

```
// Calls lists destructor when it moves out of scope
ENTITY_LIST list;
list.add(body);
FILE* save_file = fopen("save.sat","w");
api_save_entity_list(save_file, TRUE, list);
```

By constructing the ENTITY_LIST in this way, its destructor will be called automatically when the **save_ent** function exits.

4.3.3 Tombstones

Because **ENTITY_LISTs** are implemented as hash tables the function **remove** leaves an entry in the list with a **NULL** pointer, known as a *tombstone*. These can be accommodated in three ways:

1. If traversing a list using a counter and subscript operator, NULL entries can be identified using the **LIST_ENTRY_DELETED** variable.

    ```
    // Check for tombstones in ENTITY_LIST list
        int i = list.count();
        for (int c = 0; c < i; c++){
            if (list[c] != LIST_ENTRY_DELETED)
                printf("Entry %d is OK \n",c);
        }
    ```

2. The iterator methods **init** and **next** can be used to traverse the list, automatically skipping tombstones as they go.

    ```
    // For fast processing of long lists
        list.init();  // Where list is an ENTITY_LIST
        ENTITY *thing;
        while((thing = list.next()) != NULL)
            printf("This thing is a %s \n",thing->type_name());
    ```

3. A tided-up list can be reconstructed without the tombstones:

    ```
    // List copy to remove tombstones
        ENTITY_LIST list;         // list to be rebuilt
        ENTITY_LIST temp_list;    // temporary list
        list.init();              // prepare to iterate
        ENTITY *thing;
        while((thing = list.next()) != NULL)
            temp_list.add(thing); // fill the temporary list
        list.clear();             // wipe the original list
        temp_list.init();         // prepare to iterate
        while((thing = temp_list.next()) != NULL)
            list.add(thing);      // refill the original list
    ```

4.3.4 ENTITY_LIST Arguments

When ENTITY_LISTs are used as arguments in a function several problems can arise because they have no copy constructor. Take, for example, the following user-defined function:

4.3 ENTITY_LISTs

```
// Add the edge to the list if the start and end vertex are the same.
void add_if_closed(ENTITY_LIST list, EDGE *e)
{
    if(e->start() == e->end())
        list.add(e);
}
```

This code has a deadly flaw. Because there is no copy constructor, the **list** variable receives a bitwise **copy** when passed by value to the function. Because of this the local ENTITY_LIST *in* the function and the one *outside* it point to the same list data. This doesn't sound so bad until you realize that when the ENTITY_LIST (and its corresponding list data) *in* the function are destroyed, by scoping before return, the ENTITY_LIST *outside* the function will contain pointers to corrupt data.

The solution to this problem is *always* to pass ENTITY_LISTs by reference:

```
// Add the edge to the list if the start and end vertex are the same.
void add_if_closed(ENTITY_LIST& list, EDGE *e)
{
    if(e->start() == e->end())
        list.add(e);
}
```

This is the way to do it even when you want to make changes to a copy of the list in a function which will be distinct from the list outside. In such cases the outside list should be duplicated and then passed by reference.

4.3.5 Save and Restore using the Scheme AIDE

Again the most painless way of creating a sat file is to use Scheme's AIDE:

save.scm

```
; Make a block
(solid:block (position -5 -5 -5) (position 10 10 10))
; Save it to a file called savebk.sat
(part:save "savebk.sat")
; delete the block
(part:clear)
; get it back
(part:load "savebk.sat")
; make a window
(view:gl)
; make a shaded picture of it
(render)
```

The Scheme AIDE holds a number of ENTITYs in a single container class called a PART (hence the extension names **part:save** etc.). However, despite the different terminology, PARTs are stored in standard sat files along with attributes for any materials, lighting or working coordinate systems (WCS) associated with them.

4.4 Adding and Subtracting Shapes

Because B-rep data structures are complex, it has long been recognized that automated ways of constructing them are required. The primitive creation functions go some way towards this but can never hope to cover all the shapes one might wish to generate. A better approach is to provide ways of cutting and joining *simple* shapes together to form more complex ones. These sorts of operation can be expressed in terms of a mathematical formalism known as **set theory** and are generically referred to as Boolean operations.

The three basic operations **Unite**, **Subtract** and **Intersect** are simple to use and frequently generate objects and geometry far more complex than the original BODYs. The following program, **unite.cxx**, illustrates this point by creating a cross from two cylinders (referred to as frustums). The resulting BODY contains several **elliptical curves** between the two cylindrical faces and also demonstrates how the location of ENTITYs can be changed.

unite.cxx

```
// This program creates two cylinders, rotates one and then
// unites them before the resulting BODY is saved to a file.
#include <stdio.h>                              // Declares I/O functions
#include "boolean/kernapi/api/boolapi.hxx"     // Declares boolean api
#include "constrct/kernapi/api/cstrapi.hxx"    // Declares constructor api
#include "kernel/kernapi/api/kernapi.hxx"      // Declares kernel api
#include "kernel/kerndata/top/alltop.hxx"      // Declares BODY class
#include "kernel/kerndata/lists/lists.hxx"     // Declare ENTITY_LIST class for save
#include "baseutil/vector/transf.hxx"          // Declares the transform class
#include "baseutil/vector/vector.hxx"          // Declares the vector class
#include "kernel/kerndata/savres/fileinfo.hxx" // Declares fileinfo class

void save_ent(char*, ENTITY*);   // Defined in save.cxx, page 72)

void main()
{
    api_start_modeller(0);

    api_initialize_booleans();   // Initializes the boolean component

    BODY *cyl1, *cyl2; // Make two cylinders

    api_make_frustum(100,20,20,20, cyl1);
```

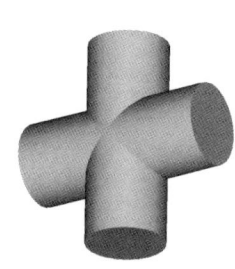

4.4 Adding and Subtracting Shapes

```
    api_make_frustum(100,20,20,20, cyl2);

    //Rotate one cylinder 90 degrees about X axis
    transf rotX = rotate_transf(M_PI/2,vector(1,0,0));
    api_apply_transf(cyl1,rotX);

    //Unite the two cylinders
    api_unite(cyl1,cyl2); // cyl2 is the result

    //Save the result
    save_ent("unite.sat",cyl2);

    api_terminate_booleans();
    api_stop_modeller();
}
```

When compiled and executed this program saves the resulting BODY to a file called **unite.sat**. There are two important points to notice about the call to **api_unite** which lies at the heart of this program:

1. The resulting BODY is assigned to the second argument (i.e. cyl2, known as the blank).

2. The BODY associated with the first argument (i.e. cyl1, known as the tool) is deleted.

This loss of the so-called tool BODY can be avoided either by copying it before the operation or, if the resulting BODY is not required, by using the rollback system (see Chapter 9) to restore it after the function has finished.

To view **unite.sat** using the Scheme AIDE type:
(view:dl)
(load:part 'unite.sat)

4.4.1 Mathematical Classes

The program **unite.cxx** (page 80) uses two of ACIS's mathematical utility classes. A **vector** object is used to define the axis about which rotation is performed and a **transf** object to specify a rotational transformation. A summary of all ACIS's mathematical utility classes is given in Table 4.3.

Table 4.3 Mathematical utility classes

Class	Description	Comments
position	A position vector (or point) in 3D Cartesian space	Can be subjected to certain vector and transformation operations
parameter	A curve parameter value	A floating point number
vector	A displacement vector in 3D Cartesian space	Has infix dot and cross product operators
unit_vector	A direction in 3D Cartesian space that has unit length	A sub-class of vector
vector	A displacement vector in 3D Cartesian space	Has infix dot and cross product operators
transf	A general 3D affine transformation	A 4×3 matrix for use with homogeneous vectors
tensor	A 3×3 tensor	Returned by mass-property and second-order gradient functions
symtensor	A symmetric 3×3 tensor	A sub-class type of tensor
interval	An interval between two real numbers	Class defines overloaded operators (e.g. $+, -, *, \&$) for interval arithmetic
matrix	3×3 affine transform	For use with vectors
par_xxx	Mathematical ENTITYs for use in (u,v) parameter space	xxx = class name such as **box, pos, vec, dir**
box	Axis-aligned rectangular box which encloses an ENTITY	Used to speed up intersection calculations

4.4.2 Set Operations using the Scheme AIDE

The Scheme AIDE's basic set operations are almost a direct mapping from the API calls except for the fact that they allow any number of tool BODYs to be given as arguments. Consequently a number of BODYs may be united, intersected or subtracted from the blank BODY (the argument) in a single call. For example, in the program **unite.scm** three cylinders are added together by a single call to **(solid:unite)**.

unite.scm

```
; Unite three orthogonal cylinders
(define c1 (solid:cylinder (position 0 0 -50)
                           (position 0 0 50) 20))
(define c2 (solid:cylinder (position 0 0 -50)
                           (position 0 0 50) 20))
(define c3 (solid:cylinder (position 0 0 -50)
                           (position 0 0 50) 20))
(define t1 (transform:rotation (position 0 0 0)
                               (gvector 1 0 0) 90))
(define t2 (transform:rotation (position 0 0 0)
                               (gvector 0 1 0) 90))
(entity:transform c1 t1)  ;Rotate about the x-axis
(entity:transform c2 t2)  ;Rotate about the y-axis
(define cross (solid:unite c1 c2 c3)) ; Result is c1, c2 and c3 are deleted
```

> As well as being some of the most useful modeling procedures, Boolean operations are also amongst the hardest to implement. Problems stem from both the vast number of cases which have to be considered and fundamental problems of numerical accuracy. The large number of cases arises from the fact that to be useful the Boolean operations must work for *every* kind of intersection which might appear between *every* type of FACE, EDGE and VERTEX supported by the modeler. These problems are compounded by the fact that many of these cases will require the detection of overlapping and coincident geometry which is always difficult in the presence of rounding and other numerical errors. Figure 4.1 illustrates these points by enumerating the six possible outcomes for the intersection of two orthogonal cylinders.

About Booleans

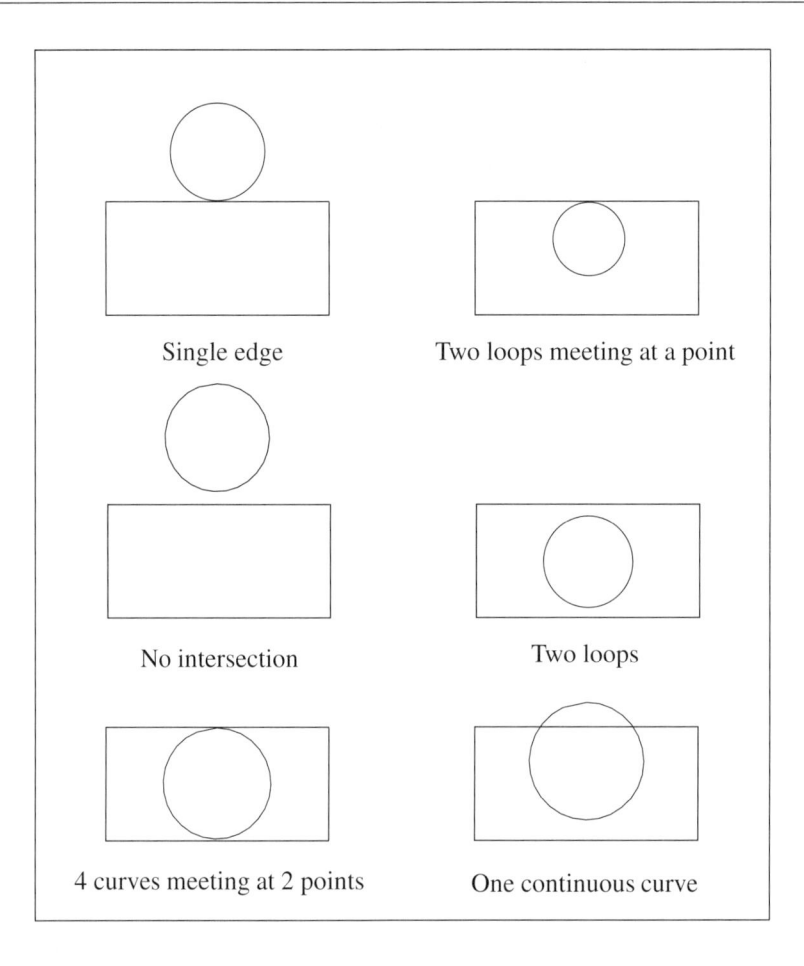

Figure 4.1: Six cases of orthogonal **cylinder-cylinder** intersection (viewed along one cylinder's axis). From Bowyer and Woodwark, "Introduction to Computing with Geometry".

With these powerful operators you can start to explore some classical geometry. For example the program **twocones.scm** generates the cross-section of a double cone forming a hyperbola.

twocones.scm

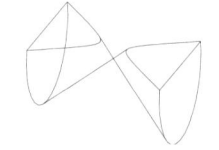

Boolean operations are not restricted to solids.

```
; Unite two cones at their tips and then section them
(define cone1 (solid:cone (position 40 0 0) (position 0 0 0) 25 0))
(define cone2 (solid:cone (position -40 0 0) (position 0 0 0) 25 0))
(define cone3 (solid:unite cone1 cone2))
(define plane (face:plane (position -100 100 5) 200 200 (gvector 0 0 -1)))
(define cut (sheet:face plane))
(bool:subtract cone3 cut )   ; Result is cone3, cut is deleted
```

Notice that because the subtraction operation was between a **3D solid** and a **2D sheet** BODY (effectively a half-space) the program used the more general **bool:subtract** rather than **solid:subtract** (which is restricted to 3D solids). For more details on set operations and intersections see Chapter 6.

4.5 Calculating Mass Properties

One thing that distinguishes solid modelers from the plethora of approximate representations is their ability to accurately calculate the mass properties (e.g. inertia and volume) of objects. The program **volume.cxx** demonstrates how the function **api_body_mass_pr** can be used to determine the intersecting volume of three identical cylinders each with its axis at right angles to the other two. This shape is used because it has an exact analytical answer against which the results can be compared. To make things slightly more interesting one of the cylinders is converted to a spline representation before the intersection is carried out.

volume.cxx

```
// This program calculates the volume of a solid defined by
// the intersection of three cylinders
#include <stdio.h>                              // File I/O functions
#include "boolean/kernapi/api/boolapi.hxx"      // Declares boolean API's
#include "constrct/kernapi/api/cstrapi.hxx"     // Declares constructor API's
#include "kernel/kernapi/api/kernapi.hxx"       // Declares kernel API's
#include "kernel/kerndata/top/alltop.hxx"       // Declares the body class
#include "operator/kernapi/api/operapi.hxx"     // Declares convert_to_spline API
#include "baseutil/vector/transf.hxx"           // Declares the transform class
#include "baseutil/vector/vector.hxx"           // Declares a vector class
#include "baseutil/vector/unitvec.hxx"          // Declares unit-vector class
#include "baseutil/vector/position.hxx"         // Declares position class
#include "kernel/kernutil/tensor/tensor.hxx"    // Declares tensor class
#include "kernel/kerndata/lists/lists.hxx"      // Declares ENTITY_LIST class for save

void volume( BODY *b, double &v);

// Defines an inline function (i.e. an abbreveation) to save some typing
  inline position p( double p1, double p2, double p3)
                      { return position(p1,p2,p3);}

  void main()
  {
     api_start_modeller(0);
     api_initialize_booleans();

     BODY *cyl1, *cyl2, *cyl3;   // Analytical Cylinders
     ENTITY *scyl1;              // Spline Cylinder
     api_solid_cylinder_cone(p(0,0,100), p(0,0,-100), 20, 20, 20, NULL, cyl1);
```

```
    api_solid_cylinder_cone(p(0,100,0), p(0,-100,0), 20, 20, 20, NULL, cyl2);
    api_solid_cylinder_cone(p(100,0,0), p(-100,0,0), 20, 20, 20, NULL, cyl3);

    // Create cylinder with spline surface
    api_convert_to_spline((ENTITY*)cyl1, scyl1);

    api_intersect((BODY*)scyl1,cyl2); // cyl2 is the result
    api_intersect(cyl2,cyl3); // cyl3 is the result

    double vol = 0;
    volume(cyl3, vol);

    printf("Volume = %f \n",vol);
    api_terminate_booleans();
    api_stop_modeller();
}

// Volume calculation routine
void volume(BODY* part, double &volume)
{
    position root = position(0.0,0.0,0.0);
    unit_vector root_vec = unit_vector(1.0,0.0,0.0);
    position cofg(0.0,0.0,0.0);
    tensor inertia; inertia.zero();
    double p_moments[3];
    unit_vector p_axes[3];
    double accuracy_got = 0;

    api_body_mass_pr(part, root,
            root_vec, 2, 0.001,
            volume, cofg, inertia,
            p_moments, p_axes, accuracy_got);
}
```

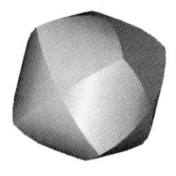

Intersection volume of three cylinders.

When compiled and executed the program should print out:

```
Volume = 37491.47
```

Since the three cylinders have equal radii (20 units) and normal axes to one another, their intersection forms a solid with 12 curved faces. It can be shown by direct integration that its volume V is $(2 - \sqrt{2})d^3$, so for diameter= 40, volume= 37490.33.

The program **volume.cxx** uses several of the mathematical utility classes. A **unit_vector** object is used to define the axis along which the BODY is projected for the mass properties calculation. A **tensor** object is created to hold the principal axes and a position object used to hold the centre of gravity. Although the function **api_body_mass_pr** gives you the option of specifying a desired accuracy and a projection plane, they make little difference to the results in the vast majority of cases.

4.5 Calculating Mass Properties

The program **vol.scm** shows how physical properties can also be calculated using the Scheme AIDE's extensions.

vol.scm

```
; This program intersects 3 cylinders and calculates the resulting volume.
; Make a window
(define mywin (view:gl))
; Make a cylinder called cyl1
(define  cyl1 (solid:cylinder(position  0 0 -50) (position  0 0 50) 20))
; Make a cylinder called cyl2
(define  cyl2 (solid:cylinder(position  0 -50 0) (position  0 50 0) 20))
;  Make a cylinder called cyl3
(define  cyl3 (solid:cylinder  (position  -50 0 0) (position  50 0 0) 20))
; Intersect cyl1 and cyl2 and cyl3
(solid:intersect  cyl1 cyl2 cyl3)
; Change the viewpoint
(view:set  (position  300 300 300) (position  0 0 0) (gvector  1 0 0))
; Have a look at the intersection
(option:set "cone_param_lines" #t)
(option:set "u_param_lines" 5)
(option:set "v_param_lines" 5)
;  Find out its mass
(solid:massprop  cyl1)
; Result is (("volume" . 37490.3513788031)
; and ("accuracy achieved" . 1.59770766448665e-005))
```

The resulting shape of the intersection is analogous to a semi-regular polyhedron called a **tetrakis hexahedron** which has been *tweaked* (see page 287) to have cylindrical rather than planar faces. Depending on the direction from which it is viewed, the intersection volume can appear to be cylindrical or cuboidal.

In addition to a BODY, the function **solid:massprop** accepts four other, optional, arguments. Two of these allow specification of the information returned (see Table 4.4 on page 88) and the **tolerance** or accuracy of the result required. The last two arguments allow the projection plane (used within the mass property analysis routine) to be specified by means of a point (i.e. a position) and a surface normal (i.e. a gvector). In **vol.scm** none of these are given, causing the default values [(0, 0.01, (0,0,0) and (0,0,1)] to be used.

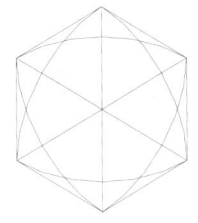

Table 4.4   solid:massprop arguments.

2nd Argument	Mass properties returned
0 (default)	Volume and tolerance
1	Volume, tolerance, center of mass, principal moments, principal axes
2	All of the above plus inertia tensors

About mass properties

> The mass properties (volume, inertia and center of gravity) of boundary models can be calculated either by **direct integration** or by numerical techniques based on the **divergence theorem**.
>
> Although direct integration provides exact results, the equations of curves and faces are not always available or easy to solve. However, a more general method exploits the fact that both surface area and volume can be computed using the (u,v) parameterization associated with each FACE. For example, the area of a FACE can be expressed as:
>
> $$Area = \int_{u_1}^{u_2} \int_{v_1}^{v_2} f(u,v)\, du\, dv$$
>
> where the limits of integration cover an entire face. By carrying out this integration (using, say, Gaussian quadrature) for every FACE on a BODY the total surface area can be calculated.

4.6 Extruding Shapes

A large number of interesting shapes can be created by ACIS's sweeping API. The simplest form of sweep involves the formation of a solid by translating a 2D planar **profile** along a **path**. If the plane of the profile is *perpendicular* to the path it is being swept along, then the operation is referred to as a **sweep**. If this is not the case it is called a rigid sweep or **rsweep**.

Any planar WIRE or a FACE can be used as a **profile**, and a **path** can be:

- along a vector
- around an axis
- along a WIRE

Sweeps can also have **draft** angles associated with them which taper, or expand, any profile formed by circular arcs and straight lines,[5] as it is moved along the path. It

---

[5] If the swept profile includes a circular arc the EDGEs adjacent to it must be tangent continuous with it (i.e. smoothly join it).

4.6 Extruding Shapes

is also possible to specify a degree of **twist** which rotates a profile as it is translated along the path.

Not surprisingly there are some restrictions on the types of profiles and paths that can be used in sweeping operations. For example:

- If the sweep path contains a spline curve (i.e. intcurve), or noncircular ellipse, then the profile must be perpendicular to the path at the beginning of the NURB curve.

- A WIRE path must start in the plane of the profile to be swept.

Although it is very easy to generate some spectacular shapes, care must be taken because (by default) no check is made to determine if a sweep results in interpenetrating FACEs (i.e. a self-intersecting BODY).

In the example **prism_sw.cxx** a standard modeling primitive (i.e. a prism) has two of its FACEs swept in different directions. The first is swept along a vector at right angles to the surface; the second is extruded along a wire in the shape of a quarter circle.

This is done by calling the following API functions:

api_make_prism: creates a BODY with 10 planar FACEs.

api_find_face: gets a pointer to the rectangular FACE aligned with the x-axis.

api_sweep_with_options: sweeps it in the x-direction 10 units, creating two new FACEs.

api_find_face: gets a pointer to the now 10-sided FACE aligned with the z-axis.

api_mk_ed_ellipse: makes a circular EDGE that turns through a quarter circle.

api_make_ewire: makes a WIRE BODY from an array of EDGEs.

api_reverse_wire: swaps the start and end VERTEXs on the WIRE.

api_sweep_with_options: carries out the sweeping operation using the default values of the **sweep_options** class.

api_save_entity_list: saves the swept BODY to file.

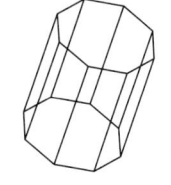

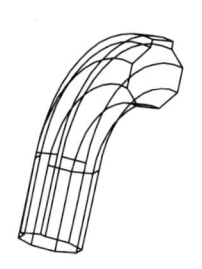

prism_sw.cxx

```
// This program makes a Prism and then sweeps one face in the
// direction of its surface normal and another along a circular arc.
#include <stdio.h>                          // File I/O functions
#include "kernel/acis.hxx"                  // System Definitions
#include "constrct/kernapi/api/cstrapi.hxx" // Declares constructor api
#include "kernel/kernapi/api/kernapi.hxx"   // Declares kernel api
```

```cpp
#include "sweep/kernapi/api/sweepapi.hxx"    // Declares sweeping api
#include "kernel/kerndata/top/alltop.hxx"    // Declares Topology classes
#include "kernel/kerndata/lists/lists.hxx"   // Declares ENTITY_LIST class
#include "baseutil/vector/unitvec.hxx"       // Defines unit-vector class
#include "baseutil/vector/position.hxx"      // Defines position class
#include "kernel/kerndata/savres/fileinfo.hxx"  // Declares fileinfo class
#include "sweep/sg_husk/sweep/swp_opts.hxx"   // Declares sweep_options class

void main()
{
    api_start_modeller(0);
    api_initialize_sweeping(); // Initialize highest level lib
    BODY* hook;
        // Create an 8 sided prism, axis along z, one face along x
    api_make_prism(100,40,40,8,hook);
    FACE *xface;
        // Get the face aligned with the x-axis
    api_find_face(hook,unit_vector(1,0,0),xface);
        // Sweep it ten units in the direction of its surface normal

    sweep_options *ops = new sweep_options;
    api_sweep_with_options(xface, 10, ops, hook);

    FACE *zface;
        // Get the face aligned with the z-axis
    api_find_face(hook,unit_vector(0,0,1),zface);
        // Make an EDGE for sweeping along
    position cen(-150,0,50);
    unit_vector norm(0,1,0);
    vector major(0,0,150);
    EDGE *sweep_edge;
    api_mk_ed_ellipse(cen,norm,major,1.0,0,pi/2,sweep_edge);
        // Make a WIRE BODY with the EDGE for sweeping .

    EDGE* e_list[1]; //Make an array of EDGE pointers, one long
    e_list[0] = sweep_edge;
    BODY *wb;
    api_make_ewire(1,e_list,wb);

        // Its sense is wrong so reverse it
    api_reverse_wire(wb->wire());

        // Sweep the face along the arc
    api_sweep_with_options(zface,wb,ops, hook);

        // Save the final BODY in a SAT file
    FILE *fp = fopen("hook.sat", "w");
    ENTITY_LIST save_list;
    save_list.add(hook);
```

4.7 Blending

```
    FileInfo info;  // Create FileInfo Object
    info.set_product_id("HW-University");  // set info's data
    api_set_file_info(FileId, info);
    api_save_entity_list(fp, TRUE, save_list);
    fclose(fp);

    api_terminate_sweeping();
    api_stop_modeller();
}
```

The program **prism_sw.cxx** is interesting because, for the first time in this chapter, it makes direct reference to FACE and EDGE ENTITYs.

Many more sophisticated sweeping operations are explored in Section 10.5 on page 249.

4.7 Blending

Very few engineering components actually have sharp corners or edges. In practice transitions between FACEs are, more often than not, rounded or smoothed to make manufacture and assembly easier. Although important, the blending of EDGEs is usually left until last, and only done after all the larger features and dimensions of the shape have been created.

The program **blend.cxx** demonstrates how a simple blending operation can be carried out using the **API** interface. A block and cylinder with coplanar top FACEs are united so that the resulting top FACE has three straight and one circular EDGEs. The cylinder is positioned so that the circular EDGE meets two of the straight EDGEs tangentially. The circular EDGE is used to seed a list looking for EDGEs that meet other EDGEs with continuity of slope. These are then blended.

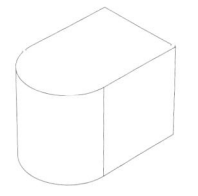

blend.cxx

```
// This program blends a sequence of three edges, two straight and one round.
#include <stdio.h>                              // File I/O Functions
#include "kernel/acis.hxx"                      // Declares system-wide information
#include "blend/kernapi/api/blendapi.hxx"       // Declares blending API's
#include "boolean/kernapi/api/boolapi.hxx"      // Declares boolean API's
#include "constrct/kernapi/api/cstrapi.hxx"     // Construction API's
#include "kernel/kernapi/api/kernapi.hxx"       // Declares kernel API's
#include "kernel/kerndata/top/alltop.hxx"       // Declare Topology Classes
#include "kernel/kerndata/geom/ellipse.hxx"     //Declare ellipse class
#include "kernel/kerndata/lists/lists.hxx"      //Declare ENTITY_LIST class
#include "baseutil/vector/transf.hxx"           //Declare transf class
#include "kernel/kerndata/savres/fileinfo.hxx"  // Declares fileinfo class

void main()
{
```

```
api_start_modeller(0);

api_initialize_blending();  // Initialize highest level lib

BODY* block;
BODY* cylinder;

// Create a block
api_make_cuboid(100,100,100,block);

// Move block around
api_apply_transf(block,translate_transf(vector(-50,0,0)));
// Create a cylinder
api_make_frustum(100,50,50,50,cylinder);
// Unite block and cylinder
api_unite(cylinder,block);
    // Find the VERTEX at the start of the cylinder
VERTEX *v1;
api_find_vertex(block, position(0,50,50),v1);
// Get a list of all (3) EDGEs which meet at the VERTEX
ENTITY_LIST vedge_list;
api_q_edges_around_vertex(v1, &vedge_list);
// Find the circular edge
EDGE *cir_edge = NULL;
for(int i = 0; i < vedge_list.count(); i++){
    EDGE *ed = (EDGE*)vedge_list[i];
    CURVE *cur = ed->geometry();
    if (cur->identity() == ELLIPSE_TYPE)
    { cir_edge = ed;
      break; }
};

ENTITY_LIST edges_to_blend;
    // Constructs a list of adjacent edges which meet smoothly
    // (i.e. with tangent continuity). The list will include the
    // seed EDGE given as argument.
api_smooth_edge_seq(cir_edge,edges_to_blend);
    // Create implicit blends by attaching
    // a Blend Attribute (specifying a rounding radius
    // of 10) to each edge in the list.
api_set_const_rounds(edges_to_blend,10.0,0,0);
    // Create explicit blends by replacing each edge
    // and vertex in the list with a new face.
api_fix_blends(edges_to_blend);
FILE *save_file = fopen("blend.sat", "w");

ENTITY_LIST save_list;
save_list.add(block);
FileInfo info;  // Create FileInfo Object
info.set_product_id("HW-University");  // set info's data
```

4.8 Blending in the Scheme AIDE

```
        api_set_file_info(FileId, info);
        api_save_entity_list(save_file, TRUE, save_list);
        fclose(save_file);

        api_terminate_blending();
        api_stop_modeller();
}
```

Blending is a two-step process that involves first marking which ENTITYs of the model are to be blended and then doing it. ACIS implements these processes in the following way:

1. Identifying a **blendable network** of EDGEs and VERTEXs by the attachment of **blend attributes**.
2. Modification of the solid is carried out by **api_fix_blends**.

In the program **blend.cxx** a **blendable network** is created using **api_smooth_edge_seq** which collects a number of tangent continuous EDGEs together in an ENTITY_LIST. This is only one form of **blendable network**; more generally they can be constructed from sequences of EDGEs and VERTEXs which either already (or would after blending) meet with tangent plane continuity. The constant radius blend, used in the example, is not the only type available; **fillets**, **chamfers**, and **variable radius** blends can also be created (see page 300).

4.8 Blending in the Scheme AIDE

The **solid:blend-edges** command takes a single EDGE or a list of EDGEs, and puts a constant radius blend on them.

blend.scm

```
; This program creates a test object and then blends all its edges
(define bldemo
  (lambda()
   (let*(
   ( tube (solid:cylinder (position 0 0 -40) (position 0 0 40) 10))
   ( box (solid:block (position -20 -20 -20) (position 20 20 20)))
   (union (solid:unite tube box)))
      (solid:blend-edges (entity:edges union) 7))))

(bldemo)  ; Execute the program
```

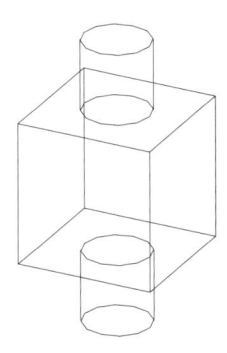

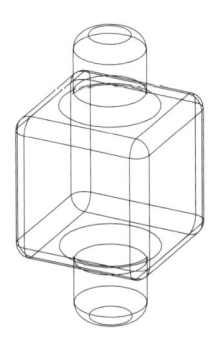

Results of blend.scm

The illustrations in the margin show what interesting things can happen when blends start to overlap as the radius is increased.

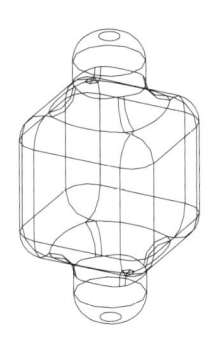

4.9 Sectioning

It is common practice in engineering (or technical) drawing systems to use cross-sections through a component, or assembly, for the display of internal details. With this application in mind the program **section.cxx** sections BODY along a plane and saves the two resulting BODYs in separate files.

The program's algorithm can be explained in terms of the API calls:

api_wiggle: Creates a block with a "wiggly" spline surface for testing purposes.

get_body_box: Returns a box object, aligned with the axes whose member functions **low** and **high** are used to extract the coordinates of two opposite corners. The distance between these points is used to give a rough measure of the size of the object being sliced.

api_mk_ed_ellipse: Is the start of a three step process which creates a circular BODY to do the sectioning. This first step creates a circular EDGE with a normal in the direction (1,1,0). Later this vector is used to determine which side of the plane represents material and which fresh air.

api_make_ewire: Creates a WIRE BODY from the circular EDGE. WIRE BODYs have no LUMPs, SHELLs or FACEs, only EDGES.

api_cover_wires: Covers the circular EDGE with a planar FACE. This is called with a NULL reference argument, ***(surface*)NULL** (see Section 2.4.5).

api_copy_body: The sectioning of the body will cause both the original and the cutter to be destroyed. Thus without copying, only one "half" of the sectioned object could be generated. So it is necessary to copy both objects in order to generate the other "half".

api_reverse_face: Switch the side representing material by reversing the sense of the FACE (see Section 1.7).

api_boolean: Subtract the plane from the body.

section.cxx

```
//This program sections a BODY with a plane
#include <stdio.h>                              // File I/O functions
#include "boolean/kernapi/api/boolapi.hxx"      // Declares boolean API's
#include "constrct/kernapi/api/cstrapi.hxx"     // Construction API's
#include "cover/kernapi/api/coverapi.hxx"       // Declares api_cover_wires
#include "kernel/kernapi/api/kernapi.hxx"       // Declares kernel API's
#include "kernel/kerndata/top/alltop.hxx"       // Declares topology classes
#include "kernel/kerndata/lists/lists.hxx"      // Declares ENTITY_LIST class
#include "kernel/kerndata/geometry/getbox.hxx"  // Declares get_body_box
#include "baseutil/vector/box.hxx"              // Declares box class
```

4.9 Sectioning

```cpp
#include "baseutil/vector/position.hxx"    // Declares position class
#include "baseutil/vector/vector.hxx"      // Declares vector class
#include "baseutil/vector/unitvec.hxx"     // Declares unit vector class
#include "kernel/kerndata/savres/fileinfo.hxx" // Declares fileinfo class

void save_ent(char* file_name, ENTITY *ent);  // Defined in save.cxx

void main()
{
    // Initialization of the modeller must be done before any other calls.
    api_start_modeller(0);
    api_initialize_covering();

    // Create a body to section
    BODY* wig;
    api_wiggle(100,100,30,1,2,-1,2,TRUE,wig);

    // Find out the length of its diagonal
    position bottom = wig_size.low();
    position top = wig_size.high();
    vector diagonal = top - bottom;
    double len = diagonal.len();

    // Create a circular edge
    position cen(30,14,20); // point on sectioning plane
    vector direction(1,1,0); // material side of sectioning plane
    unit_vector norm = normalise(direction);
    vector radius(0,0,len*0.6); // Box size plus a bit
    EDGE *slicer_e;
    api_mk_ed_ellipse(cen,norm,radius,1,0,0,slicer_e);

    // Turn the edge into a WIRE Body
    EDGE *e_array[1];
    e_array[0] = slicer_e;
    BODY *slicer_wb;
    api_make_ewire(1,e_array,slicer_wb);

    // Cover the WIRE to make a planar FACE
    ENTITY_LIST slicer_faces;
    api_cover_wires(slicer_wb,*(surface*)NULL,slicer_faces);

    // Back up some copies of the BODIES
    BODY *bk_wig, *bk_slicer_wb;
    api_copy_body(wig,bk_wig);
    api_copy_body(slicer_wb, bk_slicer_wb);

    // Reverse the direction of the cutter
    FACE *sf = (FACE*)slicer_faces[0];
    api_reverse_face(sf);
```

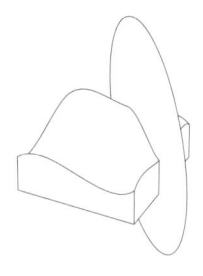

```
// Carry-out the section
api_boolean(slicer_wb,wig,SUBTRACTION);
save_ent("left-section.sat", wig);

api_boolean(bk_slicer_wb,bk_wig,SUBTRACTION);
save_ent("right-section.sat", bk_wig);

api_terminate_covering();
api_stop_modeller();
}
```

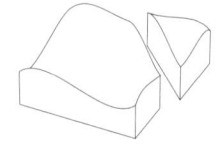

One might wonder why it is neccessary to create two seperate lamina BODYs with the same geometry but opposite orientations. Why not simply create one, double-sided lamina (by applying **api_body_to_2d**) and do a single subtraction?

Unfortunately this approach will not work because the Boolean subtraction of a double sided lamina (i.e. a sheet BODY) will not cause the blank BODY to separate (see Table 4.5). Instead the resulting BODY simply gains an internal FACE that partitions its volume.

Table 4.5   Sheet/Solid Booleans.

Operation	Result
Subtracting a Sheet from Solid	Areas of sheet within the solid are embedded in BODY.
Uniting a Solid with Sheet	Any areas of sheet exterior to solid region added to the BODY.
Intersecting a Solid with Sheet	Areas of the sheet within the solid remain.

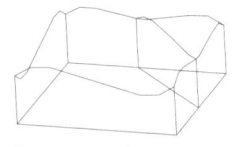

Results of a sheet boolean operation

In contrast a single-sided FACE acts like "half-spaces" during Boolean operations with everything "behind" them being treated as an infinite volume of solid material.

4.10   Slicing with the Scheme AIDE

The program **section.scm** takes a slightly different approach to sectioning a cone. A 2D BODY is created by covering a planar wire (generated with the **solid:slice** command) to form a single-sided FACE. This 2D BODY is then subtracted from the cone to create a cross-section.

4.11 Coordinate Systems

section.scm

```
; This program renders a picture of a cone sliced down the middle
(define ConicSection
  (lambda ()
   (let*(
    (cone (solid:cone (position 0 0 0) (position 0 0 40) 25 0))
    (slice-wire (solid:slice cone (position 0 0 10) (gvector 1 0 0)))
    (cutter (sheet:planar-wire slice-wire)))
     (begin
      (sheet:2d cutter )
      (solid:sweep-face (car (entity:faces cutter)) 0.01)
      (let*(
         (two-halfs  (solid:subtract cone cutter) )
         (blist (body:separate two-halfs))
         (fill (filter:type "planar-face?"))
         (pface (car (filter:apply fill (entity:faces (car blist)))))
         (dir (planar-face:normal  pface))
         (transdir (gvector:scale (gvector:reverse dir) 10))
         )
     (entity:transform (car blist) (transform:translation transdir))
     (render))))))
```

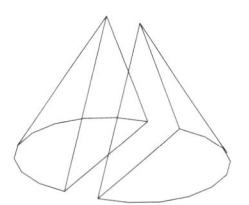

The sheet BODY is then covered again (using **sheet:2d**) to add a second FACE.[6] The sheet BODY's FACEs are then extracted in a list (using **entity:faces**). The first one in the list is swept along its normal by an arbitrarily small amount (using **solid:sweep-face**). The thin solid created by the sweep is then subtracted from the cone BODY, creating two LUMPs.

The LUMPs are then formed into separate BODYs (using **body:separate**) and placed in a list. A planar FACE on the first BODY in this list is found and the BODY translated 10 units in the opposite direction to its FACE normal.

Visually this looks fine but an obvious problem with this method is that the cross-sections of the two "halves" will be slightly different (because of the thickness of the cutter).

4.11 Coordinate Systems

Implicit in almost all the programs discussed in this book is the assumption that there is a single **origin** and a world coordinate frame relative to which everything is created and transformed. The Scheme AIDE extends this idea by allowing multiple frames

---

[6]In practice **sheet:2d** simply sets the "sided" flag to double-sided. No new FACE ENTITY is created.

of reference, so-called **Working Coordinate Systems** (**wcs**), to be specified. Two **wcs** exist by default: the **view** and the **model** coordinate systems. These have the following properties:

view Is an unchangeable wcs associated with each view created. The **view-wcs** has its origin in the center of the window, the z-axis is orientated out of the screen, y is vertical and x is positive going from left to right.

model Is the frame of reference in which objects are defined.

By default the **view-wcs** and the **model-wcs** are aligned. Furthermore all implicit references to a coordinate frame, say via the creation of a position, are by default assumed to refer to the model-wcs. So the command:

>   (**view:set** (**position** 20 -80 20)
>   (**position** 0 0 0) (**gvector** 0 0 1))

will set the position of the view port *in terms of the model's wcs*. Effectively what is being created is a transform between the viewing window's frame and the model frame.

positions and **gvectors** can be specified in other coordinate frames using a fourth argument, which can also specify the *type* of coordinates as well as the frame of reference. For example:

>   (**position** 20 -80 20) ; *Cartesian coordinates relative to the active*
>   ; *coordinate system.*
>   (**position** 20 -80 20 "model") ; *Cartesian coordinates relative*
>   ; *to the model coordinate system.*
>   (**position** 20 -80 20 "spherical") ; *Spherical coordinates relative*
>   ; *to the active coordinate system.*
>   (**position** 20 -80 -20 "polar") ; *Polar coordinates relative*
>   ; *to the active coordinate system.*
>   (**position** 20 30 78 "cylindrical") ; *Cylindrical coordinates relative*
>   ; *to the active coordinate system.*
>   (**gvector** 0 0 1 "mydatum")) ; *Cartesian coordinates relative to a*
>   ; *coordinate system called mydatum.*

In the above code **mydatum** is the name of a default wcs created by the user. Other wcs can be constructed (and automatically displayed) using the following commands:

>   ; *Create a wcs called myframe*
>   ; *Origin at (5,5,5), x-axis direction (1,1,0), y-axis direction (0,0,1)*
>   (**define** myframe (**wcs** (**position** 5 5 5)
>   (**gvector** 1 1 0) (**gvector** 0 0 1)))
>   ; *Display it in a green color*
>   (**env:set-active-wcs-colour** GREEN)

4.11 Coordinate Systems

```
; Make it the default coordinate system
(wcs:set-active myframe)
; Create a block relative to it
(solid:block (position 7 7 7) (position 14 14 14))
```

Although this is useful the real power of wcs is the way they can be used to generate *transforms between locations*. Given two wcs this is done in a three-step process using the following Scheme extensions:

wcs:to-model-transform Creates a transform between the given wcs (or active if no argument is given) and the model wcs.

wcs:to-wcs-transform Creates a transform between the active wcs and the one given.

transform:compose Combines two transforms into one.

The program **wcs.scm** illustrates the use of these in the construction of a mace head. Figure 4.2 shows the six wcs used.

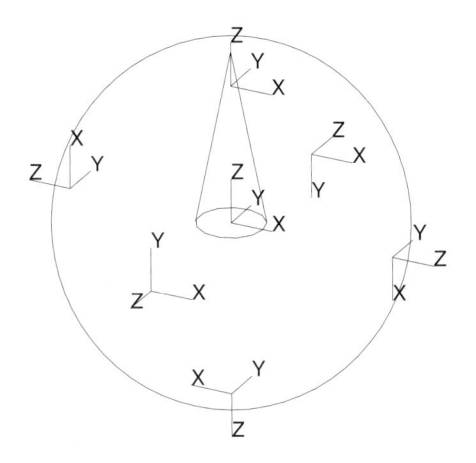

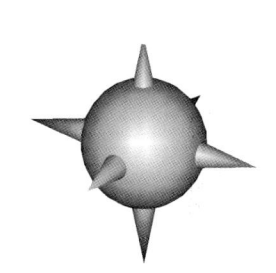

Figure 4.2: Six working coordinate systems.

wcs.scm

```
; This program locates six working coordinate systems (wcs)
; on the surface of a sphere and then moves and unites a cone
; at the origin of each one, creating a mace head

; First define a wcs at the origin
(define origin (wcs (position 0 0 0) (gvector 1 0 0) (gvector 0 1 0)))
(wcs:set-active origin)
```

```
(define ball (solid:sphere (position 0 0 0) 50))
(define cone (solid:cone (position 0 0 0) ( position  0 0 50) 10 0))

; specify six working coordinate systems in terms of spherical coordinates
; relative to the active coordinate system (i.e. "origin")
(define top (wcs (position 0 0 40)
   (gvector 1 0 0) (gvector 0 1 0)))
(define bottom (wcs (position 50 180 -0 "spherical")
   (gvector -1 0 0) (gvector 0 1 0)))
(define left (wcs (position 50 90 270 "spherical")
   (gvector 1 0 0 ) (gvector  0 0 1)))
(define right (wcs (position 50 90 90 "spherical")
   (gvector 1 0 0 ) (gvector   0 0 -1)))
(define back (wcs (position 50 90 180 "spherical")
   (gvector  0 0 1 ) (gvector 0 1 0)))
(define front (wcs (position 50 90 0 "spherical")
   (gvector  0 0 -1) (gvector 0 1 0)))
(define wcs-list (list top bottom left right back front))

; Move an entity from one working coordinate system to another
(define move
  (lambda (body from to)
    (let* (
     (part1 (wcs:to-wcs-transform from))
     (part2 (wcs:to-model-transform to))
     (trans (transform:compose part1 part2)))
      (entity:transform body trans))))

; Takes a list of wcs, and two BODYs
(define make
  (lambda(loc-list spike-body ball)
   (if (null? loc-list) (render)
      (let ((spike (entity:copy spike-body)))
       (move spike origin (car loc-list))
       (solid:unite ball spike)
       (make  (cdr loc-list) spike-body ball)))))

(make) ; Execute the program
```

Working coordinate systems can be used to support many assembly and construction tasks. The API functions underlying the Scheme extensions described here can also be accessed directly via the C++ interface.

4.12 Part Manager

It is impossible to use the Scheme AIDE without having some contact (albeit unwittingly) with the **Part Manager** as it provides default names for any ENTITYs created.

A **part** is a collection of ENTITYs which can be saved and restored *en măsse*. Parts can also associate attributes like wcs, materials and colors with specific ENTITYs.

Every ENTITY created with a Scheme Extension is assigned an identification number by the Part Manager and, by default, added to the **active part**. A default part (known as **part 1**) is created when the Scheme AIDE is initialized. To see the current part type:

```
acis>(env:active-part)
#[part 1]
acis>(define another-part (part:new))
acis>(env:set-active-part another-part)
acis>(solid:sphere (position 67 56 45) 80)
#[entity 1 2]  ;Entity Number 1 of Part 2
```

In the C++ code of a Scheme extension any ENTITY created can be registered with the Part Manager simply by bracketing the functions concerned, say **api_make_cuboid**, between the BEGIN_CREATE and END_CREATE macros (see Section 13.3).

4.13 Exercises

1. The program **ball.cxx**, on page 67, produces a debug file detailing the data structure required to represent a sphere. Test your understanding of Boundary Representations by writing down what you expect the file **ball.dbg** contains *before* checking your answer.

2. Write a Scheme program that calculates the *symmetric difference* between an arbitrary number of BODYs in a list (see Section 13.2).

3. It is possible for the intersection volume of four identical cylinders to be symmetrical if their axes have the symmetry of a regular tetrahedron. Create this intersection volume (which should have 24 FACEs) and check that its volume is $\frac{3}{2}\sqrt{2}(2-\sqrt{3})d^3$ (where d is the cylinder's diameter). How many SURFACE objects are required to represent this BODY?

4. An interesting comparison to the total storage figure for a block is generated by the **sp_block.cxx** program (see ftp site) which uses **api_convert_to_spline** to change the simple representation of the block's surfaces to a NURBS one. The debug output from this new model is shown below.

```
sp_cube.dbg
1 body record,        32 bytes
1 lump record,        32 bytes
1 shell record,       40 bytes
6 face records,      264 bytes
6 loop records,      192 bytes
6 surface records,   528 bytes
24 coedge records,  1056 bytes
12 edge records,     528 bytes
24 pcurve records,  1920 bytes
8 vertex records,    192 bytes
12 curve records,   1344 bytes
8 point records,     384 bytes
Total storage 6512 bytes
```

Explain the differences in storage requirements between the simple and NURBS surface representations.

Using the Direct Interface

5

In contrast to the high-level API commands in the previous sections the examples in this chapter show how programs can exploit the ACIS data structure's direct interface (i.e. the access methods of the classes). Although it is possible to construct and modify objects via the direct interface, its most common use is to provide arguments for API routines.

For example an application might wish to highlight all the cylindrical FACEs aligned with the x-axis. Because no API exists for locating this specific group of FACEs the direct interface could be used to create a list of them before calling, say, **api_gi_highlight_entity** for each one in turn.

However this increased flexibility requires more careful programming because (unlike the API routines) the direct interface functions do not return outcome objects for error checking.

To understand fully the C++ examples in this chapter requires a knowledge of each class's member functions (which are summarized in Appendix B). At first glance the number of methods might appear overwhelming. However, for the most common use (i.e. the **reading** of the data structure) only a small subset of the **access methods** are required to travel down the B-rep hierarchy from the BODY class to the ENTITY of interest.

5.1 Counting Faces

The program **f_count.cxx** shows how the number of faces on a model can be counted using the public access methods of the topological classes.

f_count.cxx

```
// This program creates a block and counts the number of faces on it
#include <stdio.h>                      // Declares Input/output functions
#include "constrct/kernapi/api/cstrapi.hxx"  // Declares Constructor API's
```

```
#include "kernel/kernapi/api/apimsc.hxx"    // Declares API function prototypes
#include "kernel/kerndata/top/alltop.hxx"   // Declares all topological classes

void main()
{
    api_start_modeller(0);
    api_initialize_constructors(); // Initialize component
    BODY* block;
    api_make_cuboid(100,150,200, block);
    FACE *ff = block−> lump()−> shell()−> face();
    int count = 0;
    while(ff != NULL)
      {
            count++;
            ff = ff−> next();
      }
    printf("The block has %d faces\n",count);
    api_terminate_constructors();
    api_stop_modeller();
}
```

When compiled and run this program will print the following on the screen:

        The block has 6 faces

This example demonstrates how the **direct interface** can be used to traverse the model's internal data structure. The line:

> **FACE** *ff = block−> **lump**()−> **shell**()−> **face**();

makes a series of calls to member functions of different classes. The same line could have been written in a more verbose way as:

> **LUMP** *ll = block−> **lump**();
> **SHELL** *ss = ll−> **shell**();
> **FACE** *ff = ss−> **face**();

> This example assumes the BODY does not have multiple LUMPs or SHELLs.

the function **lump()** is a method of the BODY class which returns a pointer to the first LUMP in the list of LUMPs associated with it. Similarly the function **shell()** is a method of the LUMP class and returns a pointer to the first in the list of SHELLs associated with it, and so on.

In the Scheme AIDE the same operation can be done in a couple of lines:

5.2 Accessing FACE Details

```
acis>(define b (solid:block (position 10 10 10) (position 50 50 50)))
b
acis>(define face-list (entity:faces b))
face-list
acis>(length face-list)
6
```

The Scheme extension **entity:faces** returns a list and the built-in procedure **length** returns the number of elements in it.

In both examples the list of FACEs obtained from the model *is not guaranteed* to tell you anything other than the identity and number of FACEs. However, in practice the sequence of ENTITYs in the list often reflects the *order* in which they were created, with the most recent first.

5.2 Accessing FACE Details

It is often the case that an application is only interested in FACEs with certain properties (e.g. those that are planar). The program **f_detail.cxx** illustrates how the **type** and number of LOOPs on a given FACE can be determined.

f_detail.cxx

```cpp
// This program creates a truncated cone and identifies the type of each face on it
#include <stdio.h>                                  // Declares Input/Output functions
#include "constrct/kernapi/api/cstrapi.hxx"         // Declares constructor API's
#include "kernel/kernapi/api/kernapi.hxx"           // Declares kernel API's
#include "kernel/kerndata/top/alltop.hxx"           // Declares Topological Classes
#include "kernel/kerngeom/surface/allsfdef.hxx"     // Topological Surface Classes
#include "kernel/kerndata/geom/allsurf.hxx"         // Geometric Surface Classes

void face_details(FACE*,int);    //Prints number of LOOPs and geometry type

void main()
{
    api_start_modeller(0);
    api_initialize_constructors();  //Initialize highest component
    BODY* truncated_cone;
    api_make_frustum(50,20,30,10,truncated_cone);
    FACE *ff = truncated_cone-> lump()-> shell()-> face_list();
    int count = 0;
    while(ff != NULL)
        {
            count++;
            face_details(ff,count);
            ff = ff-> next();
```

```
                }
                api_terminate_constructors();
                api_stop_modeller();
        }

        void face_details( FACE* f, int f_num)
        {
                int loopc = 0;
                for(LOOP *lp = f–>loop(); lp != NULL; lp = lp–>next(), loopc++);

                printf("face %d has %d loops of edges\n", f_num, loopc);

                SURFACE *fs = f–>geometry();
                printf("Face %d, is a %s type \n", f_num, fs–>type_name());
        }
```

Once compiled and run the following text is printed to the screen:

```
Face 1 has 2 loops of edges
Face 1 is a cone type
Face 2 has 1 loops of edges
Face 2 is a plane type
Face 3 has 1 loops of edges
Face 3 is a plane type
```

There are several other ways in which a FACE's geometric type can be established. For instance, the default behavior of the **identity** method could be used to return the most specific type of the SURFACE in the following way:

```
// Check ENTITY type
        if(fs–>identity() == CONE_TYPE)
                printf("Face %d is a cone \n",f_num);
        else if(fs–>identity() == PLANE_TYPE)
                printf("Face %d is a plane \n",f_num);
```

A second, and more obtuse way would be to extract the lowercase geometry class, used to record a surface's parameters, and interrogate it. The following code fragment demonstrates this approach:

```
// Check the geometry type
        const surface& sfs = fs–>equation();
        printf("Face %d is a %s type \n", f_num, sfs.type_name());
```

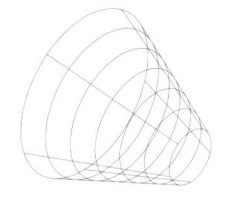

Lines of constant parameter on the surface of a truncated cone.

The result can be verified using the Scheme AIDE.

5.2 Accessing FACE Details

f_detail.scm

```scheme
; This program identifies the type and number of loops on each face of a body
; Create a test body
(define cone (solid:cone (position 0 0 0) (position 0 0 50) 30 10))
; Get a list of the faces
(define get-faces
 (lambda (body)
   ( let*(
   (face-list (entity:faces body))
   (number (length face-list)))
    (begin
      (display "Number of faces= ")
      (display number)
      (newline)
      (for-each  face-type face-list)))))

; Get a list of the loops on a face
(define face-type (lambda (face)(
    let( (loopn (length (entity:loops face))))
    (begin
      (display " The ") (print-face-type face)
      (display " face has ")(display loopn)
      (display " loops ")(newline)))))

; Print out the type of surface associated with the face
(define print-face-type
 (lambda (face)
  (cond
    ((face:planar? face)    (display "planar"))
    ((face:conical? face)   (display "conical"))
    ((face:spline? face)    (display "spline"))
    ((face:toroidal? face)  (display "toroidal"))
    ((face:spherical? face) (display "spherical"))
    (else (display "unknown")))))
```

To run type:
(get-faces cone)

Given the truncated cone the program **get-faces** outputs the following when executed in the Scheme AIDE's Interpreter:

```
Number of faces= 3
The conical face has 2 loops
The planar face has 1 loops
The planar face has 1 loops
()
```

5.3 Counting EDGEs

Below the level of FACE the ACIS data structure becomes a **graph** (or **network**) rather than the **hierarchical** relationships of the BODY, LUMP and SHELL objects. This introduces a new twist to the problem of traversing the lower parts of the data structure since now, rather than simply working through a neat linear list of, say, FACE objects, the programmer has to traverse a web of interconnections. One of the classic problems of B-rep navigation involves counting every EDGE of a model. The difficulty lies not so much in visiting every EDGE (which we can do using every LOOP of every FACE in a SHELL's list of FACEs) but in ensuring that no EDGE is counted more than once. The program **e_count.cxx** illustrates one solution that exploits the fact that each COEDGE has a direction, or **sense**, which allows a LOOP to be traversed with the material of the FACE on the left.

e_count.cxx

```cpp
// This program creates an ENTITY list of the edges on a body
#include <stdio.h>                              // Input/Output functions
#include "constrct/kernapi/api/cstrapi.hxx"     // Declares constructor API's
#include "kernel/kernapi/api/kernapi.hxx"       // Declares kernel APIs
#include "kernel/kerndata/top/alltop.hxx"       // Topological Classes
#include "kernel/kerndata/lists/lists.hxx"      // Entity List Class

// Overload Function Prototypes
void get_edges(ENTITY_LIST&, BODY*);
void get_edges(ENTITY_LIST&, LOOP*);

void main(){
    api_start_modeller(0);
    api_initialize_constructors();
    BODY *block;
    api_make_cuboid(100,150,200, block);
    ENTITY_LIST edge_list;
    get_edges(edge_list,block);
    printf("The block has %d edges \n", edge_list.count());
    api_terminate_constructors();
    api_stop_modeller();
}

void get_edges( ENTITY_LIST& edge_list, BODY *bod)
{ // Assume "bod" has one lump and one shell
    FACE *ff = bod->lump()->shell()->face_list();
    while(ff != NULL)
      {
        for(LOOP *lp = ff->loop(); lp !=NULL; lp = lp->next())
            get_edges(edge_list,lp);
        ff = ff->next_in_list();
      }
}
```

5.3 Counting EDGEs

```
void get_edges( ENTITY_LIST& edge_list, LOOP *lp)
{
    COEDGE *ce = lp->start();
    do {
     if (ce->sense() == FORWARD)
         edge_list.add(ce->edge());
     ce = ce->next();
    } while (ce != lp->start());
}
```

Once compiled and run the following text is printed to the screen:

```
The block has 12 edges
```

The problem here is to make sure EDGEs are added to the tally only once because every EDGE can be reached from *at least* two different COEDGEs, each lying on different LOOPs associated with different FACEs of the BODY. So, to stop EDGEs being counted twice the program includes the following check:

> **if**(ce-> **sense**() == **FORWARD**)
>   edge_list.**add**(ce-> **edge**());

This test is only applicable to manifold solids.

This line ensures that the EDGE count is only incremented when a COEDGE which points in the same direction as its parent EDGE is encountered (i.e. its start and end points are the same as its parent's). However, in actual fact this program would give the same result if the "if" statement was left out and the inner loop simply read:

> **do**{
>   edge_list.**add**(ce-> **edge**());
>   ce = ce-> **next**();
> } **while**(ce != lp-> **start**());

Although easier to program, this is probably a less efficient way of making the count and relies on the fact that ENTITY_LISTs do not keep duplicate copies. So, although the program will try to add a pointer to every EDGE twice, the list will in practice only record one copy.

In Scheme the whole operation can be carried out in one line as follows:

> *; Count the number of EDGEs in a BODY*
>   (**define** number (**length** (**entity:edges** body)))

At first glance the Scheme extension **entity:edges** appears to do exactly the same as the **e_count.cxx** program. However, it has a far more general definition, accepting any type of ENTITY as its argument. This allows EDGE counts to be made on individual FACEs, LOOPs, LUMPs or SHELLs, not to mention nonmanifold, mixed dimension and incomplete BODYs.

5.4 Getting Vertex Coordinates

The access methods used in the previous examples have all returned pointers to other classes. This *pointer chasing* stops when lowercase geometry or utility classes are reached. Because these hold the actual numerical values of the various geometric variables they have no need to refer to other ENTITYs and are therefore *not* returned as pointers by the access methods of the ENTITY classes. This can be seen in the program **v_coords.cxx** where position classes are returned from the APOINT class.

v_coords.cxx

```
// This program prints out the start and end coordinates
// of every EDGE on a solid, manifold BODY
#include <stdio.h>                              // Input/Output functions
#include "constrct/kernapi/api/cstrapi.hxx"    // Construction API's
#include "kernel/kernapi/api/kernapi.hxx"      // Declares kernel API's
#include "kernel/kerndata/top/alltop.hxx"      // Declares Topological Classes
#include "kernel/kerndata/lists/lists.hxx"     // Declares Entity_list Class
#include "baseutil/vector/position.hxx"        // Declares Position Class
#include "kernel/kerndata/geom/point.hxx"      // Declares Point Class
#include "baseutil/vector/transf.hxx"          // Declares the transform class
#include "baseutil/vector/vector.hxx"          // Declares the vector class
#include "kernel/kerndata/geom/transfrm.hxx"   // Declares the TRANSFORM class


// Function Prototypes
void get_edges(ENTITY_LIST&, BODY*);  // As defined in e_count.cxx
void get_edges(ENTITY_LIST&, LOOP*);  // As defined in e_count.cxx

void main()
{
    api_start_modeller(0);
    api_initialize_constructors();
    BODY *block;
    api_make_cuboid(100,150,200, block);

    // Translate block 23 along x-axis
    transf moveX = translate_transf(vector(23,0,0));
    api_apply_transf(block,moveX);

    ENTITY_LIST edge_list;
```

5.4 Getting Vertex Coordinates

```
    get_edges(edge_list, block);
    printf("\nThe block has %d edges\n", edge_list.count());

    // Get a transformation to map between local and global coordinate system
    transf bod_tran = block-> transform()-> transform();

    for (int i = 0; i<edge_list.count(); i++)
    {
        EDGE *e = (EDGE*)edge_list[i];

        // Get the untransform coordinates from each VERTEX
        position untransform_start = e-> start()-> geometry()-> coords();
        position untransform_end = e-> end()-> geometry()-> coords();

        // Apply the BODY's transform to each coordinate
        position start = untransform_start * bod_tran;
        position end = untransform_end * bod_tran;

        printf("edge %d: \n", i+1 );
        printf("start coord = %f,%f,%f\n",start.x(),start.y(),start.z());
        printf("end coord   = %f,%f,%f\n",end.x(),end.y(),end.z());
    }

    api_terminate_constructors();
    api_stop_modeller();
}
```

Once compiled and run the following text is printed to the screen:

```
The block has 12 edges
edge 1:
start coord = 73.000000,-75.000000,100.000000
end coord   = 73.000000,75.000000,100.000000
edge 2:
start coord = 73.000000,75.000000,100.000000
end coord   = -73.000000,75.000000,100.000000
        :
        :
        :
        :
start coord = 73.000000,-75.000000,100.000000
end coord   = 73.000000,-75.000000,-100.000000
```

This example takes the list created by the function **get_edges** (described earlier in this chapter) and steps through it printing out the start and end coordinates of each edge. Perhaps the most interesting line of the program is:

EDGE *e = (**EDGE***)edge_list[i];

The above line of code is doing two things :

- **edge_list[i]** Get the *i*th element out of the list.
- **(EDGE*)edge_list[i]** Cast the *i*th element to be a pointer to an EDGE (rather than an ENTITY).

The line:

$$\textbf{position } \text{untransform\_start} = e\text{->}\textbf{start}()\text{->}\textbf{geometry}()\text{->}\textbf{coords}();$$

also condenses several steps that could be more clearly seen if they were written:

VERTEX *sv = e−> **start**();
APOINT *sp = sv−> **geometry**();
position untransform_start = sp−> **coords**();

The geometry obtained using the direct interface takes no account of any transforms that have been applied to the BODY (see section 1.12, page 15). So before printing out the coordinates, each position is transformed using the infix operator * to apply the BODY's transform **bod_tran**.

In this example the ENTITY_LIST contains only 12 items so speed is not really an issue. If, however, the list contained several thousand ENTITYs it would be much more efficient to use the **init()** and **next()** functions (see page 76).

In the Scheme AIDE the user can operate above this level of detail by using the Scheme extension **entity:vertices** which returns a list of VERTEX objects for any ENTITY.

v_coords.scm

```
; This function prints out the coordinates of all an entity's vertices
(define vcoords (lambda(body)
    (let*( (vertex-list (entity:vertices body)))
        (begin
            (display "Total number of vertices =")
            (display (length vertex-list))
            (newline)
            (for-each (lambda (vertex)
                (display (vertex:position vertex))
                (newline))
                vertex-list)))))
```

5.5 Creating Primitives from ENTITYs

In the majority of cases objects can be created using API functions, though it is also possible to create models using the direct interface by calling the constructor functions of each **ENTITY** required. The program **tetra.cxx** provides an example of this (laborious) process by using a number of ENTITY constructors to create a tetrahedron. Because of its length the program is presented in a number of fragments.

tetra.cxx

```
// Create a tetrahedron using the direct interface
#include <stdio.h>                            // File I/O functions
#include "kernel/acis.hxx"                    // System wide Definitions
#include "kernel/kernapi/api/api.hxx"         // Declares API macros
#include "kernel/kernapi/api/kernapi.hxx"     // Declares kernel API's
#include "kernel/kerndata/top/alltop.hxx"     // Declare Topology classes
#include "kernel/kerndata/geom/point.hxx"     // Declare point class
#include "kernel/kerndata/geom/allcurve.hxx"  // Declare curve classes
#include "kernel/kerndata/geom/allsurf.hxx"   // Declare surface classes
#include "kernel/kerndata/lists/lists.hxx"    // Declare ENTITY_LIST class
#include "baseutil/vector/position.hxx"       // Declare position class
#include "baseutil/vector/vector.hxx"         // Declare vector class
#include "baseutil/vector/unitvec.hxx"        // Declare unit_vector class
#include "kernel/kerndata/savres/fileinfo.hxx" // Declares fileinfo class

void tetra(BODY*&);          // Function prototype
void save_ent(char*, ENTITY*);  // See page 72)

void main()
{

    api_start_modeller(0);
    api_initialize_kernel();

    BODY* body;
    tetra(body);

    save_ent("tetra.sat",body);

    api_terminate_kernel();
    api_stop_modeller()
}

void tetra(BODY*& body)
{
    // Macro to ensure current BULLETIN_BOARD has been created
    API_BEGIN
    // create points
```

```
APOINT* p0 = new APOINT(0.0, 0.0, 0.0);
APOINT* p1 = new APOINT(10.0, 0.0, 0.0);
APOINT* p2 = new APOINT(0.0, 10.0, 0.0);
APOINT* p3 = new APOINT(0.0, 0.0, 10.0);

// create straights
STRAIGHT* s0 = new STRAIGHT( p0->coords(),
                normalise(p1->coords() - p0->coords()));
STRAIGHT* s1 = new STRAIGHT( p1->coords(),
                normalise(p2->coords() - p1->coords()));
STRAIGHT* s2 = new STRAIGHT( p2->coords(),
                normalise(p0->coords() - p2->coords()));
STRAIGHT* s3 = new STRAIGHT( p2->coords(),
                normalise(p3->coords() - p2->coords()));
STRAIGHT* s4 = new STRAIGHT( p0->coords(),
                normalise(p3->coords() - p0->coords()));
STRAIGHT* s5 = new STRAIGHT( p3->coords(),
                normalise(p1->coords() - p3->coords()));
```

Note: All changes to ENTITYs via the direct interface should be enclosed between calls to **API_BEGIN** and **API_END**. See Appendix C, page 383.

The program starts by declaring the necessary header files, and a **main** function which calls the function **tetra** to create a BODY which is then saved for later viewing and verification. The function **tetra** starts its creation of a tetrahedron by creating four pointers to APOINT[1] objects, one at the origin and one 10 units along each of the x, y and z axes. Having created the simplest ENTITYs the procedure continues by constructing six **STRAIGHT** CURVEs, one to provide the shape of each of the tetrahedron's EDGEs. The STRAIGHT constructor creates an object that represents an infinite line through the given position and in the direction of the unit_vector given as the second argument.

The positions are extracted from the APOINT objects using their **coords** member function. The unit_vectors are generated by subtracting two point objects, forming a vector which is then **normalized**.

Although this procedure appears straightforward enough, it should be remembered that the **direction** of the STRAIGHTs created here will determine the arguments of the constructor calls later in the program.

tetra.cxx (continued)

```
// create surfaces
PLANE* pln0 = new PLANE(p0->coords(),unit_vector(0.0, 0.0, -1.0));
PLANE* pln1 = new PLANE(p0->coords(),unit_vector(-1.0, 0.0, 0.0));
PLANE* pln2 = new PLANE(p0->coords(),unit_vector(0.0, -1.0, 0.0));
PLANE* pln3 = new PLANE(p1->coords(),normalise(vector(1.0, 1.0, 1.0)));

// create vertices
```

---

[1] In early versions of ACIS the APOINT class was known simply as POINT. The name was changed because it conflicted with a Microsoft structure of the same name.

5.5 Creating Primitives from ENTITYs

```
VERTEX* v0 = new VERTEX(p0);
VERTEX* v1 = new VERTEX(p1);
VERTEX* v2 = new VERTEX(p2);
VERTEX* v3 = new VERTEX(p3);

// create edges
EDGE* e0 = new EDGE(v0, v1, s0, FORWARD);
EDGE* e1 = new EDGE(v1, v2, s1, FORWARD);
EDGE* e2 = new EDGE(v2, v0, s2, FORWARD);
EDGE* e3 = new EDGE(v2, v3, s3, FORWARD);
EDGE* e4 = new EDGE(v0, v3, s4, FORWARD);
EDGE* e5 = new EDGE(v3, v1, s5, FORWARD);
```

Having constructed APOINT and STRAIGHT objects the program continues by creating four **PLANE** SURFACEs to underlie (i.e. provide geometry for) the object's FACEs. The PLANE's constructor defines an infinite plane through the given point and normal to the specified unit_vector. Again the choice of direction is critical because the unit_vector given here will be used later to determine which side of the material represents solid and which fresh air. Four VERTEX objects are then constructed to define the boundary (or ends) of the STRAIGHT objects created earlier. Having done this the EDGE constructors' arguments specify the EDGE's two bounding VERTEXs, its CURVE and its **sense** relative to its underlying geometry (i.e. the STRAIGHT CURVE). Notice that all the EDGE objects are created with a **FORWARD** sense reflecting the careful choice of direction made when the STRAIGHTs were being constructed.

tetra.cxx (continued)

```
// create loops of coedges
    // xy plane
    COEDGE* c0 = new COEDGE(e0, REVERSED, NULL, NULL);
    COEDGE* c2 = new COEDGE(e2, REVERSED, c0, NULL);
    COEDGE* c1 = new COEDGE(e1, REVERSED, c2, c0);
    c0->set_previous(c1);
    LOOP* l0 = new LOOP(c0, NULL);

    // yz plane
    c2 = new COEDGE(e2, FORWARD, NULL, NULL);
    COEDGE* c4 = new COEDGE(e4, FORWARD, c2, NULL);
    COEDGE* c3 = new COEDGE(e3, REVERSED, c4, c2);
    c2->set_previous(c3);
    LOOP* l1 = new LOOP(c2, NULL);

    // xz plane
    c0 = new COEDGE(e0, FORWARD, NULL, NULL);
    COEDGE* c5 = new COEDGE(e5, REVERSED, c0, NULL);
```

```
        c4 = new COEDGE(e4, REVERSED, c5, c0);
        c0->set_previous(c4);
        LOOP* l2 = new LOOP(c0, NULL);

        // canted plane
        c1 = new COEDGE(e1, FORWARD, NULL, NULL);
        c3 = new COEDGE(e3, FORWARD, c1, NULL);
        c5 = new COEDGE(e5, FORWARD, c3, c1);
        c1->set_previous(c5);
        LOOP* l3 = new LOOP(c1, NULL);
```

The construction of the COEDGE objects is probably the most complex step in the whole process. Each COEDGE object has to be located in a linked list of other COEDGEs which bound the same FACE. The constructor takes four arguments: the "parent" EDGE, a direction or sense relative to the given EDGE and the **previous** and **next** COEDGE objects in the LOOP. Having constructed the linked lists of COEDGEs the LOOP objects can be created. The LOOPs constructor arguments initialize the first COEDGE on the LOOP and the **next** LOOP in a list of LOOPs bounding the FACE. Since each FACE has only one LOOP, the second argument is NULL.

tetra.cxx (continued)

```
        // create faces
            FACE* f0 = new FACE(l0, NULL, pln0, FORWARD);
            FACE* f1 = new FACE(l1, f0, pln1, FORWARD);
            FACE* f2 = new FACE(l2, f1, pln2, FORWARD);
            FACE* f3 = new FACE(l3, f2, pln3, FORWARD);
        // create shells
            SHELL* sh0 = new SHELL(f3, NULL, NULL);
            LUMP* lu0 = new LUMP(sh0, NULL);
        // create body
            body = new BODY(lu0);
        // close current BULLETIN_BOARD
            API_END
        }
```

The use of **new**, or **ACIS_NEW**, is mandatory. ENTITYs are not allowed to be created on the stack.

The process of tetrahedron creation ends with the construction of the high level ENTITYs FACE, SHELL, LUMP and lastly BODY. All the ENTITYs have been created using the **new** operator which prevents the memory used being deallocated when the function terminates and ensures that **create** BULLETINs are posted.

The program **tetra.cxx** shows what hard work it is creating a solid from scratch and how the **constructor** functions of the different classes can be used to create a complete data structure. Implicit in the reading of this example should be the realization that it is *very* rare to see objects created in this way because it is so easy to make errors.

5.6 Parametric Surface Inquiries

Of course, not all objects have such complex data structures; a sphere, for example, can be defined by creating only a handful of ENTITYs.

mk_sph.scm

```
; This function creates a sphere from a single face, with no edges or vertices
(define mk_sphere (lambda(radius)
    (let((fs (face:sphere (position 0 0 0) radius)))
      (sheet:face fs))))

(mk_sphere 25) ; To execute the function
```

The program **mk_sph.scm** exploits the fact that spheres (like tori) have data structures identical to the **sheet** BODYs constructed by the Scheme extension **sheet:face**. The following commands show this:

```
acis> (define test (mk_sphere 20))
acis> (entity:debug test 3)
    1 body record,           32 bytes
    4 attribute records,   1284 bytes
    1 lump record,           32 bytes
    1 shell record,          40 bytes
    1 face record,           44 bytes
    1 surface record,       168 bytes
Total storage 440 bytes
"solid body"
```

The four attribute records have been created by default display operations.

The function **entity:debug**[2] reveals how the original data structure of the single FACE has been developed into a BODY. Generally the Scheme AIDE offers higher-level functions for creating objects (such as sweeping and Booleans), though parts can also be built from individual FACEs.

5.6 Parametric Surface Inquiries

All ACIS surfaces have a parametric interface that allows points on them to be defined in terms of an embedded (u, v) coordinate frame. This allows generic applications to be written in a manner which is independent of the underlying geometry.

To demonstrate this the program **wiggle.cxx** creates a cylinder whose axis is *normal* to a surface at an arbitrary location. In this case the example works on a doubly curved surface created with **api_wiggle** but if, say, **api_make_frustum** were used instead the program would still function. The example starts by locating the spline FACE (i.e. obtaining a pointer to it) in the BODY's list of FACEs. Methods

---
[2]Debug levels go from a basic type check (0) to full print out of the data-structure (4).

from the **surface** class are then used to create a cylinder *normal* to the spline FACE at an arbitrary location. The surface class methods used are:

param which returns a parameter position, known as a **par_pos** (i.e. $u - v$ coordinates), of a point on the surface close to a given Cartesian position.

eval_position which returns a Cartesian position (for the end of the cylinder) given a **par_pos**.

eval_normal which returns the surface normal at a given **par_pos**.

The program ends with the creation of a cylinder with the center of one end lying on the spline surface.

wiggle.cxx

```
// This program makes a block with one spline face and places
// a cylinder on, and normal to, the surface at a given point
#include <stdio.h>                                // I/O functions
#include "constrct/kernapi/api/cstrapi.hxx"       // Construction API's
#include "kernel/kernapi/api/kernapi.hxx"         // Declares kernel API's
#include "kernel/kerndata/top/alltop.hxx"         // Defines Topology classes
#include "kernel/kerndata/lists/lists.hxx"        // ENTITY_LIST class for save
#include "baseutil/logical.h"                     // Defines TRUE/FALSE
#include "baseutil/vector/vector.hxx"             // Defines the vector class
#include "kernel/kerndata/geom/allsurf.hxx"       // SURFACE classes
#include "kernel/kerngeom/surface/allsfdef.hxx"   // surface classes
#include "kernel/kerndata/savres/fileinfo.hxx"    // Declares fileinfo class

void main()
{
    api_start_modeller(0);
    api_initialize_constructors();
    BODY *block, *pointer;
    double width = 50, depth = 50, height = 10;
    int low_v = 1;  // S shaped curve on each v side
    int high_v = 1;
    int low_u = 2; // Double hump curve on each u side
    int high_u = 2;
    api_wiggle(width,depth,height,
            low_v, high_v,
            low_u, high_u, TRUE, block);
    SHELL *eg= block-> lump()-> shell();
    FACE *ff;
    for(ff = eg-> face_list();   // Find the spline FACE
          ff != NULL && ff-> geometry()-> identity() != SPLINE_TYPE;
          ff=ff-> next());
    SURFACE *fs = ff-> geometry();
    const surface& sfs=fs-> equation();
```

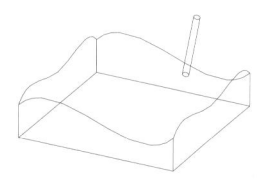

```
        // Find u-v params closest to point
    par_pos spp = sfs.param(position(10,15,10));
        // Get point on surface at u-v params
    position start = sfs.eval_position(spp);
        // Get normal on surface at u-v params
    unit_vector pn = sfs.eval_normal(spp);
        // Make a cylinder lying along the normal
    position end = start + pn*20;
    api_solid_cylinder_cone(start, end, 1, 1, 1, NULL, pointer);

    FileInfo info;  // Create FileInfo Object
    info.set_product_id("HW-University");  // set info's data
    info.set_units(1.0);  // Millimeters

    // Sets header info to be written to sat file
    api_set_file_info(FileId | FileUnits, info);

    FILE *fp = fopen("wiggle.sat", "w");
    ENTITY_LIST savelist;
    savelist.add(block);
    savelist.add(pointer);  // Save the wiggle and the cylinder
    api_save_entity_list(fp,TRUE,savelist);
    fclose(fp);

    api_terminate_constructors();
    api_stop_modeller();
}
```

The functions used in this program are not special to spline surfaces and can be applied to any type of ACIS surface.

5.7 An EDGE Marching Algorithm

There are many modeling operations which require an EDGE to be "marched" along and some form of analysis carried out at each step. This type of procedure is easy to implement in a naïve way, where a fixed step size can be used, but in the general case (where the increment has to be chosen so that no detail is missed) they can be very hard to implement.

The program **vexity.scm** demonstrates the application of some ACIS functions which use marching procedures to determine the vexity of an EDGE (i.e. convex or concave).

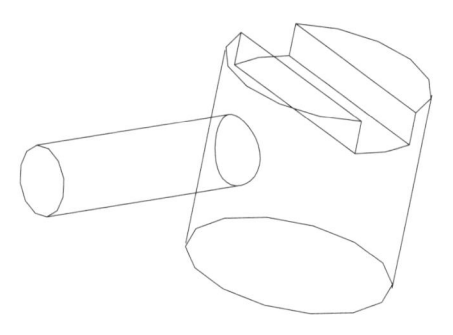

Figure 5.1: The vexity.scm program test component.

vexity.scm

```
; This program unites two orthogonal cylinders (one with a slot)
; and then marches along each edge testing if it has acute angled
; concavity along its entire length

(define vexity
 (lambda ()
  (let*    ;Create a test object
    ((cyl1 (solid:cylinder  (position 0 0 0) (position 0 0 30) 15))
     (cyl2 (solid:cylinder  (position 0 0 0) (position 0 0 50) 5))
     (slot (solid:block (position -25 -5 25) (position 25 5 40)))
     (rot (transform:rotation  (position 0 0 0) (gvector 1 0 0) 90))
     (up (transform:translation ( gvector 0 0 20)))
     (move (transform:compose  rot up)))
   (begin
    (entity:transform cyl2 move)
    (solid:unite cyl1 cyl2)
    (solid:subtract cyl1 slot))  ; Finished test part
    (let
      ((faces (entity:faces cyl1))) ;List of FACEs
      (do ((index (length faces) (- index 1)))
        ((< index 1) 'Done)  ; do for each face
        (newline)(display "Face ") (display  index)(display " has:" )(newline)
        (walk-loops (entity:loops (list-ref faces (- index 1)))) )))))
```

Having created a test component and extracted a list of FACEs the Scheme procedure **vexity** ends with a **do** loop which calls **walk-loops** for each FACE in turn. **Walk-loops** gets a list of each FACE's LOOPs and then calculates the vexity of each EDGE on each LOOP in turn.

5.7 An EDGE Marching Algorithm

vexity.scm (continued)

```
; Calculate the vexity of each EDGE of each LOOP in the given list
(define walk-loops
  (lambda(loops)
    (let* ((no_loops (length loops))) ; find the number of loops in the list
      (begin
        (do ((no 0 (+ no 1)))
            ((>= no no_loops) 'finish) ; do for each loop
          (let* ( (results (list )) ; create a list to hold the results
                  (edges (entity:edges (list-ref loops no))) ; create list of edges in a loop
                  (no_edges (length edges))) ; find the number of edges in the list
            (do ((eno 0 (+ eno 1))) ; do for each edge
                ((>= eno no_edges) 'finish)
              (let* ((cvty (ed-cvty-info:instantiate ; call edge vexity routines for each edge
                            (edge:ed-cvty-info (list-ref edges eno) 0.01) 0.01)))
                (cond   ; Test the type of edge convexity found
                  ((cvty:inflect cvty) (set! type 'inflect))
                  ((cvty:convex cvty) (set! type 'convex))
                  ((cvty:concave cvty) (set! type 'concave))
                  ((cvty:tangent cvty) (set! type 'tangent))
                  ((cvty:knife cvty) (set! type 'knife))
                  ((cvty:mixed cvty) (set! type 'mixed))
                  ((cvty:unknown cvty) (set! type 'unknown))))
              (set! results (cons type results)))
            (display "   Loop Vexity = ")(print results)))))))

(vexity) ; To execute the program
```

The program calls the Scheme extension **edge:ed-cvty-info** for each EDGE of each LOOP. It is this routine that marches along the EDGE and computes convexity information which includes the minimum and maximum angles along the EDGE (+ve for convex, -ve for concave).

The routine **ed-cvty-info:instantiate** is called to classify the angle information with respect to a given tolerance. When instantiated the convexity information can be classified as convex or concave etc. When executed the program prints out the vexity information for each LOOP on each FACE in the following format:

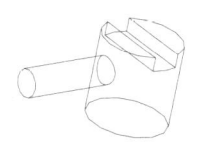

```
Face 9 has:
    Loop Vexity = (convex convex)

Face 8 has:
    Loop Vexity = (convex)

Face 7 has:
```

```
        Loop Vexity = (convex)

Face 6 has:
    Loop Vexity = (convex)
    Loop Vexity = (concave)

Face 5 has:
    Loop Vexity = (convex)
    Loop Vexity = (concave)
    Loop Vexity = (convex convex convex convex con-
vex convex convex convex)

Face 4 has:
    Loop Vexity = (convex convex)

Face 3 has:
    Loop Vexity = (concave convex convex convex)

Face 2 has:
    Loop Vexity = (concave convex concave convex)

Face 1 has:
    Loop Vexity = (convex convex concave convex)
Done
```

Visual inspection of these results shows they are correct for the component shown. For example the largest cylindrical FACE is easily identifiable as FACE number 5 in the output by the presence of three LOOPs (two entirely convex and one concave).

The functions for assessing point and edge vexity in C++ can be found in the file kern/kernel/kerndata/top/edge.hxx.

5.8 Sense and Sharing of FACE Geometry

This chapter ends with a program that demonstrates two important features of the ACIS data structure:

- The sharing of geometry (i.e. points, curves and surfaces).

- The orientation of FACEs: each FACE has associated with it a **sense** that defines if its **outside** is in the same direction as its SURFACE geometry.

The program **use_cont.cxx** illustrates these two aspects of the data structure via the creation of a "slotted block". Both the number of SURFACEs and the sense of each FACE are output before and after the Boolean operation used to create the slot. The sense of a FACE is determined by a flag, known as a REVBIT, which indicates if the surface normal at any point on the FACE is opposite to (i.e. REVBIT == REVERSED) or the same as (i.e. REVBIT == FORWARD) the underlying SURFACE.

5.8 Sense and Sharing of FACE Geometry

use_cont.cxx

```cpp
// This program makes a block with a slot through it and compares
// the values of each FACE's use-counts and revbits before and after the
// Boolean. Any FORWARD normals are also printed out.
#include <stdio.h>                              // Defines file I/O functions
#include "boolean/kernapi/api/boolapi.hxx"      // Declares boolean API's
#include "constrct/kernapi/api/cstrapi.hxx"     // Constructor API's
#include "kernel/kernapi/api/kernapi.hxx"       // Declares kernel API's
#include "kernel/kerndata/top/alltop.hxx"       // Declares the body class
#include "kernel/kerndata/geom/allsurf.hxx"     // Declares SURFACE Classes
#include "kernel/kerngeom/surface/allsfdef.hxx" // Declares surface classes
#include "kernel/kerndata/lists/lists.hxx"      // Declares ENTITY_LIST class
#include "baseutil/vector/position.hxx"         // Declares position class
#include "baseutil/vector/unitvec.hxx"          // Declares unit_vector class

inline position p(double p1, double p2, double p3){return position(p1,p2,p3);}
void output_details(BODY*,char*);

void main()
{
    api_start_modeller(0);
    api_initialize_booleans();
    BODY *block, *slot;
    api_solid_block(p(-50,-50,0), p(50,50,50), block);
    api_solid_block(p(-60,-25,25), p(60,25,55), slot);
    output_details(block,"Block before subtract");
    output_details(slot,"Slot before subtract");
    api_subtract(slot,block);
    output_details(block,"Block after subtract");
    api_terminate_booleans();
    api_stop_modeller();
}

void output_details(BODY* part, char* comment)
{
    printf("%s\n", comment);
    FACE *ff=part->lump()->shell()->face();
    int nof = 0, rvb_count = 0;
    ENTITY_LIST surf_list;
    while(ff != NULL)
       {
       nof++;
       surf_list.add((ENTITY*)(ff->geometry()));
       if (ff-> sense() == REVERSED)
            rvb_count++;
       else {
             const surface& suf = ff->geometry()->equation();
             if (suf.type() == plane_type){
```

```
                        plane forw_pl = (plane&)suf;
                        unit_vector norm = forw_pl.normal;
                        printf("Forward Surface Normal = (%f,%f,%f)\n",
                                norm.x(),norm.y(),norm.z());
                    }
                }
                ff = ff-> next();
            }
            printf("%d FACEs (%d with REVERSED sense) and
                    %d SURFACEs \n \n",
                    nof, rvb_count, surf_list.iteration_count());
    }
```

When run this program should produce the following output:

```
Block before subtract
Forward Surface Normal = (0.0,0.0,1.0)
6 FACEs (5 with REVERSED sense) and 6 SURFACEs

Slot before subtract
Forward Surface Normal = (0.0,0.0,1.0)
6 FACEs (5 with REVERSED sense) and 6 SURFACEs

Block after subtract
Forward Surface Normal = (0.0,1.0,0.0)
Forward Surface Normal = (0.0,0.0,1.0)
Forward Surface Normal = (0.0,-1.0,0.0)
Forward Surface Normal = (0.0,0.0,1.0)
Forward Surface Normal = (0.0,0.0,1.0)
10 FACEs (5 with REVERSED sense) and 9 SURFACEs
```

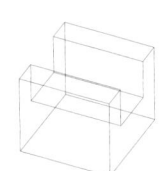

What's going on? We can see that when first created the two blocks have only one surface which has the same orientation as the FACE it supports. However, after the subtraction of the slot, three new FACEs are formed on the BODY, each of which has opposite orientation to the FACE it was formed from on the tool BODY. In other words the Boolean operation creates the three new FACEs on the blank from the SURFACEs of the tool BODY. So although the new FACEs have different boundaries from their originals they share the same SURFACE, but with the orientation reversed. Hence we see that the number of FORWARD orientated FACEs on the block increases from one to five. One is created because of the division of the original FORWARD face; the other three are the new slot FACEs.

The division of the "top" FACE by the slot also accounts for the difference between the numbers of FACEs and SURFACEs. The two top FACEs share the same SURFACE.

5.9 Exercises

1. Discuss how the program **e_count.cxx** would behave if the function **get_edges** were presented with a nonmanifold **BODY**.

2. Rewrite the **get_edges** functions of the program **e_count.cxx** so that they will work on a BODY which has multiple LUMPs and SHELLs.

3. The first step in most hidden-line algorithms is to remove those FACEs completely invisible to the user. Write a program that, given a vector and a planar faced BODY, creates an ENTITY_LIST containing only FACEs orientated "towards" the given vector.

4. Archimedes, allegedly, investigated a class of polyhedra all of whose VERTEXs[3] are identical but whose FACEs are of two different types, each type being a regular polygon. Figure 5.2 and Table 5.1 detail three semi-regular polyhedra for which sat files are available from this book's web site.

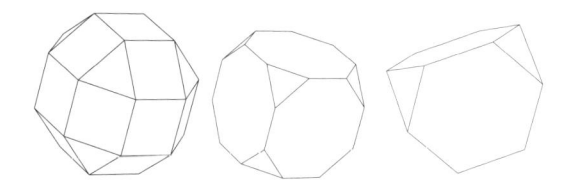

Figure 5.2:   Three semi-regular polyhedra.

Table 5.1   Three semiregular polyhedra.

Name	VERTEXs	EDGEs	FACE type 1	FACE type 2
Truncated cube	24	36	6 octagons	8 triangles
Small rhombicuboctahedron	24	48	8 triangles	18 squares
Truncated tetrahedron	12	18	4 triangles	4 hexagons

Write a program which, given one of these three models, can name it.

---

[3] Identical because the number and angle of EDGEs meeting at a given VERTEX are the same.

ENTITY Intersections and Booleans 6

This chapter explores the different ways in which ACIS supports the calculation of intersections between ENTITYs (i.e. EDGE-FACE, FACE-FACE etc.) and set (or Boolean) operations, such as unite, intersect and subtract.

Intersection and set operations are the bread and butter of geometric modeling, being fundamental to many common applications, such as:

- Ray-tracing: where the path of light is followed until it **intersects** the surface of a BODY.

- Blending: where material is added to or **subtracted** from a BODY.

- Clash detection: where potential collisions between BODYs are checked for by calculating the **intersection** volume.

Intersections arise when two ENTITYs both occupy the same region of space. This chapter starts by looking at how intersections can be calculated for geometry of differing dimensions (i.e. points, curves, surfaces and finally 3D BODYs). The focus towards the end of the chapter is on the way in which these procedures are used in combination to carry out Boolean (or set) operations. These later sections also illustrate how WIRE and nonmanifold BODYs arise quite naturally during the calculation of a Boolean operation.

6.1 Comparisons of Points

The simplest intersections occur in one dimension, where testing for coincidence simply means checking if two VERTEXs (or points) lie at the same position. What constitutes the *same* is generally determined by the modeler's own internal tolerances.

Because computers have a fixed number of digits available to represent numbers, accuracy is a trade-off between **precision** (i.e. the number of decimal places) and **range** (i.e. the difference between the largest and smallest number representable).

For example, imagine a computer that can use only five digits to represent numbers. Such a system could be visualized as using five sequential boxes, each representing one of the digits available; thus the number 65591 could be envisaged as:

| 6 | 5 | 5 | 9 | 1 |

Likewise the range of numbers representable with five boxes if a precision of one decimal place were used could be envisaged as:

| 0 | 0 | 0 | 0 | . | 1 | → | 9 | 9 | 9 | 9 | . | 9 | (i.e. $10^{-1} \to 10^4$)

Obviously the precision of the system can only be increased by reducing the range. For example if numbers are held to three decimal places of accuracy:

| 0 | 0 | . | 0 | 0 | 1 | → | 9 | 9 | . | 9 | 9 | 9 | (i.e. $10^{-3} \to 10^2$)

Similarly it is no surprise that if the system precision is set to four decimal places the range decreases still further:

| 0 | . | 0 | 0 | 0 | 1 | → | 9 | . | 9 | 9 | 9 | 9 | (i.e. $10^{-4} \to 10^1$)

Notice in each case that the ratio (which ACIS refers to as **resnor**) between the biggest and smallest numbers that can be represented *remains constant*. In other words in all cases:

$$\text{resnor} = \frac{\text{smallest}}{\text{largest}} = \text{constant } (10^{-5} \text{ in this example})$$

The ratio reflects the number of digits available to represent a number. In ACIS the smallest number representable is known as **resabs**.

Table 6.1 gives the default values for resnor and resabs. Internally ACIS is unitless; however, in order to make the sharing of data between applications easy, a unit (e.g. mm, inches, km, etc.) is held in the ACIS model[1]. If one assumes that one unit is equivalent to a millimeter (mm) the range of dimensions available in the default modeling world would be

$$10^{-6} \text{mm} \to 10^4 \text{mm}$$

or

$$0.000001 \text{ mm} \to 10 \text{ meters}$$

Only the value of resabs can be modified by an application (Spatial Corp advises that resnor should not be changed). For example, if you wanted to change the internal units from mm to inches resabs could be set to:

$$\text{resabs} = \frac{10^{-6}}{25.4} = 3.937 \times 10^{-8}$$

---

[1] See the **UNITS_SCALE** class.

6.1 Comparisons of Points

Table 6.1 ACIS tolerance variables.

Name	Description	Default value
resabs	**Smallest number representable**: If two points are separated by less than **resabs** they are considered coincident.	10^{-6}
resnor	**Ratio between the biggest and the smallest values representable**.	10^{-10}
resfit	**Tolerance applied to approximate geometric comparisons**: Every point on an approximating curve or surface should be within **resfit** of the exact geometry to which it approximates.	10^{-3}
resmch	**Computer precision**: The biggest small number a given computer considers to be zero (i.e. 1 + resmch != 1). Reflects the precision to which numerical values are calculated.	10^{-11}

giving a modeling world of up to 393.7 inches. Values of resnor and resabs are recorded in when models are saved to disk (see page 72). However, it is the responsibility of the application to scale incoming models with different tolerances. No action is taken automatically.

Although these variables allow an application to modify the modeler tolerance, much bigger changes in scale can be made simply by assuming a different value for the modeler units. For example, a group of astrophysicists intent on representing the Milky Way, which spans some 10^{14} kilometers (or 100,000 light years), could simply assume that one ACIS unit was equivalent to one light year and start modeling (spheres mostly). As long as everyone using ACIS to model galaxies agreed this was the scale (and didn't independently choose parsecs or astronomical units) then there would be no need to adjust resabs.

Common uses of resabs range from determining when points lie inside or outside a volume through to checking the **geometric integrity**[2] of a model.

The tolerance variables are declared in the file **base.hxx**. Although they can be used directly the following routines (declared in **acistol.hxx**) are provided for making comparisons against these values:

is_equal: Tests if pairs of **doubles**, **positions**, **vectors**, and **unit_vectors** are within **resabs** of each other.

---

[2] The geometric integrity of a model can be checked by ensuring that the surfaces of its FACEs intersect *along* the curves of the EDGEs on their common boundary and that the points of the VERTEXs lie *on* the curves of the EDGEs they bound (see page 11.8).

are_parallel: Tests if pairs of **unit_vectors** or **vectors** are parallel and pointing in the same direction.

are_perpendicular: Tests if pairs of **unit_vectors** or **vectors** are perpendicular to within a tolerance.

is_zero: Tests if individual **doubles**, **positions**, **vectors** and **unit_vectors** are within **resabs** of zero.

6.2 Point Intersection and Containment

In addition to testing pairs of positions for coincidence, ACIS provides functions for determining the relationship between individual points and other geometric entities of higher dimension, such as curves and surfaces. For example, to determine if a given position lies *on* a piece of geometry, the following **position-based inquiry functions** defined in both the curve and surface classes can be used:

test_point: Tests whether a point lies on a curve or surface, to resabs.

test_point_tol: Tests whether a point lies on a curve or surface to a given precision or resabs (whichever is larger).

The use of a similar virtual function (**param**, which gets the u-v coordinates of a point on a surface closest to a given point) was shown in Section 5.6.

Point containment can be determined by using **api_point_in_body** which checks whether a given point lies inside, outside, or on the boundary of a given BODY. This function can also be called via the Scheme extension **solid:classify-position**.

6.3 Lowercase Geometry Intersections

In addition to calculating the intersection of the **bounded** pieces of geometry that represent the FACEs and EDGEs of a BODY, ACIS provides functions for analyzing the interactions of unbounded[3] curves and surfaces.

The reader will recall that unbounded geometry is represented by a number of classes, such as **straight**, **sphere** and **spline** which are distinguished from the associated ENTITY classes (**STRAIGHT**, **SPHERE** and **SPLINE**)[4] by their lowercase names.

Intersections between the elements of lowercase geometry can be calculated by a number of routines not found in the API interface. The routines fall into three categories:

---

[3]Unbounded in the sense that its extent is not restricted by LOOPs of EDGEs. Spheres and circles are unbounded but finite.

[4]The uppercase geometry classes (i.e. **PLANE**) are derived from ENTITY and record model information, such as how many FACEs lie on a given PLANE SURFACE. The lowercase geometry objects (i.e. **plane**) represent only geometry and contain no method for supporting, say, rollback etc.

6.3 Lowercase Geometry Intersections

curve–curve intersections, which return a linked list of **curve_curve_int** objects each of which details an individual intersection.

surface–curve intersections, which return a linked list of **curve_surf_int** objects each of which details an individual intersection.

surface–surface intersections, which return a linked list of **surf_surf_int** objects each of which details an individual intersection.

The program **surfint.cxx** illustrates the simple application of a specific surface–surface intersection routine, namely **int_plane_torus**. After creating a torus lying on the x-y plane, a while-loop is used to create a number of (infinite) plane surfaces each at a different location on the x-axis. The intersection curve between each of these plane surfaces and the torus is calculated and turned into an EDGE (or EDGEs) by an API routine. This transformation from curve to EDGE is done so the resulting ENTITYs can be written to file using the **save_ent** routine defined in Section 4.2.

surfint.cxx

```
// This program creates two lowercase surfaces (a torus and a
// plane) and calculates a number of intersections between them
#include <stdio.h>
#include "constrct/kernapi/api/cstrapi.hxx"   //Construction API's
#include "kernel/kernapi/api/kernapi.hxx"     //Declares kernel API's
#include "kernel/kerndata/top/alltop.hxx"     //Declares upper case topology
#include "kernel/kerngeom/surface/allsfdef.hxx" //Declares non-ENTITY surfaces
#include "kernel/kerndata/geom/allcurve.hxx"  //Declares upper case geometry
#include "kernel/kerngeom/curve/allcudef.hxx" //Declares non-ENTITY curve class
#include "kernel/kerndata/lists/lists.hxx"    //Declares ENTITY_LIST class
#include "baseutil/vector/position.hxx"       //Declares position class
#include "baseutil/vector/unitvec.hxx"        //Declares unit_vector class
#include "intersct/kernint/intsfsf/intsfsf.hxx" //surface/surface intersect classes
#include "kernel/kernint/intsfsf/sfsfint.hxx"   //surface/surface intersect classes
#include "kernel/kerndata/savres/fileinfo.hxx" // Declares fileinfo class

void save_ent(char*, ENTITY_LIST&); // Function prototype; see save.cxx

void main()
{
    api_start_modeller(0);
    api_initialize_constructors();
    ENTITY_LIST int_edges;
    double x_coord = 0.5;
    // make a torodial surface centered on the origin
    torus hoop(position(0,0,0), unit_vector(0,0,1), 40, 15);
    while (x_coord < 50){
        // make a plane surface
        plane slice(position(x_coord,0,0), unit_vector(1,0,0));
```

```
// do a specific surface-surface intersection
surf_surf_int *spiras = int_plane_torus(slice,hoop);
while(spiras != NULL) {
    curve *spira = spiras-> cur;
    EDGE *spiraE;
    if (spira-> type() == ellipse_type){
        ellipse *cur = (ellipse*)spira;
        api_mk_ed_ellipse(cur-> centre, cur-> normal,
                        cur-> major_axis, cur-> radius_ratio,
                        0, 360, spiraE);
    } else if (spira-> type() == intcurve_type){
        intcurve *cur = (intcurve*)spira;
        api_mk_ed_bs3_curve(cur-> cur(),spiraE);
    } else {
      printf("A surprise type\n");
      exit(0);
    }
   int_edges.add(spiraE);
   spiras = spiras-> next;
  }
  x_coord = x_coord + 2.0;
} // end while
save_ent("spira.sat", int_edges);
api_terminate_constructors();
api_stop_modeller();
}
```

Cross-Sections of a Torus

The **surfint.cxx** program is "hard-wired" to test for only two types of resulting curve (i.e. an ellipse or an intcurve, see page 9). Intersections between other combinations of geometry might require more interrogation to determine the nature of the result. Usually surfaces are intersected as part of a Boolean operation when they are associated with a FACE. Because the FACEs have boundaries an optional argument (not shown in the **surfint.cxx** example) can be passed to the intersection routines indicating that intersection need only be calculated within a specific bound (i.e. box).

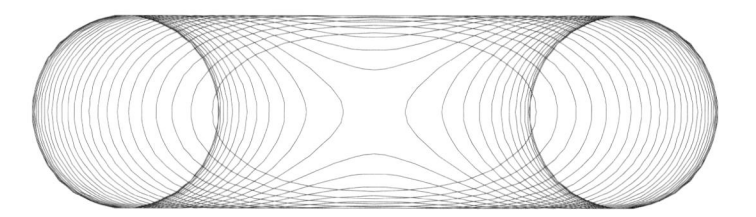

Figure 6.1: The spiric sections of Perseus (x-sections of a torus).

The results of the program (shown in Figure 6.1) have an interesting history. Possibly, one of the first people to wish he had a torus/plane intersection routine was the Greek mathematician Perseus, whose name is now associated with the resulting curves or *spiric sections* (the Greek for torus is spira). The curves themselves are also known as **Cassini's ovals** and have been used (since the 1680s) to model planetary motions.

> About curve/curve intersections
>
> For pairs of analytical curves specific solutions can be used (e.g. circle/circle intersections). However, in the general case intersections between two curves can be found using their parametric representation. The procedure is easily outlined for two parametric curves $s(w)$ and $t(u)$:
>
> 1. Use projection to establish if it is possible for an intersection point to exist between the two curves. This first step exploits the observation that if two curves intersect then their projections must also intersect. In the case of B-spline curves the projection of the control points can be used to test for non-intersection.
>
> 2. Note that at each intersection point $s(w) = t(u)$, implying that
>
> $$x(w) = x(u)$$
> $$y(w) = y(u)$$
> $$z(w) = z(u)$$
>
> Because there are three equations and only two unknowns a solution can be found using some form of numerical analysis. However, the process is complicated by the possibility of more than one intersection point and the accuracy with which digital computations can be performed.

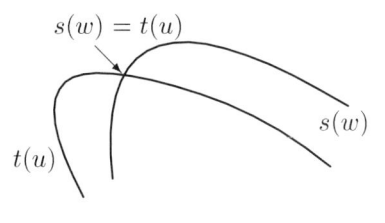

6.4 EDGE–FACE Intersection

Having looked at the intersection of points and unbounded geometry in previous sections the focus now changes to **bounded geometry**. ACIS supports this sort of calculation at the API level and provides routines for the calculation of EDGE–EDGE, EDGE–FACE and FACE–FACE intersections.

In all these routines there is uncertainty about the exact nature of the results. Take, for example, the intersection of an EDGE with a FACE. In simple cases this gives rise to a single VERTEX. However, a number of VERTEXs can result if the EDGE intersects the FACE several times or touches it in more than one place. EDGEs can also arise in places where the curve of an EDGE lies at least in part on a FACE (i.e. it is coincident with the surface). Because of this the API routine used to compute such intersections returns an ENTITY_LIST containing all the VERTEXs and EDGEs required to represent the intersection. The program **edfaint.cxx** demonstrates how this ENTITY_LIST is generated, and its contents analyzed, when an EDGE is intersected with FACE of a sphere.

edfaint.cxx

```
// This program intersects a straight EDGE with the FACE of a sphere
#include <stdio.h>                              // Input/Output functions
#include "constrct/kernapi/api/cstrapi.hxx"     // Construction API's
#include "intersct/kernapi/api/intrapi.hxx"     // Declares intersector API's
#include "kernel/kernapi/api/kernapi.hxx"       // Declares kernel API's
#include "kernel/kerndata/top/alltop.hxx"       // Topological Classes
#include "kernel/kerndata/lists/lists.hxx"      // Entity List Class
#include "kernel/kerngeom/surface/allsfdef.hxx" // surface definitions
#include "kernel/kerndata/geom/surface.hxx"     // SURFACE definitions
#include "kernel/kerndata/geom/point.hxx"       // POINT class

void main(){
    api_start_modeller(0);
    api_initialize_intersectors()
    BODY* ball;  api_make_sphere(50.0, ball);
    FACE *ff = ball->lump()->shell()->face_list();
    position start = position(70,70,70); position end  = position(-70,-70,-70);
    EDGE *ray_edge = new EDGE;
    api_mk_ed_line(start, end, ray_edge);
    ENTITY_LIST *intPs =  new ENTITY_LIST;
    api_edfa_int(ray_edge, ff, intPs);
    printf("Number of intersections = %d \n",intPs->count());
    int num = intPs->count();
    for(int i= num-1; i>=0; i--){
      if( (*intPs)[i]->identity(VERTEX_LEVEL)==VERTEX_TYPE){
        ENTITY *pe = (*intPs)[i];
        APOINT *p = ((VERTEX*)pe)->geometry();
        printf("Intersect Point %d = (%f,%f,%f) \n",
             i,p->coords().x(), p->coords().y(), p->coords().z()); } }
    api_terminate_intersectors();
    api_stop_modeller();}
```

Once the program has been compiled and run, the following text is printed on the screen:

```
Number of intersections = 2
Intersect Point 1 = (-28.8675,-28.8675,-28.8675)
Intersect Point 0 = (28.8675,28.8675,28.8675)
```

Notice that the program outputs only isolated intersections which form VERTEXs. Any EDGE type intersections in the ENTITY_LIST are ignored.

6.5 FACE–FACE Intersection

Generally FACE–FACE intersections give rise to EDGEs and VERTEXs. The function **api_fafa_int** used in the program **fafaint.cxx** returns the results of FACE–FACE intersection calculations in the form of a WIRE BODY.

fafaint.cxx

```cpp
// This program makes a sphere and a torus and intersects their
// FACEs to create some wires and then prints out their bounding boxes
#include <stdio.h>                              // Input/Output functions
#include "boolean/kernapi/api/boolapi.hxx"      // Declares boolean API's
#include "constrct/kernapi/api/cstrapi.hxx"     // Construction API's
#include "kernel/kernapi/api/kernapi.hxx"       // Declares kernel API's
#include "kernel/kerndata/top/alltop.hxx"       // Topological Classes
#include "kernel/kerndata/geometry/getbox.hxx"  // box getting functions
#include "baseutil/vector/box.hxx"              // defines the box class
#include "baseutil/vector/transf.hxx"           // defines the transform class
#include "baseutil/vector/position.hxx"         // position class
#include "kernel/kerndata/data/debug.hxx"       // debug routines for print out

void main(){
    api_start_modeller(0);
    api_initialize_booleans();
    BODY* ball;
    api_make_sphere(50.0, ball);
    BODY* hoop;
    api_make_torus(55.0, 25.0, hoop);

    transf move = translate_transf(vector(25,25,0));
    api_apply_transf((ENTITY*)hoop,move);

    FACE *ballface = ball->lump()->shell()->face();
    FACE *hoopface = hoop->lump()->shell()->face();

    BODY *int_graph;
    api_fafa_int(ballface, hoopface, int_graph);
```

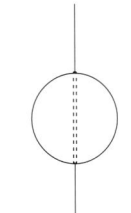

```
        WIRE *wr = int_graph−> wire();
        int count = 0;

        while(wr != NULL) {
            count++;
            box wire_frame = get_wire_box(wr);
            position bottom = wire_frame.low();
            position top = wire_frame.high();

            printf(" Wire Number %d \n ",count);
            printf("Bottom %f %f %f \n",
                   bottom.x(),bottom.y(),bottom.z());
            printf("Top %f %f %f \n",
                   top.x(),top.y(),top.z());
            wr = wr−> next();
        }
        api_terminate_booleans();
        api_stop_modeller();
    }
```

Once the program has been compiled and run, the following text is printed on the screen:

```
Wire Number 1
Bottom -3.096224, -49.904343, -25.000283
Top 49.422759, 7.577808, 25.000283
Wire Number 2
Bottom -49.904343, -3.096224, -25.000283
Top 7.577808, 49.422759, 25.000283
```

Each pair of top and bottom coordinates represent the corners of a bounding box around the WIRE BODY (shown in the marginal illustration by the next program, **fafaint.scm**).

All ACIS's geometric ENTITYs have **axis-aligned** bounding boxes associated with them. These boxes are not created by default (although since they support quick intersection tests they are created during Boolean operations). A program wishing to use a bounding box must force its calculation using one of the **get_*_box** routines (where * can be body, wire, shell, subshell, lump, face, loop or edge).

When the boxes of two ENTITYs are compared, one must be transformed into the coordinate system of the other in order for the comparison to take place. For preliminary testing, each box is transformed and then boxed in the new coordinate system. This is not optimal, but it is relatively quick.

Boxes are computed only when needed, and changed ENTITYs simply require the existing box (if any) to be deleted; however, after a box is computed it is saved for later reuse. Boxes are not logged for rollback purposes, nor are they saved to a disk file. A box pointer in a rollback record is always set to NULL. After a rollback, such boxes must be recomputed.

6.5 FACE–FACE Intersection

The box class itself has a number of useful overload operators, for example:

- `boxA >> boxB` returns TRUE if boxA encloses boxB.
- `boxA >> positionB` returns TRUE if boxA encloses positionB.
- `boxA && boxB` returns TRUE if the boxes overlap.
- `boxA & boxB` returns a box representing the overlap (i.e. the intersection) of boxA and boxB.

The process shown in the program **fafaint.cxx** can be done using Scheme extensions:

fafaint.scm

```
; This program intersects two FACEs
(define mywin (view:new))
(wcs (position 0 0 0)(gvector 1 0 0)(gvector 0 1 0))
(view:set-size 250 250)
(define ball (solid:sphere (position 0 0 0) 50))
(define hoop (solid:torus (position 0 0 0) 55 25))
(entity:transform hoop (transform:translation (gvector 25 25 0 )))
(define ballface (list-ref (entity:faces ball) 0))
(define hoopface (list-ref (entity:faces hoop) 0))
(define int-graph (face:intersect ballface hoopface))
(entity:box (entity:wires int-graph))
(render)
(view:set (position 300 250 150) (position 0 0 0) (gvector 1 0 0))
(render)
(entity:display int-graph)
```

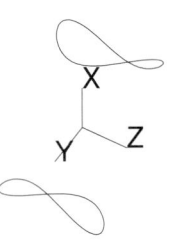

In **fafaint.scm** models of a sphere and a torus are created and the Scheme extension **entity:faces** used to extract a list of FACEs from the two shapes. Both objects are bounded by a single face so the Scheme function **list-ref** can be used to extract the single FACEs from position 0 in the lists. The intersection calculation is invoked by the **face:intersect** function which returns the curve in the form of a WIRE BODY. A list of the BODY's WIREs is created by the function **entity:wires** and used as an argument for **entity:box** which calculates axis-aligned bounding boxes around each of the ENTITYs in the given list.

About surface/surface intersections

For pairs of analytical surfaces specific solutions can be used (e.g. sphere/cone intersections). However, in the general case intersections between two surfaces must be found using only their parametric representation. Unlike the procedure outlined for curves the difficulty posed by two parametric surfaces $s(u, w)$ and $t(v, m)$ is immediately apparent. Consider that at each point along the intersection curve $s(u, w) = t(v, m)$, implying that:

$$x(u, w) = x(v, m)$$
$$y(u, w) = y(v, m)$$
$$z(u, w) = z(v, m)$$

Because there are three equations and four unknowns one is in difficulty from the outset. However, intersection curves can be found using a technique known as curve tracing. The procedure can be described in outline:

1. An initial point on each piece of the intersection curve must be established in some way. One method for doing this involves intersecting a number of isoparametric curves lying on each surface (i.e. a curve/surface intersection).

2. Guess the location of the next point on the curve, say by stepping along an approximate tangent at the last point located.

3. Refine this guess using numerical methods.

Needless to say, selecting a suitable step size for offsetting the last point is fraught with difficulties. Potential problems also exist in the first step where small pieces of curve can be easily missed.

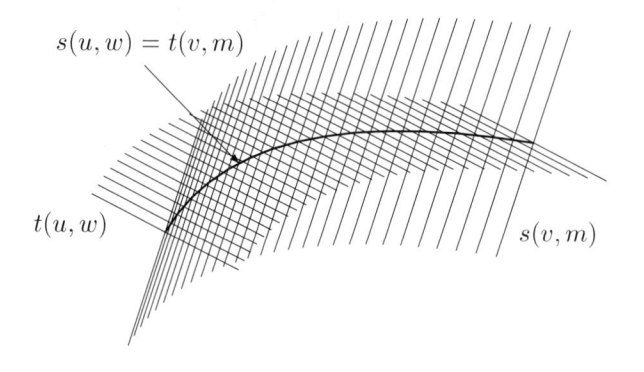

6.6 Overview of the Boolean Algorithm

The following sections give an overview of the Boolean algorithm used to **unite**,

6.6 Overview of the Boolean Algorithm

intersect or **subtract** two manifold solids. Perhaps the key to understanding this process is the realization that while new FACEs, EDGEs, VERTEXs and CURVEs can be created, **no** new SURFACEs are generated at any stage.

Tool and Blank Arguments

The order of the arguments is important in all ACIS's Boolean operations. Because of this the two BODYs given as arguments are referred to as the **TOOL** and the **BLANK** BODYs. This naming convention reflects the desired behavior of the modeler when Boolean subtractions are used to simulate manufacturing operations. In such cases the shape of the TOOL is removed from the BLANK. Similarly in intersection and union operations the resulting shape is assigned to the BLANK BODY.

Tool   Blank

Preliminaries

1. The two BODYs are positioned in the correct orientation and location relative to one another. The TOOL will be transformed into the space of the BLANK in order to perform the Boolean.

2. The Boolean process will normally change both the TOOL and BLANK BODYs.

3. The BODY which results from the Boolean process is assigned to the BLANK.

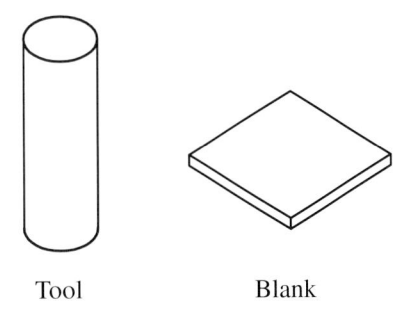

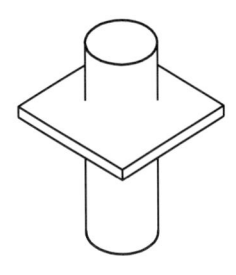

Set operations in modeling have long been referred to as Booleans because they form an algebra of the type described by George Boole (1815–1864). Boole started studying mathematics in an effort to produce better textbooks for the pupils in his school sometime around 1836. Having no clear notion of when to stop he continued studying until 1855 when he got married! The year before, he published his big idea, an algebra of logic in which certain types of reasoning were reduced to the simple manipulation of symbols. From this came the abstract idea of a *Boolean Algebra* which has two operations and must satisfy the laws of closure, commutativity, associativity, distributivity, identity and complement. The systems of logic, set theory and switching networks each form a Boolean algebra. Because the operations of union, subtraction and intersection can be defined in terms of operations on sets of points they are naturally referred to as Booleans.

About Booleans and George Boole

Imprinting

The first step determines the intersection curves formed by intersecting pairs of FACEs. This process is known as *imprinting* and can be envisaged as a line drawn round the profile that each BODY's boundary makes on the boundary of the other.

The result takes the form of a WIRE BODY known as an *intersection graph*. Each EDGE in an intersection graph's WIRE has four COEDGEs, two of which lie on the BLANK and two of which lie on the TOOL.

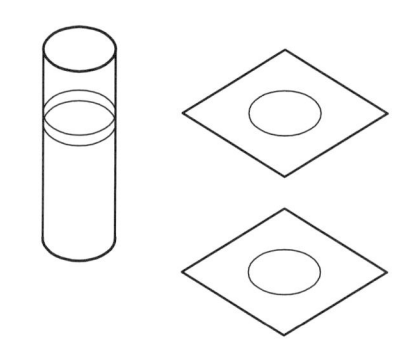

Each COEDGE in the WIRE also has an attribute attached to it which describes its relationship with its owning FACE (see Section 6.7). The geometry of the EDGE represents the intersection curve between two FACEs (in cases where point contact exists this EDGE might degenerate to a VERTEX).

In principle the process of generating an intersection graph requires every FACE on one object to be intersected with every FACE on the other object. Comparing every FACE pair in this way can take a long time so in practice the efficiency of the process is improved by initially testing only the bounding boxes.

About Euler Operators

> The low level details of splitting and stitching are often described and implemented in a stepwise manner which changes the model incrementally. This approach:
>
> - preserves the *validity* of the model's data structure,
> - makes the algorithm more comprehensible.
>
> The functions used for carrying out the individual steps of this process are known as **Euler Operators** because they leave the model conforming to Euler's equation at the end of each step (see page 19). The operators themselves are defined in terms of six topological parameters and have names like:
>
> **MEV**: Make an Edge and a Vertex
>
> Euler operators are not used in ACIS Booleans (where nonmanifold operations need to be carried out) but are used in blending and sweeping.

6.6 Overview of the Boolean Algorithm

Splitting and Labeling of FACEs

Intersecting FACEs on both TOOL and BLANK are subdivided by the addition of EDGEs from the intersection graph to form new FACEs. The new and existing FACEs of each BODY are classified as being on, inside or outside the boundary of the other BODY. In other words the FACEs of the TOOL are classified as lying on, inside or outside the BLANK and vice versa.

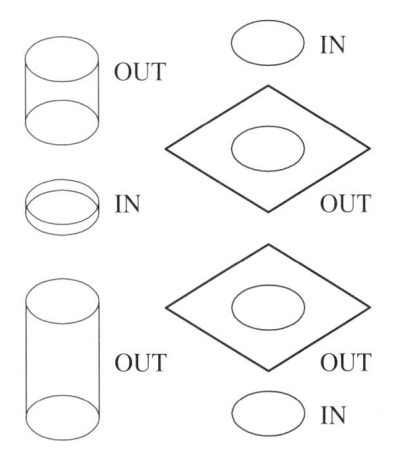

This process illustrates why the sharing of surfaces and curves is commonplace in complex models. For example, if the cylinder is united with the plate then both the upper and lower cylindrical FACEs will share the *same* surface, because of their common origin.

It is at this point that any attributes attached to the FACEs being divided will have their **split** methods invoked (see Chapter 12). These user-defined methods specify the manner in which a FACE's existing attributes are assigned (e.g. copied, transferred, deleted etc.) to a new FACEs.

Stitching for Union

The FACEs of the two BODYs can be *stitched* back together in several different ways depending on the desired outcome. For a union of the two BODYs all the FACEs labeled OUT are stitched together (i.e. their respective COEDGEs will be associated with a common EDGE). All the FACEs labeled IN are deleted and any attributes attached to them will have their **delete** method called.

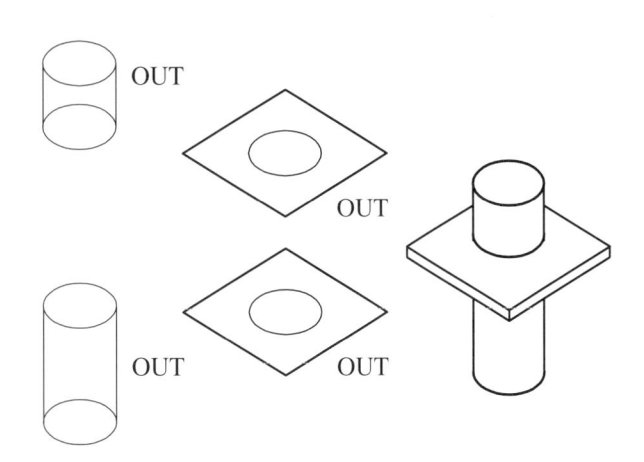

It is also possible at this point in the Boolean process for two FACEs to **merge** during the stitching process. Any attributes attached to such FACEs will have their **merge** methods invoked at this point.

Stitching for Intersection

In contrast to the union operation the intersection algorithm joins all the FACEs labeled IN along common EDGEs. The resulting volume represents the overlap between the two BODYs. The remaining FACEs labeled OUT are deleted. A NULL intersection indicates that the two BODYs do not touch at any point.

Once again the appropriate methods will be called to handle the deleting or merging of any FACE attributes. Intersection volumes are useful for detecting if any interference exists between complex BODYs. The calculation of the intersection BODY is often far more exact than many applications need and the **qdint.cxx** program later in this chapter shows how a simple TRUE/FALSE collision test can be implemented.

Stitching for Subtraction (Tool from Blank)

Subtraction of the TOOL from the BLANK is achieved by:

1. Reversing the labels of the BLANK's FACEs, so that all the OUTs become INs.

2. Reversing the sense of the TOOL's FACEs, so the material lies on the other side of the SURFACE (i.e. the SURFACE normal is flipped).

3. Stitching all the IN FACEs together as in the union operation.

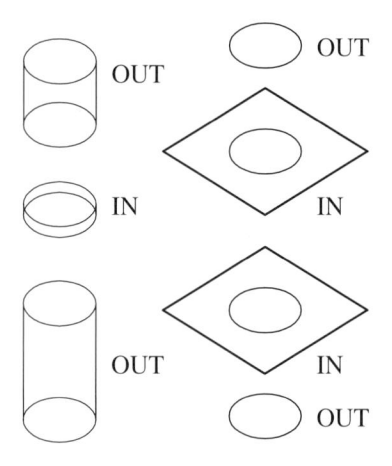

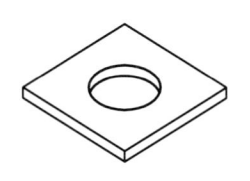

In practice ACIS does not reverse the SURFACE normals but allows each FACE to hold a flag known as a REVBIT which indicates if the sense of the FACE is opposite to (i.e. REVBIT == REVERSED) or the same as (i.e. REVBIT == FORWARD) the underlying SURFACE.

Stitching for Subtraction (Blank from Tool)

Subtraction of the BLANK from the TOOL is achieved by:

1. Reversing the labels of the TOOL's FACEs, so that all the OUTs become INs.

2. Reversing the sense of the BLANK's FACEs, so the material lies on the other side of the surface (i.e. the REVBIT is flipped).

3. Stitching all the IN FACEs together as in the union operation.

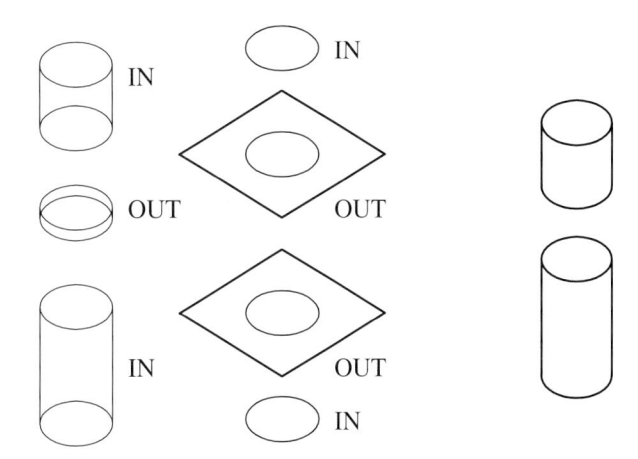

ACIS provides API level access to several of the intermediate steps in the Boolean process. The following three programs, **abebool1–3.cxx**, follow the process of uniting an eight-sided prism with an orthogonally oriented cylinder.

This description of the Boolean process has assumed that both the TOOL and the BLANK are manifold solids. However, ACIS can also perform Booleans on other combinations of ENTITYs (e.g. WIREs and sheets).

6.7 Generating an Intersection Graph

The intersection graph between two BODYs provides valuable information about which elements of an object's boundary will be modified by a Boolean operation. Such information could be used by applications, such as feature modeling, to determine which areas of a model's surface will be changed by a Boolean operation. The program **abebool1.cxx** starts the Boolean process by creating an intersection graph, in the form of a WIRE BODY, whose EDGEs represent the intersection curves between intersecting FACEs on the two BODYs. Unlike standard WIRE BODYs those representing intersection graphs have two COEDGEs associated with every EDGE (note that **api_slice** kills the COEDGEs which lie on the TOOL BODY). Each COEDGE has an attribute attached to it, known as **ATTRIB_INTCOED**, which records intersection data.

abebool1.cxx

```cpp
// This program creates an intersection graph between a prism and a cylinder.
// The attribute attached to one COEDGE of the WIRE's EDGEs is also examined
#include <stdio.h>         // Declares file I/O functions
#include "boolean/kernapi/api/boolapi.hxx"  // Declare boolean API's
#include "constrct/kernapi/api/cstrapi.hxx" // Construction API's
#include "kernel/kernapi/api/kernapi.hxx"   // Declare kernel API's
#include "kernel/kerndata/top/alltop.hxx"   // Declares topology classes
#include "boolean/kernbool/boolean/at_bool.hxx" // Declares attribute classes
#include "baseutil/vector/transf.hxx"       // Declares the transform class
#include "baseutil/vector/vector.hxx"       // Declares a vector class
#include "kernel/kerndata/lists/lists.hxx"  // Declares the ENTITY_LIST class

void main()
{
    api_start_modeller(0);
    api_initialize_booleans();
    BODY *tool_cyl, *blank_prism, *intgraph;
      // z-axis aligned prism
    api_make_prism(120,20,30,8,blank_prism);
      // y-axis aligned prism
    api_make_frustum(200,20,20,20, tool_cyl);
    transf rotX = rotate_transf(pi/2,vector(1,0,0));
    api_apply_transf(tool_cyl,rotX);
      // create the intersection graph
    api_slice(tool_cyl,blank_prism, *(unit_vector*)NULL, intgraph);

      // count the number of WIREs in the intersection graph
    int wc = 0;
    for(WIRE *iw = intgraph->wire(); iw != NULL;
                            iw=iw->next(),wc++);
    printf("Intersection graph has %d WIREs\n",wc);
      // Count the number of COEDGEs in first WIRE of the graph
    int eco = 0;
    COEDGE *st_coedge = intgraph->wire()->coedge();
    COEDGE *int_coedge = st_coedge;
    do{
        int_coedge = int_coedge->next();
        eco++;
        }while(int_coedge != st_coedge && int_coedge != NULL);
    printf("First WIRE has %d COEDGEs\n",eco);
    // examine the ATTRIB_INTCOEDGE attribute of the first COEDGE

    ATTRIB *att = ((ENTITY*)st_coedge)->attrib();
    ATTRIB_INTCOED *incoed_at = (ATTRIB_INTCOED *)att;

    if (incoed_at != NULL){
        printf("Attribute name %s\n", att->type_name());
```

6.7 Generating an Intersection Graph 145

```
            printf("This intersection COEDGE lies ");
            switch(incoed_at−>type()){
              case edge_class:
                printf("wholly on the boundary of a FACE\n");
                break;
              case boundary_class:
                printf("within a face, ");
                printf("but has a VERTEX on its boundary\n");
                break;
              case face_class:
                printf("wholly within the boundary of a FACE\n");
                break;
            };
        }
        api_terminate_booleans();
        api_stop_modeller();
    }
```

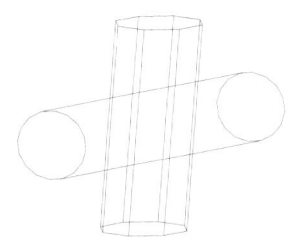

Prism and cylinder

When compiled and run this program prints out:

```
Intersection graph has 2 WIREs
First WIRE has 8 COEDGEs
Attribute named intcoed
This intersection COEDGE lies within a FACE, but has a
VERTEX on its boundary
```

Intersection graph

The **ATTRIB_INTCOED** holds the following information:

- A pointer to the FACE on which it lies.

- The relationship between material on the left of the COEDGE (as it traverses in a FORWARD direction) and the other BODY (i.e. inside, outside and several other special cases).

- The relationship between the COEDGE and the boundary of the FACE it lies on.

Note: **abebool1.cxx** gets a pointer to the **ATTRIB_INTCOED** class by casting the COEDGE object to the type **ENTITY** and then using the member function **attrib()** to return a pointer. Since this implicitly assumes that the EDGE will have only one attribute attached to it the operation is unlikely to work in the context of a larger program. A safer way to get hold of an ATTRIB class, irrespective of the number of attributes in the ENTITY's list, is to use the function **find_named_attribute** discussed in Chapter 12.

6.8 Imprinting and Division of FACEs

Having established the mutual intersection curves formed by the inter-penetrating boundaries of the two objects involved in the Boolean, the next step is to **imprint** these curves onto the FACEs of both objects. This process divides up the existing FACEs of a model so that portions of them can be retained or discarded in the latter stages of the process. The program **abebool2.cxx** takes the uniting of the eight-sided prism and the cylinder one step further by dividing up the FACEs in this way using the **api_imprint** function.

abebool2.cxx

```
// This program imprints a prism on a cylinder and so creates some new FACEs
#include <stdio.h>                              // Declares file I/O functions
#include "boolean/kernapi/api/boolapi.hxx"      // Declare boolean API's
#include "constrct/kernapi/api/cstrapi.hxx"     // Construction API's
#include "kernel/kernapi/api/kernapi.hxx"       // Declares kernel API's
#include "kernel/kerndata/top/alltop.hxx"       // Declares topology classes
#include "baseutil/vector/transf.hxx"           // Declares the transform class
#include "baseutil/vector/vector.hxx"           // Declares a vector class

void main()
{
    api_start_modeller(0);
    api_initialize_booleans();
    BODY *tool_cyl, *blank_prism;
    api_make_prism(120,20,30,8,blank_prism);    // z-axis aligned prism
    api_make_frustum(200,20,20,20, tool_cyl);   // y-axis aligned prism
    transf rotX = rotate_transf(pi/2,vector(1,0,0));
    api_apply_transf(tool_cyl,rotX);
    api_imprint(tool_cyl,blank_prism);          // imprint them
    FACE *ff = tool_cyl->lump()->shell()->face_list();
    int ffc = 0;
    while(ff != 0){
        ffc++;
        ff = ff->next_in_list();
    }
    printf("Cylinder now has %d FACEs\n",ffc);
    api_terminate_booleans();
    api_stop_modeller();
}
```

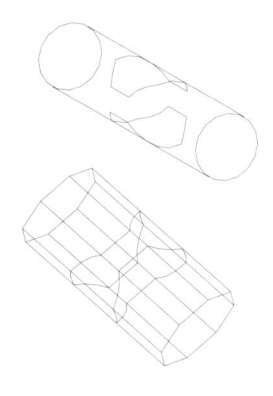

Imprinted BODYs

When this program is compiled and run the following text is displayed:

```
Cylinder now has 5 FACEs
```

The two new FACEs have been formed by the loops of EDGEs imprinted onto the cylindrical FACE of the frustum.

6.9 Stitching and Separation

Having imprinted the two BODYs the final step of the union process involves the **stitching** together of their two boundaries. This is done by forming **nonmanifold** EDGEs (i.e. adjacent to more than two FACEs) where there are coincident EDGEs or VERTEXs (i.e. along the intersection curves).

In the real Boolean algorithm parts of the boundary would be marked for retention or disposal depending on whether a union, subtraction or intersection operation was being carried out. However, in the program **abebool3.cxx** the process is completed by testing points on each FACE to see if they lie inside either of the original TOOL or BLANK BODYs. Any inside FACEs are then removed along with their COEDGEs and LOOPs.

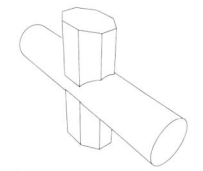

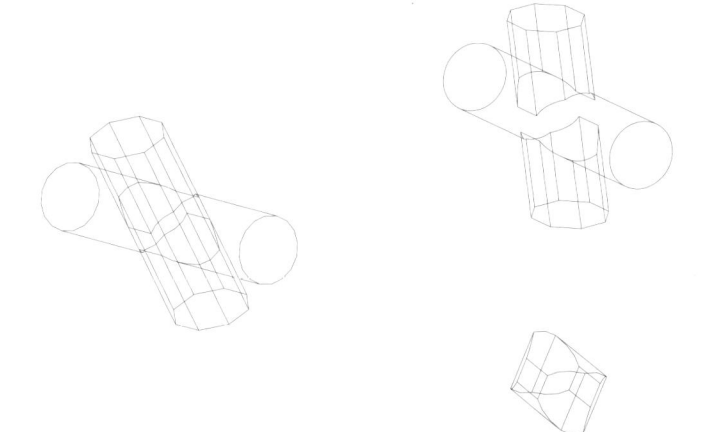

Figure 6.2: The nonmanifold BODY (left) resulting from **api_imprint_stitch** and the manifold BODY (top right) which results from the removal of the internal FACEs (bottom right).

abebool3.cxx

```
// This program imprints a prism and a cylinder on each other
// and then stitches them together to create a nonmanifold object.
// The removal of all internal FACEs results in a manifold model of a
// united prism and cylinder.
#include <stdio.h>                          // Declares file I/O functions
#include "boolean/kernapi/api/boolapi.hxx"  // Declares boolean API's
#include "constrct/kernapi/api/cstrapi.hxx" // Declares Construction API's
#include "intersct/kernapi/api/intrapi.hxx" // Declares intersector API's
#include "kernel/kernapi/api/kernapi.hxx"   // Declares kernel API's
#include "kernel/kerndata/top/alltop.hxx"   // Declares topology classes
#include "baseutil/vector/transf.hxx"       // Declares the transform class
```

```cpp
#include "baseutil/vector/vector.hxx"      // Declares the vector class
#include "baseutil/vector/position.hxx"    // Declares the position class
#include "kernel/kerndata/lists/lists.hxx" // Declares the ENTITY_LIST class
#include "kernel/kerndata/data/debug.hxx"  // Declares debug functions
#include "kernel/kerndata/savres/fileinfo.hxx" // Declares fileinfo class

void save_ent(char*, ENTITY*); // Function prototype; see save.cxx
logical check_face(FACE*, position, BODY*, BODY*); // Check FACE function

void main()
{
    api_start_modeller(0);
    api_initialize_booleans();
    BODY *tool_cyl,*tool_copy, *blank_prism, *blank_copy;
        // z-axis aligned prism
    api_make_prism(120,20,30,8,blank_prism);
    api_make_prism(120,20,30,8,blank_copy);
        // y-axis aligned cylinder
    api_make_frustum(200,20,20,20, tool_cyl);
    api_make_frustum(200,20,20,20, tool_copy);
    transf rotX = rotate_transf(pi/2,vector(1,0,0));
    api_apply_transf(tool_cyl,rotX);
    api_apply_transf(tool_copy,rotX);
        // imprint them
    api_imprint_stitch(blank_prism, tool_cyl);
    FILE *output1 = fopen("abebo1.dbg","w");
    FILE *output2 = fopen("abebo2.dbg","w");
    debug_size(blank_prism,output1);

    ENTITY_LIST face_list;
    api_get_faces(blank_prism, face_list);
    face_list.init();
    FACE *ff = NULL;
    int count = 0;

    // Define three positions close to the CofG for inside/outside test
    position cofg = position(0,0,0); // At the center
    position up_cofg = position(0,0,5); // Slightly above
    position dw_cofg = position(0,0,-5); // Slightly below

    while((ff = (FACE*)face_list.next()) != NULL)
      {
        if ( check_face(ff, cofg, blank_copy, tool_copy) ||
             check_face(ff, up_cofg, blank_copy, tool_copy) ||
             check_face(ff, dw_cofg, blank_copy, tool_copy) )
          {
            count++;
            printf("Face %d removed\n",count);
            api_remove_face(ff); // Remove all inside FACEs
```

6.9 Stitching and Separation 149

```
            }
          }

        debug_size(blank_prism,output2);
        fclose(output1); fclose(output2);
        save_ent("abebool3.sat",blank_prism);
        api_terminate_booleans();
        api_stop_modeller();
    }

    // Returns TRUE if the closest point to the given position on
    // the given FACE lies inside either of the two BODYs
    logical check_face(FACE *tf, position tp1, BODY *b1, BODY *b2)
    {
        position test_pt; // Position on FACE tf
        // Find closest position on tf to tp1
        api_find_cls_ptto_face(tp1, tf, test_pt);

        point_containment pc1, pc2;
        api_point_in_body (test_pt, b1, pc1);
        api_point_in_body (test_pt, b2, pc2);

        if(pc1 == point_inside || pc2 == point_inside)
          return TRUE;
        else
          return FALSE;
    }
```

The internal ENTITYs of the nonmanifold BODY are identified by testing each FACE to see if a point on it lies inside either the tool or the blank BODYs. In addition to **point_inside**, **point_outside** and **point_unknown**, **api_point_in_body** can also return **point_boundary**. Consequently it is important that the points being tested lie clearly within the boundary of the FACE. Algorithms for generating such points on an arbitrary FACE are rather involved so **abebool3.cxx** cheats and uses the function **api_find_cls_ptto_face** to calculate a position on a FACE closest to a given position known to generate the correct result. Inspection of the stitched BODY shows that there is no single point guaranteed to generate non-boundary test points on every internal FACE (i.e. on the two cylindrical FACEs the closest point to the origin could lie on their boundary). Because of this, three different positions are used to generate FACE internal test points. The creation of a more general procedure is left as an exercise for the reader!

Table 6.2 shows the number and type of ENTITYs recorded in each of the debug files output by the **abebool3.cxx** example. Notice that in the BODY resulting from the **api_imprint_stitch** operation (the stitched body) are 16 nonmanifold EDGEs. These are the imprinted EDGEs from the intersection graph which are connected to the FACEs of both the prism and the cylinder, which means that each EDGE has a ring of four COEDGEs. Thus the 16 nonmanifold EDGEs support 64 COEDGEs, while the remaining 42 manifold EDGEs are associated with 84 COEDGEs. After the internal FACEs have been removed the resulting manifold BODY has 21 FACEs.

Table 6.2   ENTITY counts before and after decomposing the BODY.

ENTITY	Stitched BODY	Manifold BODY
BODY	1	1
LUMP	1	1
SHELL	1	1
FACE	31	21
LOOP	34	24
SURFACE	13	13
COEDGE	148	100
EDGE	58	50
VERTEX	34	34
CURVE	36	36
POINT	34	34

6.10 Quick Intersection Tests

The intermediate stages of Boolean operations can be used for many purposes. The program **qdint.cxx** illustrates one obvious, non-Boolean, application of the intersection graph. Using **api_boolean** to test for intersections puts unnecessary load on your computer if all you want is a yes/no kind of answer to the question of whether two BODYs overlap. The function **api_slice**, which determines the intersection graph between two BODYs, can be used to give this sort of result.

6.10 Quick Intersection Tests

qdint.cxx

```
// This program does a cheap (TRUE or FALSE) intersection test between
// a prism and a cylinder using the api_splice function.
#include <stdio.h>                          // Declares file I/O functions
#include "boolean/kernapi/api/boolapi.hxx"  // Declares boolean API's
#include "constrct/kernapi/api/cstrapi.hxx" // Construction API's
#include "kernel/kernapi/api/api.hxx"       // Declare API_NOP macros
#include "kernel/kernapi/api/kernapi.hxx"   // Declares kernel API's
#include "kernel/kerndata/top/alltop.hxx"   // Declares topology classes
#include "baseutil/vector/transf.hxx"       // Declares the transform class
#include "baseutil/vector/vector.hxx"       // Declares a vector class

logical is_interfering(BODY*, BODY*); // Function prototype

void main()
{
    api_start_modeller(0);
    api_initialize_booleans();
    BODY *cyl, *prism;
        // z-axis aligned prism
    api_make_prism(120,20,30,8,prism);
        // y-axis aligned prism
    api_make_frustum(200,20,20,20, cyl);
    transf rotX = rotate_transf(pi/2,vector(1,0,0));
    api_apply_transf(cyl,rotX);

    if(is_interfering(cyl,prism))
            printf("Is interfering \n");
    else
            printf("Is interfering ..NOT \n");
    api_terminate_booleans();
    api_stop_modeller();
}

logical is_interfering( BODY *b1,  BODY *b2)
{
// The NOP macros ensure all changes to the model are undone

    logical interfering = FALSE;
    BODY *slice = NULL;

    API_NOP_BEGIN

    outcome res =  api_slice(b1,b2,*(unit_vector*)NULL,slice);

    if(res.ok())
        interfering = (slice!=NULL) ? TRUE:FALSE;
```

        API_NOP_END
        return interfering;
}

---

api_slice returns a WIRE BODY, but if there is no intersection the pointer returned is a NULL one. A NULL pointer is also returned when one BODY totally encloses another, so without modification the routine would fail in this special case.

6.11 Nonregularized Booleans

Boolean operations can be expressed elegantly using the notation of mathematical set theory. If you think of a three-dimensional object as being composed of a set of points then the intersection of two BODYs can be written simply as the intersection of two sets (say A and B):

$$A \cap B$$

Well, almost! The trouble with computers is that they tend to do exactly what you tell them to do, not what you meant. Since the intersection of two sets is composed of points common to both, the result should also contain any portions of coincident boundary. This is not surprising since the points on the boundary of a set are still members of that set.

However, this behavior is not always desirable; consider the intersection of a cuboid (B) and an L-shaped block (A) shown in Figure 6.3.

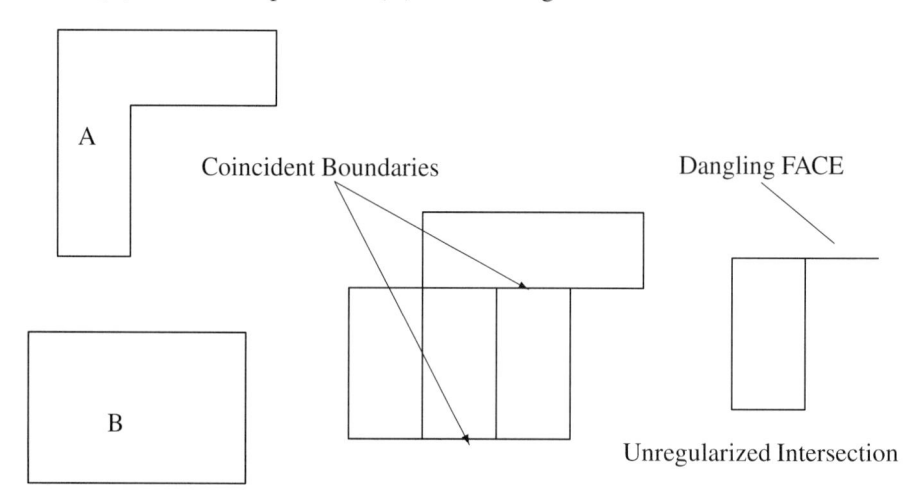

Figure 6.3: A nonregularized intersection.

The resulting volume of intersection contains two FACEs created from places where some part of a FACE on A was coincident with some part of a FACE on B. One

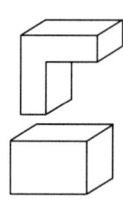

6.11 Nonregularized Booleans

bounds the solid of intersection, while the other is left *dangling* in space. More precise set-theory expressions can be written to take this sort of behavior into account by making a distinction between the set of points which form the boundary of an object and the set which constitutes its interior. Manipulation of these two sets allows expressions for *regularized Boolean* operations to be defined (written as $\cap^*, \cup^*, -^*$).

In a regular set all boundary points are adjacent to an interior point (so no dangling EDGEs or FACEs are allowed). The process of removing any isolated VERTEXs or *dangling* EDGEs and FACEs is known as **regularization**.

In practice this is done by ensuring that segments of boundary–boundary intersection, for example, are included in the regularized Boolean intersection *only if the interiors of both shapes lie on the same side* of the common boundary.

Nonregularized Booleans are used to support the modeling of multi-dimensional ENTITYs. Thus a 3D solid can have a lone EDGE emanating from the center of a FACE which won't be removed by the first Boolean the shape is subjected to. The program **noreg.scm** demonstrates this by using nonregularized Boolean operations to imprint a FACE onto the surface of a sphere and unite a WIRE with the resulting shape.

noreg.scm

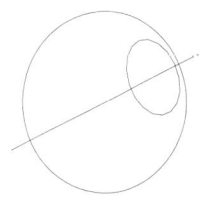

```
; This function creates a sphere and creates a circular face on it
; using some nonregularized Booleans. A wire axis is also added.

(define imprint (lambda()
    (let*
      ((ball (solid:sphere (position 0 0 0) 50))
       (cyl (solid:cylinder (position 0 0 0)(position 45 45 45) 22))
       (axis (linear-edge (position -50 -50 -50) (position 50 50 50)))
       (ballbk1 (entity:copy ball))
       (ballbk2 (entity:copy ball))
       (diff (solid:subtract cyl ballbk1))
       (patch (bool:nonreg-intersect  cyl ballbk2)))
      (begin
        (bool:nonreg-union  ball patch)
        (bool:nonreg-union  ball (wire-body axis))
        (entity:debug ball 3)))))

(imprint)   ; To execute the function
```

1 body
1 lump
1 shell
2 faces
3 wires
4 loops
1 surface
7 coedges
6 edges
5 vertex
2 curves
5 points

The output of the final debug function shows that the resulting sphere has (amongst other things) two FACE records, four LOOP records, one SURFACE record, seven COEDGE records, six EDGE records and five VERTEX records. Note that four of the five VERTEXs lie on the WIRE, one at each end and one at the entry and exit points of the sphere's surface.

6.12 Selective Booleans

Situations often arise in feature-based CAD where the effects of a Boolean operation need to be restricted. For example when adding a stiffening rib to a thin-walled component it might be desirable to ensure that no part of the rib penetrates the BODY.

The Boolean algorithm can easily accommodate this sort of demand by changing the behavior of the imprinting stage (i.e. FACE-FACE intersection, see page 140). Recall that the standard Boolean operation takes every FACE on the tool BODY and intersects it with every FACE on the blank BODY. In *selective* Booleans only certain FACEs are intersected, so that the areas changed by the Boolean operation can be limited to specific regions of a BODY.

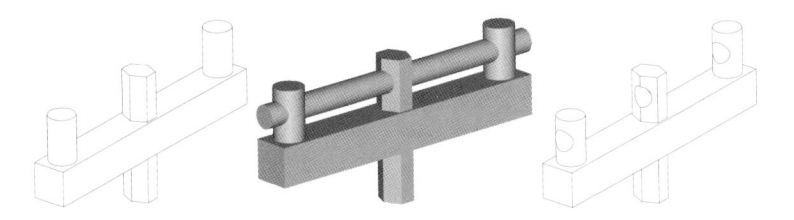

Figure 6.4: Standard Booleans subtraction.

ACIS supports selective Booleans in two ways: firstly the user can indicate explicitly which FACEs on one BODY should be intersected with which FACEs on another BODY during the Boolean process. Although effective in simpler cases the number of FACEs requiring explicit identification can quickly overwhelm the user, so in more complex situations nonmanifold cellular models are available to provide a high level structure for specifying exactly how a Boolean operation will behave. Both these approaches are demonstrated in the examples **sel_bool.cxx** and **cell_bool.cxx**.

First **sel_bool.cxx** demonstrates the explicit selection of FACEs to ensure that a subtraction operation only affects the cylindrical FACEs on the object shown in Figure 6.4, while the center boss is left untouched, Figure 6.5.

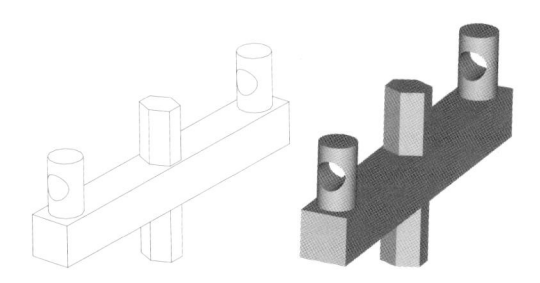

Figure 6.5: Selective Boolean subtraction.

After creating the "tool" (i.e. the cylinder) and "blank" BODYs, the program creates two ENTITY_LISTs containing pointers to each object's FACEs. Each EN-

6.12 Selective Booleans

TITY_LIST is iterated through and the function **is_cylindrical_face** used to test the geometry of each FACE. Pointers to the cylindrical FACEs found are copied to two other ENTITY_LISTs called **toolfaces** and **blankfaces**. Examination of Figure 6.4 shows that **toolfaces** will contain one FACE and **blankfaces** two FACEs.

The function **api_selectively_intersect** argument's are two equal length arrays of FACEs. These arrays have to be the same size so each FACE in one can be intersected with a corresponding entry in the other. To satisfy these requirements the program creates two arrays (**toolf** and **blankf**), each containing two FACE pointers. Because the cylindrical FACE on the "tool" BODY is required to intersect with two separate FACEs on the "blank" it appears twice in the **toolf** array.

Having identified which FACEs will be involved in the selective Boolean the process is initiated with a call to **api_boolean_start**. This function initializes several data-structures used in the Boolean process including the creation of an empty intersection graph.

The function **api_selectively_intersect** is then called with the arrays **toolf**, **blankf** and their size (i.e. 2) as arguments. The data generated by this process is attached to the intersect graph created in the previous step.

Finally the process is finished by a call to **api_boolean_complete** which carries out one of six different operations depending on its argument. Valid arguments for this function are UNION, INTERSECTION, SUBTRACTION, NONREG_UNION, NONREG_INTERSECTION, or NONREG_SUBTRACTION.

sel_bool.cxx

```
// This program starts the modeler, creates a block, with two cylindrical and one prism
// protrusions. Selective booleans are used to subtract a hole from only the cylinders
#include <stdio.h>                              // Declares I/O Functions
#include "constrct/kernapi/api/cstrapi.hxx"    // Declares Constructor APIs
#include "kernel/kernapi/api/kernapi.hxx"      // Declares Kernel APIs
#include "kernel/kerndata/top/alltop.hxx"      // Declares Topological Classes
#include "boolean/kernapi/api/boolapi.hxx"     // Declares Boolean APIs
#include "kernel/kerndata/lists/lists.hxx"     // Declares ENTITY_LIST class
#include "baseutil/vector/position.hxx"        // Declares position class
#include "kernel/geomhusk/acistype.hxx"        // Declares is_cylindrical_face function
#include "kernel/kerndata/savres/fileinfo.hxx" // Declares fileinfo class

void save_ent(char*, ENTITY*);  // Function prototype

void main()
{
    api_start_modeller(0);
    api_initialize_booleans();
    BODY *block, *p1, *p2, *p3, *tool ;

    api_make_cuboid(50,400,50,block);  // make test object
    api_make_prism(180, 25, 20, 6, p1);
```

```
        api_solid_cylinder_cone(position(0,-170,0),
                        position(0,-170,90),
                        20,20,20, NULL, p2);
        api_solid_cylinder_cone(position(0,170,0),
                        position(0,170,90),
                        20,20,20, NULL, p3);
        api_unite(p1, block);    api_unite(p2, block);    api_unite(p3, block);

        // make tool body for subtraction
        api_solid_cylinder_cone(position(0,-210,60),
                        position(0,210,60), 15,15,15, NULL, tool);

        ENTITY_LIST facelist, toolfaces, blankfaces;

        // make list of all cylindrical faces on the blank
        api_get_faces(block,facelist);
        facelist.init();
        ENTITY *face;
        while((face = facelist.next())!= NULL)
                if(is_cylindrical_face(face)) blankfaces.add(face);

        facelist.clear();  //Clear List

        // make list of all cylindrical faces on the tool
        api_get_faces(tool,facelist);
        facelist.init();
        while((face = facelist.next())!= NULL)
                if(is_cylindrical_face(face)) toolfaces.add(face);

        // create arrays of FACEs ....of matching length
        FACE* toolf[] = {(FACE*)toolfaces[0], (FACE*)toolfaces[0]};
        FACE* blankf[] = {(FACE*)blankfaces[0], (FACE*)blankfaces[1]};

        // Initializes the Boolean process and creates an empty intersection graph
        api_boolean_start(tool,block);

        // Intersects the FACEs in the two arrays and attaches
        // the results to the intersection graph
        api_selectively_intersect(2, toolf, blankf);

        // Completes remaining stages of the Boolean operation
        api_boolean_complete(SUBTRACTION);

        save_ent("sel_bool.sat",block); // save results

        api_terminate_booleans();
        api_stop_modeller();
}
```

6.12 Selective Booleans

The approach adopted in **sel_bool.cxx** is very low level and prone to error (because it requires an application to explicitly select which FACEs are going to be involved in the Boolean process). For example, if the argument UNION had been used with **api_boolean_complete** in the previous example a self-intersecting BODY would have been created.

The next example demonstrates a powerful solution to these problems that exploits the cellular topology (see page 14) formed after nonmanifold union operation between "tool" and "blank" BODYs are carried out.

Figure 6.6: A selective UNION between two orthogonal "slabs". The square hole results from the exclusion of one CELL from the UNION.

The program **cell_bool.cxx** demonstrates this approach by intersecting two orthogonal blocks. The Boolean operation halts after all the intersections are complete but before any material has been marked for removal. The CELLs of the nonmanifold BODY are then returned, in the form of a graph, and only after the application has selected which cells should be included in the final result (by deleting unwanted cells from the graph) does the Boolean process run to its conclusion.

The graph returned in **cell_bool.cxx** contains a cut-vertex (page 274) which is identified by a functions of the **generic_graph** class (see page 272). Removal of this cell results in a Boolean operation equivalent to the symmetric difference (see page 343) of the two objects.

cell_bool.cxx

```
// This program creates a nonmanifold intersection between two blocks.
// The selective Boolean API's used return a graph of the adjacency relationships
// between the objects cells. The central cell is identified. and removed, using
// operations from graph theory and a second selective Boolean API used to
// create a "hollow" cross.
#include <stdio.h>                          // Declares I/O Functions
#include "constrct/kernapi/api/cstrapi.hxx" // Declares Constructor APIs
#include "kernel/kernapi/api/kernapi.hxx"   // Declares Kernel APIs
#include "kernel/kerndata/top/alltop.hxx"   // Declares Topological Classes
#include "boolean/kernapi/api/boolapi.hxx"  // Declares Boolean APIs
```

```
#include "kernel/kerndata/lists/lists.hxx"   // Declares ENTITY_LIST class
#include "sbool/kernapi/api/sboolapi.hxx"    // Declares Selective Boolean API
#include "kernel/kernutil/law/generic_graph.hxx"  //Declares Generic Graph Class
#include "kernel/kerndata/savres/fileinfo.hxx"  // Declares fileinfo class

void save_ent(char*, ENTITY*);  // Function prototype

void main()
{
    api_start_modeller(0);
    api_initialize_sbooleans();

    BODY *block, *p1;

    // make test object
    api_make_cuboid(100,70,10,block);
    api_make_cuboid(10,70,100,p1);

    generic_graph *graph;

    // This api creates a graph structure from the cellular topology,
    // formed by the non-manifold union of block and p1. Each cell is
    // represented by a vertex in the graph and each edge defines
    // an adjacency relationship between two cells.
    api_selective_boolean_stage1(block,p1, graph);

    // create sub-graph from any cut vertices present in "graph"
    generic_graph *cuts_graph = graph->cut_vertices();

    // subtract the cut vertices from the graph
    generic_graph *cuttings = graph->subtract(cuts_graph,FALSE);

    // This api modifies the block in such a way that the
    // vertices (i.e. cells) remaining in the graph "cuttings" are kept.
    api_selective_boolean_stage2(block,cuttings);

    // This api completes the Boolean operation to create a manifold BODY
    api_boolean_complete(UNION);

    // save results
    save_ent("cell_bool.sat", block);
    api_terminate_sbooleans();
    api_stop_modeller();
}
```

The process requires calls to three API functions in strict order:

1. **api_selective_boolean_stage1**

6.13 Exercises

2. **api_selective_boolean_stage2**

3. **api_boolean_complete**

Also important is the order of the BODYs given as arguments to the stage1 and stage2 APIs; the blank must come first.

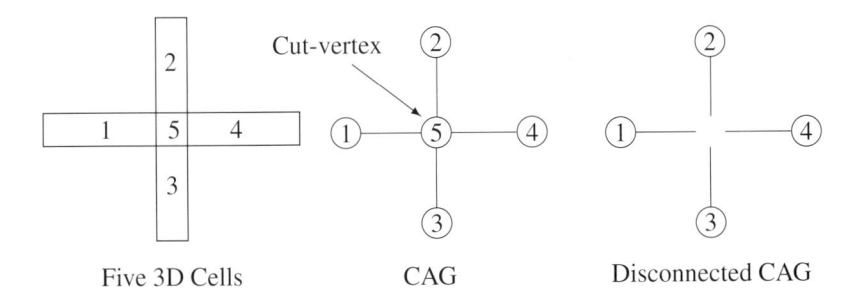

Figure 6.7: Cellular Adjacency Graph (CAG).

As in **sel_bool.cxx** the nature of the final Boolean (e.g. union, subtraction etc.) is determined by the argument given to the **api_boolean_complete** function.

In this example all the CELLs are 3D in nature and have adjacency relationships formed by the sharing of single, internal FACEs. However the approach is much more general than this and, as the example on page 277 shows, graphs can incorporate both 2D and 3D CELLs which share several FACEs.

6.13 Exercises

1. Modify **surfint.cxx** so that it trawls back and forward until it locates the point at which the intersection curve forms a figure 8.

2. Could the **abebool3.cxx** program, which produces a manifold BODY equivalent to the union of the two BODYs, be extended or modified to mimic a Boolean subtraction operation?

3. Extend the **abebool3.cxx** program so that the sheet BODY resulting from the decomposition (or unstitching) process is transformed into a manifold solid.

4. Modify the **qdint.cxx** program so that interference is detected when one BODY totally encloses the other.

5. Account for each of the FACE, LOOP, COEDGE, EDGE and VERTEX records found in the data structure of the sphere created by the **noreg.scm** program.

Rendering, Ray Firing and Faceting

7

Some aspects of solid modeling are definitely more fun than others! Whilst creating shapes and calculating intersections are undoubtedly worthy tasks, they don't have the instant gratification of **shiny computer graphics**. This chapter starts with some C++ examples which cover some of the low-level fundamentals of generating images such as:

Ray firing: used to trace the path of imaginary light rays.

Faceting: the generation of meshes of planar faces across the surface of an object.

In contrast to the detail of these first examples the later material concentrates on the facilities for display and interaction offered by the Scheme AIDE.

Computer displays have always been a trade-off between speed and quality. Because of this the Scheme AIDE offers a number of different **rendering modes** which allow the user to select just how good he or she wants the image to look. Despite the performance differences, the majority of display methods generate images using variations on two techniques:

1. **Ray tracing:** Calculates the path imaginary rays of light would take through the model world.

2. **Polygon shading:** Colors a mesh of polygons (known as facets) approximating the surface of an object.

Pure ray traced images are very realistic but can take some time to compute. Polygon shading on the other hand can be very fast and, because of some clever algorithms for smoothing out the changes in color between polygons, is surprisingly elegant. Also hardware available on most PC graphics cards can do this many times a second for large polygon meshes, allowing real-time display of complex moving models.

The pecking order of rendering methods goes like this:

Flat mode: Simply calculates a single color for each facet. Although very fast, the resulting images are angular with every facet a constant color.

Gouraud mode: This smoothes out the angular facets by interpolating the color calculated for each vertex of each facet. The resulting images miss any specular highlights that occur inside facet boundaries. This is because no interior point can be brighter than the brightest vertex.

Phong mode: This also smoothes out the angular appearance of the facets but by interpolating surface normals (rather than colors) across facets, it reproduces highlights that fall within facet boundaries.

Photo-realistic modes: Beyond the fast displays offered by Phong and Gouraud shading are a number of techniques (available in some of ACIS's optional components) which allow the passage of light rays to be **traced** through a collection of mirrored, transparent or opaque objects, which can have textures and colors "painted" onto their surfaces (see Section 7.7).

The programs **api_ray.cxx** and **facet.cxx** demonstrate the two fundamental techniques that lie behind all these different rendering algorithms. Note however, that in practice the rendering components use their own proprietary ray firing routines and not the ones found in the Kernel API.

7.1 Ray Firing

Given plenty of CPU time the highest quality images can be generated, without facets, by tracing, or following, imaginary rays of light from the viewer's eye to objects in the picture being generated, and from there towards the light sources. To do this a ray is notionally **fired** from the approximate location of a viewer's eye through each pixel of the display window and onto the virtual surfaces of the objects being modeled. If the color of the pixel was simply set to that of the ENTITY hit, only the silhouettes of objects would appear; so in order to generate more realistic images some form of **illumination model** is used to calculate the color of the surface at that particular point.

There are several different ways of calculating the color of an object at a given point but the important thing to appreciate is that they *all* require the surface normal at the point on the surface where the "ray" intersects the object in order to calculate the direction of any reflected light. The example **api_ray.cxx** shows how this vital bit of information can be obtained.

7.1 Ray Firing

api_ray.cxx

```
// This program creates a sphere, fires a ray at it and finds
// the surface normal at the point of intersection.
#include <stdio.h>                              // Input/Output functions
#include "constrct/kernapi/api/cstrapi.hxx"     // Construction API's
#include "intersct/kernapi/api/intrapi.hxx"     // Declares intersector API's
#include "kernel/kernapi/api/kernapi.hxx"       // Declares kernel API's
#include "kernel/kerndata/top/alltop.hxx"       // Topological Classes
#include "kernel/kerndata/lists/lists.hxx"      // Entity List Class
#include "kernel/kerngeom/surface/allsfdef.hxx" // surface definitions
#include "kernel/kerndata/geom/surface.hxx"     // SURFACE definitions
#include "kernel/kerndata/geom/point.hxx"       // POINT class

void main(){
   api_start_modeller(0);
   api_initialize_constructors();
   BODY* ball;
   api_make_sphere(50.0, ball);
   position ray_pos = position(70, 70, 70);
   unit_vector ray_dir = normalise(vector(-70,-70,-70));
   ENTITY_LIST results;  double* params = NULL;
   //We only want the initial intersection point, where the ray first hits the sphere
   //so ray radius = 0.1 and number of hits wanted = 1
   api_ray_test_body(ray_pos, ray_dir, 0.1, 1, ball, results, params);
   printf("Number of hits = %d\n",results.count());
   if (results.count() != 0) {
      int num = results.count();
      for(int i= num-1; i>=0; i--){
         ENTITY *hit = results[i];
         if(hit-> identity(FACE_LEVEL)== FACE_TYPE){
            position pos = ray_pos + (ray_dir * params[i]);
            FACE *fhit = (FACE*)hit;
            surface const *asf;
            asf = &(fhit-> geometry()-> equation());
            unit_vector snorm = asf-> point_normal(pos);
            if (fhit-> sense()!=0)
                   snorm = -snorm;
            printf("Surface Normal = (%f,%f,%f)\n", snorm.x(),snorm.y(),snorm.z());
            printf("Intersection Point = (%f,%f,%f)\n", pos.x(),pos.y(),pos.z());
         }
   } }
   api_terminate_constructors();
   api_stop_modeller();
}
```

When **api_ray.cxx** is compiled and run the following text is printed on the screen:

```
Number of hits = 1
Surface Normal = 0.577350269, 0.577350269, 0.57735026
Intersection Point = 28.8675134, 28.8675134, 28.867513
```

Notice that **api_ray_test_body** returns an array of parameters that indicate where along the ray the hits (or intersections) have occurred. For efficiency when a ray is intersected with a BODY its rotation, translation and scale transformations are inverted and applied to the ray so that the *intersection test* occurs in internal BODY coordinates. The results, however, are returned in world coordinates.

About illumination models

Illumination models are used to calculate the color of a point on an object's surface. One of the simplest uses **Lambertian reflection** which models the diffuse reflection of dull, matt surfaces like cardboard. Because this sort of surface reflects light equally in all directions it has approximately the same color no matter which direction you look at it from. The surface brightness depends solely on the angle α between the direction of the light source and the surface normal.

The illumination equation for diffuse reflection is simply:

$$I_{sp} = L_p k_d cos(\alpha)$$

where:
  I_{sp} is the intensity of the light reflected at the point sp,
  L_p is the point light source's intensity,
  k_d is a coefficient which varies between 1 and 0 for different materials,
  α is the angle between the surface normal at sp and the light source.

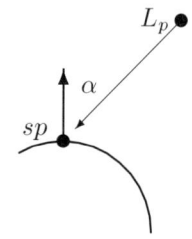

This basic model is easily extended to take the location of the viewpoint and color of the light source(s) into account.

7.2 Faceting ENTITYs

Although it is obviously possible to render any surface by firing a ray for every pixel on the screen, calculating the object's surface normal at each point hit and then

7.2 Faceting ENTITYs

applying an **illumination model** to produce a color is too expensive (in terms of computing time) to be practical for real-time displays of moving models[1]. Consequently most shading is done using a **polygon mesh** generated automatically from the model.

Unfortunately different rendering systems require different types of information from the polygons which comprise the mesh. Some applications, for example, require the u-v coordinates of each vertex in the mesh and some don't. To support these different requirements the ACIS faceter allows the programmer to specify what information is to be held at each node of the mesh and how it is to be output. This is done by deriving classes from the faceter's **MESH_MANAGER** class.

The program **facet.cxx** uses a MESH_MANAGER that attaches the facets to the FACE from which they were derived as a **POLYGON_POINT_MESH** (i.e. a singly linked, NULL-terminated list of polygons). Applications that do not want to use a POLYGON_POINT_MESH are free to derive a new class from MESH_MANAGER to handle the facet data.

facet.cxx

```
// This program facets a torus using view-independent controls based on the
// size of its bounding box. The coordinates of each node are then printed out.
#include <stdio.h>
#include "kernel/acis.hxx" // Declares system wide parameters
#include "constrct/kernapi/api/cstrapi.hxx" // Construction API's
#include "kernel/kernapi/api/kernapi.hxx" // Declares kernel API's
#include "faceter/api/af_api.hxx" // Declares faceter API's
#include "kernel/kerndata/geometry/getbox.hxx" // Declares get_*_box function
#include "baseutil/vector/box.hxx" // Declares box class
#include "kernel/kerndata/top/alltop.hxx" // Declares topology classes
#include "baseutil/vector/position.hxx" // Declares position class
#include "baseutil/vector/vector.hxx" // Declares vector class
#include "faceter/meshmgr/ppmeshmg.hxx" // Declares poly_point_mesh manager
#include "faceter/attribs/refine.hxx" // Declares refinement class
#include "faceter/attribs/af_enum.hxx" // Declares enum types

void main(){
    api_start_modeller(0);
    api_initialize_faceter();
    api_initialize_constructors();
    BODY *hoop;
    api_make_torus(50,10,hoop);
    // Set max gap between any facet and the exact surface to be 1/50 of the
    // diagonal length of the object's bounding box
    box bx = get_body_box(hoop);
    double sd = (bx.high() - bx.low()).len()/50.0;
    printf("Surface deviation to be set to %lf\n",sd);
```

---

[1] A modern display can easily contain a million pixels.

```
    // Specify the degree of refinement
    REFINEMENT *ref = new REFINEMENT();
    ref-> set_surf_mode( AF_SURF_ALL );
    ref-> set_adjust_mode( AF_ADJUST_NONE );
    ref-> set_triang_mode( AF_TRIANG_ALL );
    ref-> set_surface_tol( sd );
    api_set_default_refinement( ref );
    // Specify the information to be recorded at each vertex
    parameter_token ptoken[2];
    ptoken[0] = POSITION_TOKEN;
    ptoken[1] = NORMAL_TOKEN;
    VERTEX_TEMPLATE *vt = new VERTEX_TEMPLATE (2, ptoken);
    api_set_default_vertex_template( vt );

    FACE *f = hoop-> lump()-> shell()-> face_list();

    while(f){
       POLYGON_POINT_MESH *facets =
                (POLYGON_POINT_MESH *)NULL ;
       api_facet_face(f);
       api_get_face_facets( f, facet);

       // Print out facet data for this FACE
       POLYGON *poly;
       POLYGON_VERTEX *poly_vtx;
       position vtx_pos;
       for (poly= facets-> first(); poly != NULL;poly = poly-> next())
       {
           for (poly_vtx = poly-> first(); poly_vtx != NULL;
                   poly_vtx = poly_vtx-> next()){
              poly_vtx-> point(vtx_pos);
              printf("Position: %f %f %f\n", vtx_pos.x(), vtx_pos.y(), vtx_pos.z());
           }
       }
       printf("Face facetted with %d polygons\n",facets-> count());
       delete facets;
       f = f-> next_in_list(); }

    api_terminate_constructors();
    api_terminate_faceter();
    api_stop_modeller();
}
```

When executed this program produces an output, the first and last few lines of which are shown below:

```
Surface deviation to be set to 3.417601
Position: -40.000000 -0.000000 0.000000
```

7.2 Faceting ENTITYs

```
Position: -40.340742 0.000000 -2.588190
Position: -38.637033 10.352762 0.000000
                 ⋮
                 ⋮
                 ⋮
Position: -41.059832 -23.705905 -9.659258
Position: -38.971143 -22.500000 -8.660254
Position: -33.525212 -33.525212 -9.659258
Face facetted with 1152 polygons
```

The way in which the shape (i.e. triangular, quadrilateral or n-sided) and size of the polygons varies with curvature of the object's surface is determined by the **refinement** used. In the example program a refinement is created and four of its default parameters changed. The choice of refinement parameters used in **facet.cxx** implies the following:

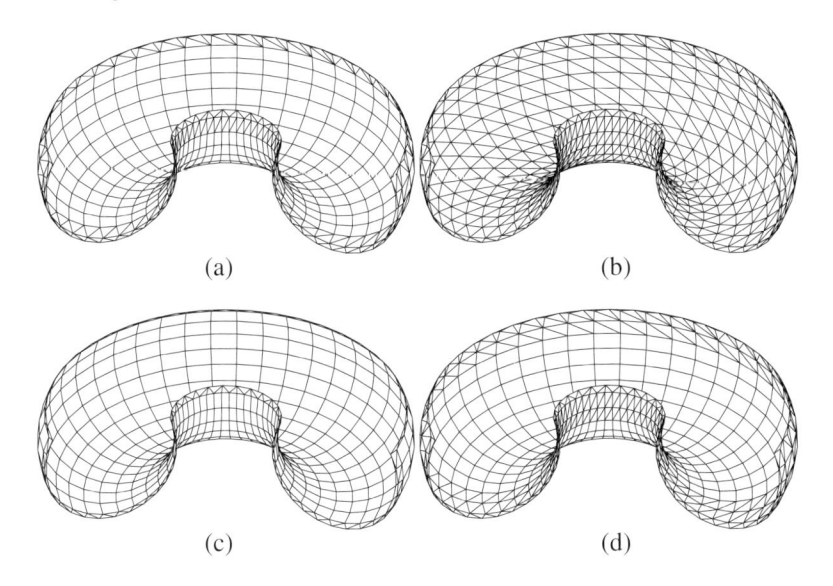

Figure 7.1: Four different faceting refinements applied to a quarter torus.

AF_SURF_ALL: Specifies that this particular instance of the **REFINEMENT** class is applicable to any type of surface. Other arguments, such as **AF_SURF_CONE**, can be used to associate a given refinement with a particular type of FACE.

AF_TRIANG_ALL: Specifies that triangulation should take place all over the FACE, not just, say, at the boundaries (see Figure 7.1(b)). Other values of this argument, such as **AF_TRIANG_FRINGE_1**, specify triangulation only along the EDGEs (Figure 7.1(c)). The default (**AF_TRIANG_FRINGE_2**) specifies that the boundaries of a FACE are triangulated two layers deep (see

Figure 7.1(a)), likewise the parameter **AF_TRIANG_FRINGE_3** causes triangulation three layers deep, (Figure 7.1(d)).

AF_ADJUST_NONE: Specifies that no second pass over the triangles generated is to be made in an attempt to increase their quality.

Surface deviation: Specifies the maximum distance between a facet and the true surface. In the example this has been made equivalent to 2% of the diagonal length across the object's bounding box. Generally between 1–5% of the model size is reckoned to be a good value.

Normal deviation: Although not explicitly shown this parameter, which specifies the maximum difference between the surface normals of adjacent facets, is set by default to be 15 degrees. Figure 7.2 shows the different facets produced by 15° and 30° differences in surface normal between facets.

Low quality facet

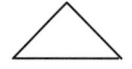

High quality facet

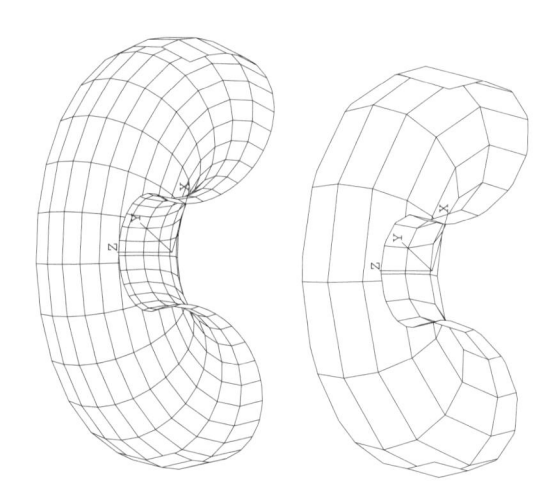

Figure 7.2: Toriodal FACE faceted with different normal deviations.

Having created a refinement, the program specifies a **VERTEX_TEMPLATE** to define what information is to be recorded at each vertex of the polygon mesh. The **facet.cxx** program specifies that the position and surface normal at each point are recorded (other possible tokens are COLOUR_TOKEN, POINTER_TOKEN, TEXTURE_COORDS_TOKEN, UV_DERIVS_TOKEN, UV_CHANGE_TOKEN, TRANSPARENCY_TOKEN, UV_TOKEN).

Facets can be used for any application (not just rendering) where the programmer wants to trade algebraic for combinatorial complexity and is willing to accept approximate answers.[2] For example, at least one commercial ACIS application uses

---

[2]The routine **make_faceted_body** (in fct/ppm_tb for creating an approximate copy of a BODY whose FACEs consist only of planar facets) has many applications other than display.

7.3 Faceted Hidden Line

facets for fast collision checking and triangular facet data in the form of STL files is also used as input for nearly all commercial rapid prototyping machines.

7.3 Faceted Hidden Line

After shading, perhaps the second most common use of facet data is the generation of hidden-line views (e.g. Figure fig:fndev). Although composed of straight line segments, images generated in this way are acceptable for all but the most precise display applications (see Section 7.4).

The API functions of the Interactive Hidden Line (IHL) Component take a previously faceted ENTITY together with a view definition and compute the silhouette lines for all its FACEs by projecting the 3D facet data on to the 2D view plane. Often the silhouette lines will lie on the existing EDGEs but sometimes (say on a cylindrical FACE) they will appear on a FACE. Consequently IHL data (in the form of attributes) is attached to a FACE only if it has a fully or partly visible silhouette line. All the other data is attached to EDGEs. The example **fhl.cxx** uses the IHL Component functions and attributes to count how many fully, or partly, visible EDGEs on a block with a through hole can be seen from a given position.

Facets are generated by a four-step process:

1. Determination of the size of grid to lay on each FACE to be faceted.

2. Discretization of EDGEs: in which the edges are divided into segments which are approximately the size of the smallest grid spacing on the adjacent FACEs.

3. FACE subdivision: in which the grid is laid out in parameter space.

4. Triangulation: in which the grid is divided into triangular cells.

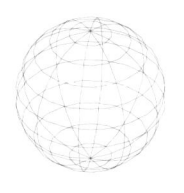

The influence of the parameter space grid can be clearly seen in the faceted sphere shown above. On the left are the u-v iso parameter lines of a sphere, on the right a rendered image of a sphere (using flat mode). Notice how the size of the facets decreases close to the poles.

About faceting

fhl.cxx

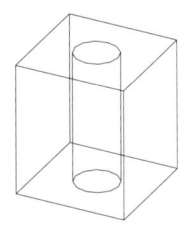

```cpp
// This program facets a block with a round hole and calculates which facets
// are visible from a given point. It then counts how many of the block's
// original EDGEs are visible
#include <stdio.h>
#include "kernel/acis.hxx"                      // Declares system wide parameters
#include "boolean/kernapi/api/boolapi.hxx"      // Declares boolean API's
#include "constrct/kernapi/api/cstrapi.hxx"     // Declares constructor API's
#include "kernel/kernapi/api/kernapi.hxx"       // Declares kernel API's
#include "kernel/kerndata/lists/lists.hxx"      // Declares ENTITY_LIST
#include "faceter/api/af_api.hxx"               // Declares faceter api's
#include "kernel/kerndata/top/alltop.hxx"       // Declares BODY,EDGE...
#include "baseutil/vector/position.hxx"         // Declares position
#include "ihl_husk/api/ihlapi.hxx"              // Declares ihl api's
#include "ihl_husk/ihl/ihl_seg.hxx"             // Declares ihl_seg
#include "faceter/api/af_api.hxx"               // Declares init_faceter api
#include "kernel/geomhusk/acistype.hxx"         // Is_edge function

void main() {
    api_start_modeller(0);
    api_initialize_constructors();
    api_initialize_booleans();
    api_initialize_faceter();
    api_initialize_interactive_hidden_line();

    BODY *block, *hole;
    api_make_frustum(70,10,10,10,hole);
    api_make_cuboid(50,50,50, block);
    api_subtract(hole,block);

    ENTITY_LIST input;  // ENTITYs in
    ENTITY_LIST output; // IHL_SEGMENTs out

    // Add block to input list
    input.add(block);
    // Specify the view
    position origin(0,0,0);
    position eye(100,100,100);

    // The optional token is an integer index to a view; the default is 0.
    // When token is 0, hidden line data is computed for drawing, but it is
    // not stored in attributes attached to the bodies. If token is nonzero,
    // the data is stored on the model as attributes and identified by token.
    // Existing attributes identified by the same token are replaced.

    api_ihl_compute (input,  // body list
```

7.3 Faceted Hidden Line

```
                      0,      // view token
                      eye,    // view eye
                      origin, // view target
                      FALSE,  // TRUE if perspective projection
                      TRUE,   // TRUE if interior segments needed
                      FALSE,  // TRUE if hidden segments needed
                      FALSE,  // TRUE to avoid refaceting body
                      FALSE,  // TRUE if you do not want to calculate
                              // hidden line but only silhouette
                      output  // segments returned
                      );

    int count = output.count(); // Number of segments in returned list
    ENTITY_LIST vis_entities;
    vis_entities.init();

    for (int i = 0; i < count; i++){
            IHL_SEGMENT *seg = (IHL_SEGMENT*)output[i];
            // For each segment check if its visible
            if (seg-> visible()){
                    ENTITY *ent = seg-> model_ent(); // Return seg's parent ENTITY
                    if (is_EDGE(ent))
                            vis_entities.add(ent); } }

    int ent_count = vis_entities.count();

    // Get a list of all the EDGEs on the block
    ENTITY_LIST edge_list;
    api_get_edges ((ENTITY*)block, edge_list);

    printf("%d EDGEs can be seen out of %d\n",ent_count,edge_list.count());

    // Clean up
    api_ihl_clean (input, 0);
    api_terminate_interactive_hidden_line();
    api_terminate_faceter();
    api_terminate_booleans();
    api_terminate_constructors();
    api_stop_modeller();
}
```

When executed **fhl.cxx** prints

```
10 EDGEs can be seen out of 14
```

The function **find_attrib** (described in Chapter 12) allows any ENTITY's associated list of ATTRIB objects to be searched for a given type of attribute. An alternative way of finding the IHL data is to use the **api_ihl_retrieve** function which returns a set of IHL attributes attached to a given ENTITY.

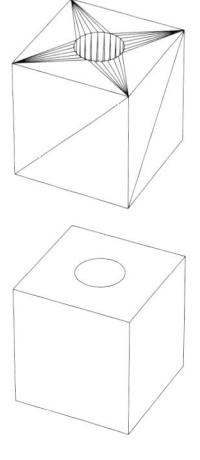

The **ATTRIB_IHL_SLIST** class returned by **find_attrib** holds a number (returned by member function **count()**) of **ihl_segments** in an array-like structure. It is these **ihl_segments** which contain the start and end points of each line (accessed through member functions **x1, y1, x2, y2**) and a flag declaring if the segment is visible or not.

The program ends by removing all the IHL data from the model and closing down the faceter.

7.4 Precise Hidden Line

Although the ACIS can generate approximate (i.e. facetted) hidden-line views some applications need exact information. For instance, if a model is going to be used to generate high quality CAD plots, the straight line segments of a Interactive Hidden Line (IHL) view would be unacceptable. The Precise Hidden Line Component (PHL) takes a view definition (i.e. a viewport) and calculates which model and silhouette EDGEs are visible from the given position. Because this information is generated as *precise* curves it can be used for applications, such as the plotting of engineering drawings, where a faceted approximation would be unsuitable.

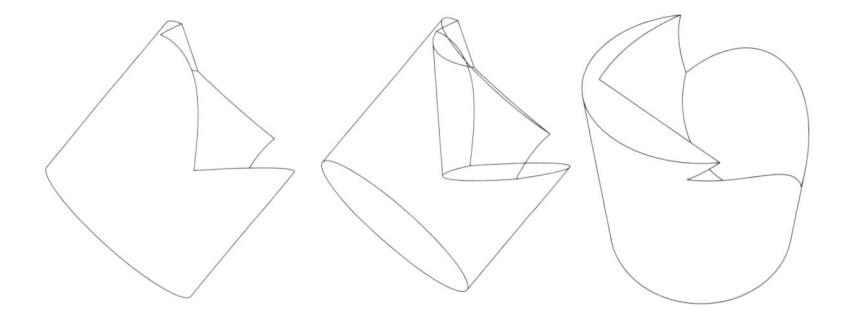

Figure 7.3: Result of **phl.scm** display as wire frame (center) and phl.

The component provides API routines for both computing and recording PHL data. The recording is optional and is implemented using ACIS's attribute mechanisms (i.e. ATTRIB_PHL class).

The program **phl.scm** demonstrates this component by subtracting two cylinders from one another and then displaying a picture of the resulting object with the hidden lines removed (Figure 7.3).

phl.scm

```
(define draw_cb ; Assume a viewport has already been created
  (lambda()(let*(
    (b (solid:cylinder (position 20 20 0) (position 0 25 35) 30))
```

```
            (c (solid:cylinder (position 0 -20 -20)(position 0 20 20) 30)))
       (bool:subtract c b)
       (phl:draw (phl:compute) #t))))
```

7.5 Rendering in C++

Having gained an insight into some of the fundamental technologies of rendering, the program **render.cxx** shows how, at a higher level, these operations are incorporated into a simple program that creates a window and displays a rendered image in it. This is done using an instance of the **view** class which provides high-level support for rendering in different Window Systems (i.e. X-Motif, MS-Windows, etc.). Using this, **render.cxx** shows the basic steps needed to create the MS window shown in Figure 7.4 (page 178) with an image of a sphere in it.

render.cxx

```
// This program makes an MS Window and displays a
// sphere in it for 8 seconds, before exiting
#include "constrct/kernapi/api/cstrapi.hxx"  // Construction API's
#include "kernel/kernapi/api/api.hxx"        // Declares outcome class
#include "kernel/kernapi/api/kernapi.hxx"    // Declares kernel API's
#include "kernel/kerndata/top/body.hxx"      // Declares BODY class
#include "kernel/kerndata/lists/lists.hxx"   // Declares ENTITY_LIST class
#include "baseutil/vector/position.hxx"      // Declares the position class
#include "faceter/api/af_api.hxx"            // Declares faceter api's
#include "faceter/meshmgr/idx_mm.hxx"        // Declares the INDEXED_MESH_MAN
#include "rnd_husk/api/rnd_api.hxx"          // Declares api_rh functions
#include "gihusk/sys_utl.hxx"                // Declares sleep_milliseconds
#include "gihusk/windows/view3dms.hxx"       // Declares view_3d_ms class
// For MACs include .../mac/view_mac.hxx"
#include "gihusk/view3d.hxx"                 // Declares view3d class
#include "gihusk/api/gi_api.hxx"             // Declares gi API's
#include "gihusk/rend.hxx"                   // Declares toolkit_rbase..
#include "br_husk/api/br_api.hxx"            // Declares BR API's

void main()
{
    api_start_modeller(0);
    api_initialize_constructors();
    api_initialize_basic_rendering();
    api_initialize_rendering();
    api_initialize_graphic_interaction();

    // sets up rendering image functions
```

Rendering, Ray Firing and Faceting

```
// rh_image_start, rh_image_scanline, rh_image_end
toolkit_rbase_app_callback* tk = NULL;
tk = new toolkit_rbase_app_callback;
add_rbase_app_cb(tk);

BODY *ball;
api_make_sphere(20,ball);

// Make a display window (platform dependent)
view3d *view = new view_3d_MS(150, 150, 450, 450, TRUE);

// For MACs
// view3d *view = new view_3d_mac(150, 150, 450, 450, TRUE);

// Define the viewpoint
view-> set_eye(position(70,70,70));
view-> set_target(position(0,0,0));

ENTITY_LIST elist;
elist.add((ENTITY*)ball);

// Sets the output window for the rendering
api_prepare_to_render(view);

// Add a white background (the default is black)
RH_BACKGROUND *background;
Render_Color white(1,1,1);

const char *name = "plain";
api_rh_create_background(name, background);
api_rh_set_background_arg(background, "color", white);
api_rh_set_background(background);

// Without lights only a black outline is displayed
ENTITY_LIST light_list;
RH_LIGHT *default_light = NULL;

api_rh_create_light("eye",default_light);
light_list.add(default_light);

api_rh_set_light_list(light_list);

// Use an INDEXED_MESH_MANAGER to hold the facets
INDEXED_MESH_MANAGER *MM = new INDEXED_MESH_MANAGER;
MM-> SetTransform(NULL);  //No transform in this case
api_set_mesh_manager(MM);
api_facet_entity((ENTITY*)ball);

// No smoothing over the facets
api_rh_set_render_mode(RENDER_MODE_FLAT);
```

```
        logical clear_screen = TRUE;

        // Render the image
        api_rh_render_entities(elist,TRUE);

        sleep_milliseconds(8000);  // Look at it in wonder
        // Clean-up in order
        api_rh_delete_background(background);
        api_rh_delete_light(default_light);
        api_rh_terminate();
        delete MM;
        api_terminate_graphic_interaction();
        api_terminate_rendering();
        api_terminate_basic_rendering();
        api_terminate_constructors();
        api_stop_modeller();
    }
```

The code needed to create a display window will vary from system to system. In X-windows, for instance, the following lines are close to the minimum required:[3]

```
        Widget toplevel;
            // X-Windows commands for creating a Scheme AIDE View Window
        toplevel = XtInitialize("acis", "ACIS", NULL, 0, &argc, argv);
        XtVaSetValues(toplevel, XtNwidth, 400, 0);
        XtVaSetValues(toplevel, XtNheight, 400, 0);
        XtRealizeWidget(toplevel);
        view3d* view = new view_3d_X(toplevel, 0, 0, 400, 400, FALSE);
```

Having switched everything on and specified a background color and a light source (so a black image will not be rendered on a black background), the program explicitly creates a mesh manager to hold the facets generated by the **api_facet_entity** function. All of the ACIS renderers (Advanced, Basic and OpenGL) require that ENTITYs are faceted with the INDEXED_MESH_MANAGER.

Having done the rendering the memory used is released by calling a number of termination functions in the same order as initialization was done.

7.6 Rendering using Scheme Extensions

The **Basic** Rendering Component creates images by specifying:

A plain uniform background: the color of which can be specified.

---

[3] See the Parted Example in scm/parted/parted.cxx for a more detailed description.

A number of (shadowless) lights: the properties of which can be specified. These are sources of illumination which do not support the computation of shadow effects.

One of three render modes: which vary how the facets are "smoothed".

A plain uniform material: the color of which can be specified.

A faceting refinement: which affects the angularity of the resulting image.

These different parameters are set using the following Scheme AIDE commands. The code fragments below can be found in the file **render.scm**.

7.6.1 Setting the Background

By default images are rendered in a black window; however, the background color can be changed by creating a **background** object of a specific type, setting its properties and making it part of the rendering context. This is done as follows:

```
; create a background
(define b (background "plain"))
; make it white
(background:set-prop b "color" (color:rgb 1 1 1))
; assign it to the renderer
(render:set-background b)
```

The Basic and Advanced Rendering Component are "plug-compatible". Consequently Advanced Rendering commands are accepted by the Basic Rendering Component. For example, the command **(background:types)** will return a list of exotic background types (e.g. "clouds" and "graduated") which can be specified by users of the Basic Rendering Component but only displayed by the Advance Rendering Component.

7.6.2 Setting the Lights

By default every object is lit by a single "eye" light, but much more interesting pictures can be generated by creating additional lights in other locations. Table 7.1 summarizes the types of lights available and their parameters.

Lights can be created and their properties set using the Scheme extensions **light** and **light:set-prop** in the following way:

```
; Light Types : "ambient", "distant", "eye", "point", "spot"
(define lite-e (light "eye"))
(light:set-prop lite-e "color" (color:rgb 0 0 1)) ; blue light
(light:set-prop lite-e "intensity" 0.3)
(define lite-p (light "point"))
(light:set-prop lite-p "location" (gvector 50 50 50))
(light:set-prop lite-p "color" (color:rgb 1 1 0)) ; yellowish light
(light:set #t (list lite-e lite-p)) ; Sets the list of currently-active lights
```

7.6 Rendering using Scheme Extensions

Table 7.1 Light type descriptions.

Light type	Description	Properties
"ambient"	Adds an equal amount of light from all directions. If used in isolation an ambiently lit object will appear as a uniform color	Only **"color"** and **"intensity"** can be varied
"distant"	Adds a source of parallel rays	In addition to **"color"** and **"intensity"**, the position and attitude of the light can be specified with two **gvectors**: **"location"** and **"to"**
"eye"	Adds parallel rays from a **viewer's** eye position	Only **"color"** and **"intensity"** can be varied
"point"	Adds light shining equally in all directions from a point	**"color"**, **"intensity"**, **"location"** and the rate of **"fall off"** can be varied
"spot"	Adds a cone of rays emanating from a point	**"color"**, **"intensity"**, **"location"**, **"cone angle"** and the rate of **"fall off"** can be varied

Be careful when specifying locations to ensure that the lights are positioned *outside* the objects and remember that the direction of the **spot** and **distant** lights is calculated by subtracting the **location** gvector from the **to** gvector.

7.6.3 Setting the Materials

While there is a great deal of choice when it comes to creating materials for use with the Advanced Rendering Component, the Basic Rendering Component lets you create only "plain" materials. Materials can be specified using the following commands:

```
;Make a block to add a material to
(define block (solid:block (position -10 -20 -30) (position 20 15 34)))
;Create a material
(define mat (material))
(material:set-color-type mat "plain" )
(material:set-color-prop mat "color" (color:rgb 1 1 1))
;Add it to a block
(entity:set-material block mat)
```

7.6.4 Setting the Facet Refinement

In addition to the lighting, background and material the quality of the rendered image will also be determined by the number and shape of facets used to approximate the model. Because of this the behavior of the faceter used to create rendered images can be changed by the user so that more, or less, triangles are generated. Although there are many parameters that can be adjusted perhaps the most effective are:

- **"normal tolerance"**: which sets the largest allowable differences between the surface normals of vertex adjacent facets (see page 168).

- **"surface tolerance"**: which sets the maximum distance allowable between a facet and the exact surface of a model.

The following code fragment demonstrates how the default values of these two parameters can be changed.

```
; Change the normal and surface tolerance faceting parameter
(define ball (solid:sphere (position 0 0 0) 50))
(entity:facet ball)              ;Create facets with default refinement
(entity:display-facets ball #t) ; Display until next input event
(entity:delete-facets ball)
(define newref (refinement))     ;Create a new refinement
(refinement:set-prop newref "normal tolerance" 50) ; change default value of 15
(refinement:set-prop newref "surface tolerance" -20) ; change default value of -1
(entity:set-refinement ball newref) ;Associate the new refinement with the sphere
(entity:facet ball)
(view:clear)
(entity:display-facets ball #t) ; Display the new blocky facets
```

The command **entity:facet** forces the computation of facet data. The command **render**, on the other hand, will only facet items that are not already faceted.

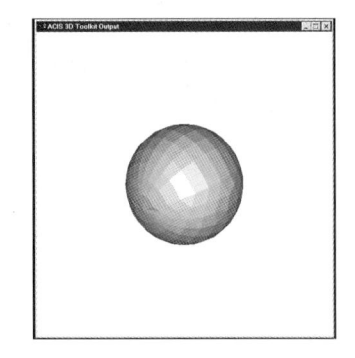

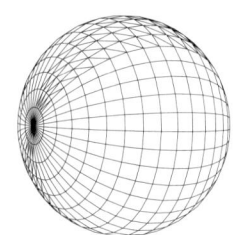

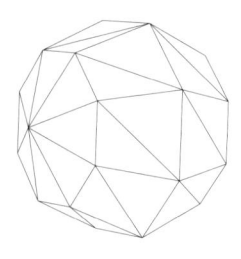

Figure 7.4: Output of the program **render.cxx**.

7.7 The Advanced Rendering Component

The **Advanced Rendering Component** provides the functions needed to generate photo-realistic images. Unlike the basic rendering this component allows materials, backgrounds and lights to be created and used in the generation of images of ACIS models. Because the pictures generated can contain shadows, reflections and refractions they have a much more authentic look about them than those produced with the basic rendering component. Amongst other things the component supports:

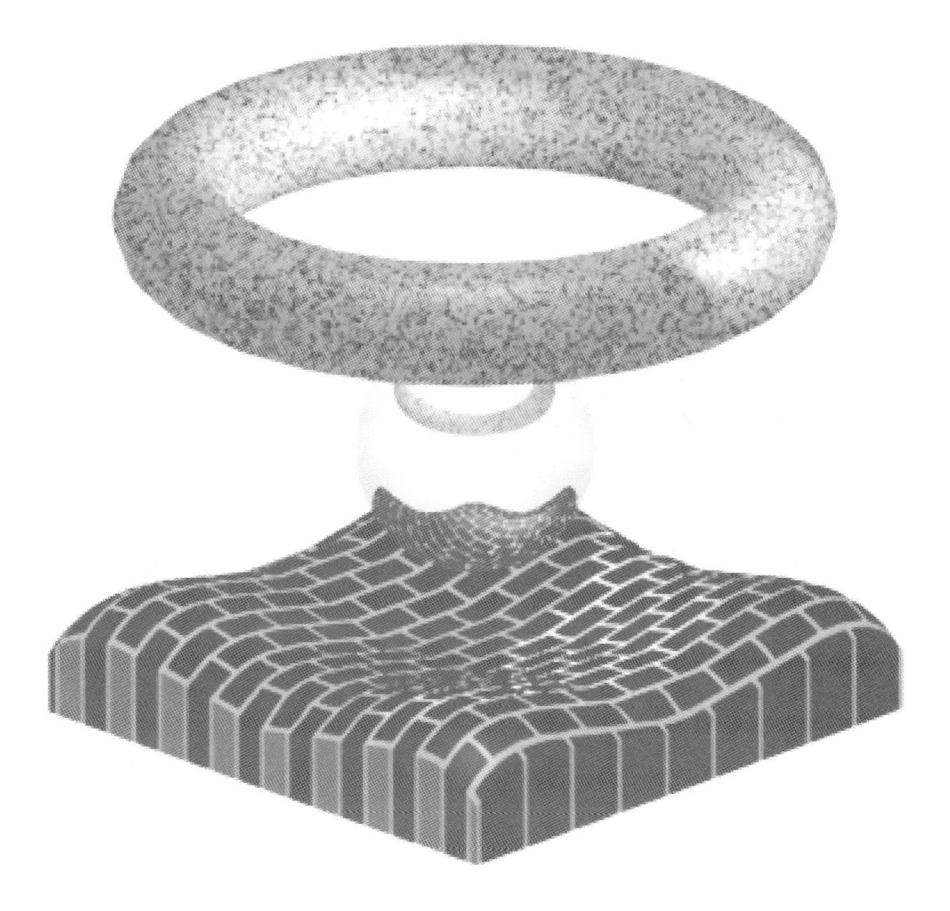

Figure 7.5: A render image.

- Color bleeding and soft-edged shadows.
- 2D and 3D tabular or procedural shading.
- Bump and environment mapping.
- Ray tracing.

adren.scm

```
; Render block, sphere and torus with materials and light sources
; Set the view for maximum effect
(view:set (position 150 150 80) (position 0 0 0) (gvector 0 0 1))
(wcs (position 0 0 0) (gvector 1 0 0) (gvector 0 1 0))
; Create a plain white background
(define b (background "plain"))
(background:set-prop b "color" (color:rgb 1 1 1))
(render:set-background b)
; Create mirrored sphere with reflected views
(define sph (solid:sphere (position 0 0 40) 20))
(define m1 (material))
(material:set-reflection-type m1 "conductor")
(material:set-reflection-prop m1 "roughness" 0.5)
(entity:set-material sph m1)
(define t1 (texture-space "spherical"))
(entity:set-texture-space sph t1)
; Create a marble doughnut
(define nut (solid:torus (position 0 0 80) 50 10))
(define m2 (material))
(material:set-color-type m2 "blue marble" )
(entity:set-material nut  m2)
(define t2 (texture-space "cylinder"))
(entity:set-texture-space nut t2)
; Create a brick covered wiggle
(define wig (solid:wiggle 100 100 10 2 2 2 2))
(define m3 (material))
(material:set-color-type m3 "wrapped brick")
(material:set-color-prop m3 "brick color" (color:rgb 1 0 0))
(material:set-color-prop m3 "scale" 10)
(entity:set-material wig m3)

; Move things around a bit
(entity:transform (list wig sph nut) (transform:scaling 0.6 0.6 0.6))
(entity:transform (list wig sph nut) (transform:translation (gvector 0 0 -30)))

; Create lights and shadows
(define l1 (light "point"))
(light:set-prop l1 "intensity" 0.3)
(light:set-prop l1 "location" (gvector 30 30 80))
(define l2 (light "ambient"))
(light:set-prop l2 "location" (gvector 180 200 50))
(define sp (light "spot"))
(light:set-prop sp "location" (gvector -20 50 40))
(light:set-prop sp "to" (gvector 0 0 0))
(light:set-prop sp "shadows" #t)
; Create a shadow map
(light:create-shadows sp (list wig sph nut))
```

```
(light:create-shadows l2 (list wig sph nut))
(light:set #t (list l1 l2 sp)) ; Sets the list of currently-active lights
(render:set-mode "raytrace-full")
(render)
```

> The Scheme AIDE allows colors to be specified with three numbers representing the proportion of red, green and blue in the color. This so-called RGB value can be visualized as the coordinates of a point inside a cube.
>
> Cyan=(0,1,1)
> Blue=(0,0,1)
> White=(1,1,1)
> Magenta=(1,0,1)
> Black = (0,0,0)
> Yellow=(1,1,0)
> Red=(1,0,0)
>
> The main diagonal of the cube represents a transition from black to white through a number of different shades of gray. Interestingly a line through RGB space does not always appear as a linear change in color because of variation in the sensitivity of the human eye.

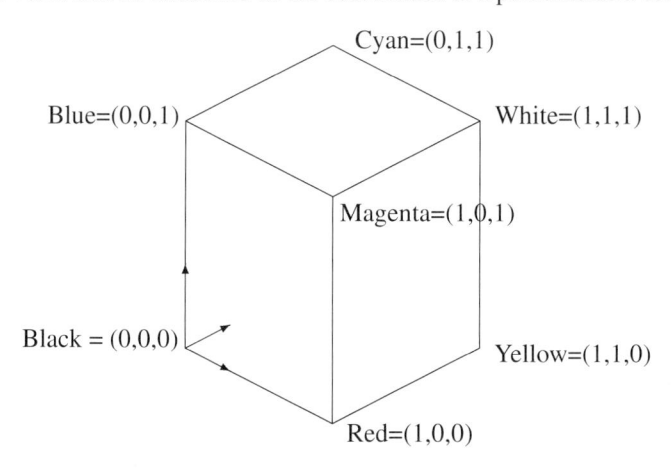

About the RGB color model

7.8 Spin!

The orientation of the image displayed in one of the Scheme AIDE's view windows is determined by two vectors. One is calculated as the difference between the **eye** and **target** positions. The other, known as the **up** vector, determines which way up an image is displayed.

Although the parameters are set when a view is created (see Section 4.1), they can also be redefined at any time. The program **spin.scm** manipulates both the **up** gvector and the **eye** position in order to rotate (i.e. spin) the displayed image about either the x, y or z axis. In this way the position of any ENTITYs in the view remain unchanged while the position (and direction) of the viewer's eye rotates through a series of small, 1 degree, steps to animate the display.

spin.scm

```
; This function spins the given view around the x (1), y (2) or z (3) axis
; by a multiple of 360 degrees, Example calls :
;   (spin (env:active-view) 1 2)  rotate about the x-axis(1) 720 (2x360) degrees
;   (spin (env:active-view) 2 1)  rotate about the y-axis(2) 360 (1x360) degrees
;   (spin (env:active-view) 3 0.2) rotate about the z-axis(3) 72 (0.2x360) degrees

(define spin
 (lambda (view-name dir num)
  (do ((ang 0 (+ ang 1)))((= ang (* 360 num))) ; Use a series of 1 degree steps
    (begin
      (if (= dir 1)(turn-about (gvector 1 0 0) view-name))
      (if (= dir 2)(turn-about (gvector 0 1 0) view-name))
      (if (= dir 3)(turn-about (gvector 0 0 1) view-name))))))

(define turn-about (lambda (axis view)
          (begin
            (view:set-eye
              (position:transform (view:eye view)
                      (transform:rotation
                        (view:target view) axis 1)) view)
            (view:set-up
              (gvector:transform (view:up view)
                      (transform:rotation
                        (position 0 0 0) axis 1)) view)
            (view:refresh view)  ; Change the view
            (system:sleep 10)))) ; Pause for 10 milliseconds
```

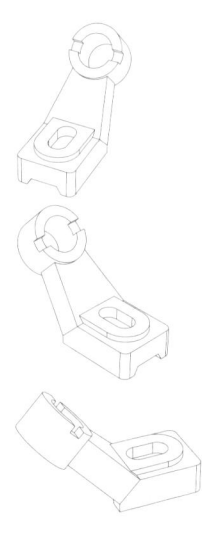

The current values of the display parameters for a given view are returned by the functions **view:eye**, **view:up** and **view:target**. In the program **spin.scm** the **eye** position is changed in the function **turn-about** by creating a rotation transform of one degree about an axis whose origin lies at the **target** position and whose direction is parallel to the given gvector (i.e. one of the principal axes).

; *Creates a rotation transform of 1 degree*
(**transform:rotation**(**view:target** view) axis 1)

This **transform** is used as an argument for the **position:transform** function which applies it to the current **eye** location. The function **set:eye** uses the resulting position to update the given view's (i.e. defined by the variable **view-name**[4]) display parameter. The process of changing the **up** vector is almost identical.

---

[4]The currently active view is returned by the function **env:active-view**.

7.9 Creating Postscript Images

To create a sharp-looking postscript picture you require a resolution of at least 600 dpi (dots per inch) on your printer. The output of the **render:postscript** command has a resolution equal to the viewport (ie window width) divided by the desired size (supplied as an argument to the function). So if a 2-inch picture requires a resolution of 600 dpi then the viewport size will have to be 1200 (assuming a square window and image, so that x and y resolutions are the same).

The function **psprint.scm** takes three arguments (an output size in inches, a resolution in dots per inch and a filename) and outputs a postscript file rendered from the active **viewport**.

psprint.scm

```
; Program for writing postscript images to file with a given dpi
; Args:- Body, Size of output (in inches), resolution in dpi, filename

(define psprint (lambda(body size resolution filename)
    (let* ((w (* size resolution))
           (orig-w (view:width))
           (op (view:viewport))
           (b (background "plain")) )
      (begin
        (background:set-prop b "color" (color:rgb 1 1 1)) ; white
        (render:set-mode "raytrace-full")
        (render:set-background b)  ; Make background white
        (view:set-viewport 0 0 w w) ; Set viewport for require dpi
        (view:set-size orig-w orig-w)
        (render:postscript  body filename (* 25.4  size)(* 25.4 size))
        (view:set-viewport (list-ref op 0) (list-ref op 1)  ; Restore original
                  (list-ref op 2) (list-ref op 3)) ; viewport
        (view:set-size orig-w orig-w) ; Restore view to original width
        (render:set-mode "gouraud") ; Restore default render mode
        (view:refresh)
    ))))
```

Notice that a **view** is really a window on a potentially much larger area known as the **viewport**. The program can be used in the following way:

```
; Create a dl display window (program assumes this is square)
(view:dl 1 1 500 500)

;create a test body
(define block (solid:block (position -10 -20 -30) (position 20 15 34)))
```

tb.ps

(**define** ball (**solid:sphere** (**position** 0 0 0) 25))
(**define** tb (**solid:unite** block ball))

; *rotate view to desired view with mouse*
(**view:compute-extrema**) ; *Fit part to window*
(**view:refresh**)

(psprint tb 1 300 "C:/temp/tb.ps") ; *Write postscript*

The program changes the background to white (it takes a long time to print black backgrounds) and scales the window to fit around the object. Although the resulting files can be very large, they reduce by up to 97% when compressed (e.g. zipped or stuffed).

7.10 Picking and Interaction

In general it is about ten times faster to write programs in Scheme than in C++. However, when it comes to writing interactive procedures that allow users to pick displayed ENTITYs via the mouse the productivity gain must be closer to twenty than ten!

The program **faceren.scm** uses the Scheme extensions **read-event**, **event:button** and **pick:face** to allow interactive selection of FACEs for highlighting.

The function **sc_group_faces** is recursive and will continue to add FACEs to a list (via left-button mouse picks) until the right-button is pressed to indicate that the selection process is finished. The procedure returns the list of selected FACEs which is used as an argument for the **face_render** function.

faceren.scm

; *This function allows the user to select a number of FACEs*
; *with the mouse and then render them.*
; *Note: Only the first (i.e. front) FACE found by each pick is selected.*

(**define** sel_faces
 (**lambda**()
  (**print** "Use left mouse button to pick faces.")
  (**print** "Use right mouse button to render picked faces.")
  (**let** ((evt (**read-event**)))
   (face_render (sc_group_face evt)))))

; *If button one is pressed pick an ENTITY and recurse*
; *else end recursion*
(**define** sc_group_face
 (**lambda**(evt)
  (**let*** ((ent (**pick:face** evt)))
   (**if** (**and** (**not** ent) (= (**event:button** evt) 1))
    (**begin**

7.11 Exercises

```
          (print "ERROR NO FACE PICKED: TRY AGAIN.")
          (sc_group_face (read-event)))
       (begin
         (if (= (event:button evt) 1)
           (begin
             (display (car ent))
             (newline)
             (entity:set-highlight (car ent) #t)
             (append (list (car ent)) (sc_group_face (read-event)))
        )))))))

(define face_render
  (lambda (faces)
    (begin
      (entity:set-render-sides (part:entities) 2)  ; Render both sides
      (env:set-highlight-color 2)
      (entity:set-highlight (part:entities) #t)
      (render faces)
     )))
```

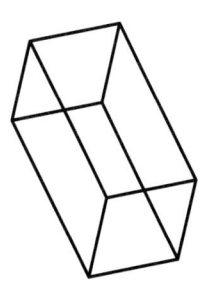

View before picking

The **face_render** function takes the list of picked FACEs and renders them before highlighting the rest of BODY. This creates the effect of several isolated FACEs sitting on a wire frame. The extension **entity:set-render-sides** defines the *sidedness* of facets lying on BODYs, LUMPs, SHELLs or FACEs. If the sidedness is set to 2 then the facets are double-sided (i.e. visible to the view on both sides). In contrast facets rendered with 1 sidedness can only be seen if their normal vector "faces" the viewer.

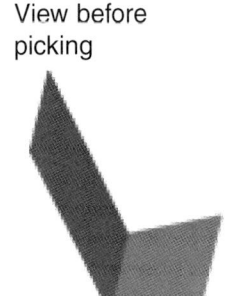

The picked FACEs

7.11 Exercises

1. Write a simple "brute force" ray tracer that assumes a default light source and uses Lambertian illumination model.

2. Modify the program **facet.cxx** (page 164) so it writes the triangle data to file in STL format. STL files are simply unordered lists of triangular facets with the following format:

```
solid ff_test01.stl
facet normal 0.000000 -1.000000 0.000000
outer loop
vertex 95.000000 75.000000 60.000000
vertex 95.000000 75.000000 0.000000
vertex 100.000000 75.000000 60.000000
end loop
end facet
```

⋮

3. Extend the **fhl.cxx** program so that the FACEs of the BODY (as well as the EDGEs) are checked for fhl_segments.

4. The program **faceren.scm** highlights only the first FACE (i.e. the front one) in the list of picked ENTITYs. Design and implement an interface that allows other FACEs (i.e. the back ones) to be picked.

Spline Surfaces

8

This chapter explores the different ways in which ACIS's APIs and the Scheme AIDE support complex, or free form, curves and surfaces. When first encountered many people find such spline geometry a difficult area with its numerous variables and unfamiliar jargon. Both of these conspire against the newcomer developing any intuitive understanding of what actually influences the final shape of the spline geometry. The unpleasant truth is that, unlike simple geometric shapes (where each parameter has a single, unique, impact on a shape) the final form adopted by a spline is determined by a complex interaction of **all** the variables (at least, locally). In other words a programmer wishing to increase the depth of a depression on a spline surface can tweak several variables, all of which will "deepen" the hole, but each in a subtly different way. Similarly the creator of spline geometry has the option of setting many parameters, whose effects are highly dependent on each other.

The chapter starts by defining some of the basic terminology used to describe splines. The generation of a cubic Bézier curve with the Scheme AIDE is demonstrated in the program **bezier.scm** which is preceded by some background theory intended to both support the example and provide an introduction to the concept of blending functions. The more general B-spline representation is then discussed and the effects of the different parameters illustrated by the creation of various curves and surfaces. Blending functions for both Bézier and B-spline curves are derived from first principles in order to make the underlying mathematics more transparent. The chapter ends with several programs that demonstrate some of the surface generation functions available (i.e. lofted, deformable and net surfaces).

8.1 Some Spline Concepts

There are many occasions when the shape of a curve is not initially specified by an analytical equation but only by a number of points that it in some sense *fits*. If the curve is required to pass through all these points, then the process used to define its shape is called *interpolation*. On the other hand if the curve needs only to be close to the points, without necessarily passing through any or all of them, then the process is known as *approximation*.

In solid modeling the need for interpolation might arise during the construction of a curve which represents the intersection of two surfaces. Such *intersection curves* might be approximated by interpolating a smooth curve through a number of points common to both surfaces. In contrast to this, curves which approximately pass through given points are more often used to produce shapes whose smooth appearance is aesthetically pleasing rather than precise.

Although the mathematics required to generate interpolating splines is much harder than that needed for approximation, the API user need not be aware of this. Indeed it is more work to *use* the API functions for approximations than interpolations.

The points used to define a spline are known as *control points* and when joined by straight lines (as in Figure 8.1) they are said to form a *control polygon*. To form a close polygon, the last control point can be linked to the first control point.

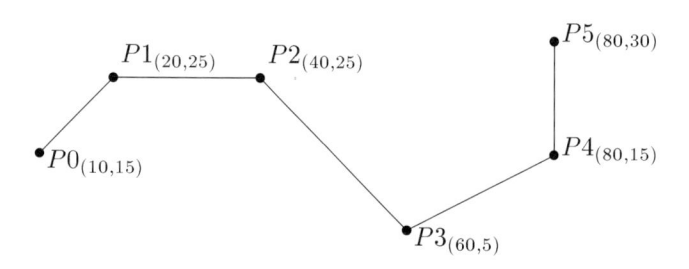

Figure 8.1: A sequence of control points forming a control polygon.

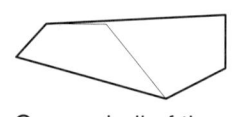

Convex hull of the control polygon

Free-form curves and surfaces are constructed by *blending* together the locations of the various control points. The control polygon, although not explicitly used in the creation of the curve, is useful as a way of visualizing the approximate final form of the spline curve. Also for Bézier, or B-Spline, curves the convex-hull[1] of the control polygon defines the boundary of the free-form shape generated. This property allows the extent of spline shapes to be *localized* and so can be used to support approximate intersection tests. The same approach can also be applied to free-form surfaces.

The control points are transformed into a smooth curve by means of a *blending function* which translates (or maps) between the normalized parametric representation of the curve (i.e. a value of u which varies between 0 and 1 along the curve) and the cartesian one (i.e. x, y, z). In other words blending functions typically calculate the (x, y, z) coordinates of a point on the curve at a given parameter value u.

The following example demonstrates how a Bézier blending function is used to form a smooth curve based on three control points.

---

[1] A minimal convex region enclosing a given shape.

8.2 Bézier Curves

Given three points $[(x_A, y_A), (x_B, y_B), (x_C, y_C)]$ a curve can be created which smoothly blends their locations by using the following expressions:

$$x(u) = x_A B_A(u) + x_B B_B(u) + x_C B_C(u) \quad (8.1)$$
$$y(u) = y_A B_A(u) + y_B B_B(u) + y_C B_C(u) \quad (8.2)$$

where:
B_k is point k's blending function
u is a parameter value between 0 and 1

Assume that the coordinates of the points and their Bézier blending functions are:

$$x_A = 0 \quad y_A = 0 \quad B_A = u^2 - 2u + 1 \quad (8.3)$$
$$x_B = 10 \quad y_B = 30 \quad B_B = 2u - 2u^2 \quad (8.4)$$
$$x_C = 50 \quad y_C = 30 \quad B_C = u^2 \quad (8.5)$$

allowing equations 8.1 and 8.2 to be written as:

$$x(u) = 0(u^2 - 2u + 1) + 10(2u - 2u^2) + 50u^2 \quad (8.6)$$
$$y(u) = 0(u^2 - 2u + 1) + 30(2u - 2u^2) + 30u^2 \quad (8.7)$$

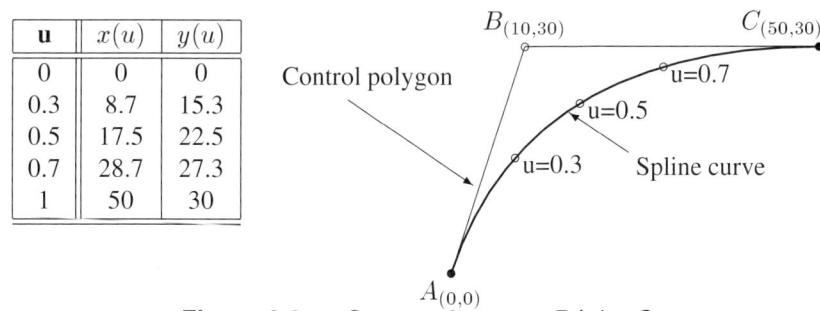

u	$x(u)$	$y(u)$
0	0	0
0.3	8.7	15.3
0.5	17.5	22.5
0.7	28.7	27.3
1	50	30

Figure 8.2: Some points on a Bézier Curve.

Before considering where the blending functions came from, notice that the curve they generate has the following properties:

1. The curve's **degree** (i.e. 2, as powers of u up to u^2 are used) is one less than the number of control points (i.e. three). It is a quadratic curve in this case.

2. The curve lies inside the convex hull of the control polygon. In this case the convex hull is a triangle defined by the points A, B and C.

3. It passes through (i.e. interpolates) the start and end points.

4. Its tangent is parallel to the control polygon at the start and end points.

These properties are true of all Bézier curves irrespective of the number or arrangement of the control points. Although quadratic and cubic Bézier curves are commonly used for the design of small segments of curve, they are rarely used with large numbers of control points. The reasons for this become apparent when the blending functions are considered in more detail.

In order to talk about blending functions in general we need to define a few expressions. The first step is to rewrite equations 8.1–8.7 so each blending function can be expressed for any number of control points. Recall that to calculate each blending function we need to know:

1. The total number of control points (n). This number will determine the degree of the curve, d (i.e. quadratic, cubic, etc.) which will be one less than the number of control points ($d = n - 1$).

2. The number of each individual control point (k) whose blending function is required.

Each blending function can now be expressed as:

$$B_{(k,d)}(u) \tag{8.8}$$

Convention dictates that the control points are numbered from zero and for ease of expression the coordinates of each point will be referred to by number (e.g. x_1 for the x-coordinate of the second point). So in the previous example where $d = 2$ (three control points) equations 8.1 and 8.2 could have been written:

$$x(u) = x_0 B_{(0,2)}(u) + x_1 B_{(1,2)}(u) + x_2 B_{(2,2)}(u) \tag{8.9}$$
$$y(u) = y_0 B_{(0,2)}(u) + y_1 B_{(1,2)}(u) + y_2 B_{(2,2)}(u) \tag{8.10}$$

or more succinctly as:

$$x(u) = \sum_{k=0}^{n} x_k B_{(k,d)}(u) \tag{8.11}$$

$$y(u) = \sum_{k=0}^{n} y_k B_{(k,d)}(u) \tag{8.12}$$

and most concisely of all the relationship can be stated in vector notation as:

$$\underline{p}(u) = \sum_{k=0}^{n} \underline{p}_k B_{(k,d)}(u) \tag{8.13}$$

The function $B_{(k,d)}$ can be expressed in several ways, each of which has its own merits.

8.2 Bézier Curves

Bernstein polynomial

This is the easiest form to compute manually:

$$B_{(k,d)} = \frac{d!}{k!(d-k)!} u^k (1-u)^{d-k} \tag{8.14}$$

So in the case of three control points ($n = 2, d = 2$):

$$B_{(0,2)} = \tfrac{2}{2} u^0 (1-u)^2 = 1 - 2u + u^2 \tag{8.15}$$
$$B_{(1,2)} = \tfrac{2}{1} u^1 (1-u)^1 = 2u - 2u^2 \tag{8.16}$$
$$B_{(2,2)} = \tfrac{2}{2} u^2 (1-u)^0 = u^2 \tag{8.17}$$

Remember:
$0! = 1$
and here we use in the limit $0^0 \to 1$.
N.B. generally 0^0 is indeterminate

Matrix form

Although the Bernstein form is easy to compute manually the symmetry of the functions is obscured by the factorial terms. The Matrix formulation makes the pattern underlying the functions clearer. This is done by reformulating equation 8.13 as:

$$p = B\,P \tag{8.18}$$

where $\quad P = \begin{bmatrix} p_0 \\ p_1 \\ p_2 \end{bmatrix}$

$$B = \begin{bmatrix} B_{(0,2)}, B_{(1,2)}, B_{(2,2)} \end{bmatrix}$$

The B matrix can be expressed as the product of two other matrices:

$$B = UBm \tag{8.19}$$

where $\quad U = [u^2, u, 1]$

$$Bm = \begin{bmatrix} 1 & -2 & 1 \\ -2 & 2 & 0 \\ 1 & 0 & 0 \end{bmatrix}$$

The symmetry becomes even more apparent when one considers the cubic Bézier blending function. For cubic blending:

$$U = [u^3, u^2, u, 1]$$

$$Bm = \begin{bmatrix} 1 & 3 & -3 & 1 \\ 3 & -6 & 3 & 0 \\ -3 & 3 & 0 & 0 \\ 1 & 0 & 0 & 0 \end{bmatrix}$$

Recursive form

Elegant, but not very accessible, is the recursive form which is stated as:

$$B_{(k,d)}(u) = (1-u)B_{(k,d-1)}(u) + uB_{(k-1,d-1)}(u) \tag{8.20}$$

The recursion terminates at $B_{(k,k)} = u^k$ or $B_{(0,k)} = (1-u)^k$. Although this form is, perhaps, the hardest to understand it is worth spending a few minutes appreciating its behavior because the B-spline blending functions (described in the next section) are expressed in this way. Also this is a numerically more stable way of computing $B_{(k,d)}$ than the matrix form or the right hand sides of 8.15-8.17.

Creating Bézier Curve

The Scheme AIDE function for creating an EDGE with a cubic Bézier curve requires four control points. This is shown in the **bezier.scm** program which creates a cubic Bézier curve and then sweeps it round an axis to create a solid body with a single complex surface.

bezier.scm

```
;This program makes a cubic Bezier curve and then rotates
;it to form a solid.

(define apple (lambda () (let*
        ((e1 (edge:bezier  (position 5 5 0)
                           (position 15 50 0)
                           (position 75 40 0)
                           (position 55 5 0)))
         (w1 (wire-body (list e1)))
         (s1 (solid:revolve-wire w1
                    (position 5 5 0)
                    (gvector:from-to
                        (position 5 5 0)
                        (position 55 5 0)) 360)))
        (render))))

(apple) ; To execute the function
```

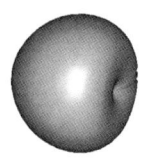

An equivalent C++ routine for the creation of cubic Bézier curves is **api_curve_bezier**.[2]

The Scheme AIDE, like many packages, provides only cubic Bézier curves because this gives reasonable flexibility of shape without the computational inefficiency

---

[2] Internally ACIS represents a Bézier curve as a special case of B-spline curve.

of higher-order polynomials. The problem with Bézier curves which are defined using more than, say, four control points is apparent in the cubic blending functions, which are:

$$\begin{align} B_{(0,3)}(u) &= (1-u)^3 \\ B_{(1,3)}(u) &= 3u(1-u)^2 \\ B_{(2,3)}(u) &= 3u^2(1-u) \\ B_{(3,3)}(u) &= u^3 \end{align}$$

Notice that all of the blending functions are non-zero across the *entire range* of u. Because of this no local control of the curve's shape is possible and a change in the location of any control point affects the geometry of the entire curve.

A second problem lies in the way that the degree of the curve is determined purely by the number of control points. Consequently a Bézier curve defined using ten control points would require a ninth degree blending function and hence a lot of computation!

About designing cubic Bézier curves

The shape of a cubic Bézier curve is determined by the location of four control points P_0, P_1, P_2 and P_3. Designing such a segment is a three-step process which determines the location of these points:

1. Choose the end-points P_0 and P_3 through which we want the curve to pass.

2. Place P_1 and P_2 on the desired tangents at P_0 and P_3.

3. Adjust the lengths of the two segments P_0P_1 and P_2P_3 simultaneously to give greater or lesser *fullness* to the curve.

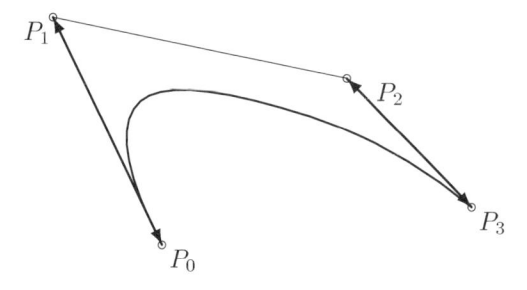

8.3 B-splines

Although ACIS provides only a couple of functions which explicitly support Bézier curves, an understanding of their theory is useful because it provides an introduction to concepts that can be developed to create a type of free-form curve known as a **B-spline**.

$P1_{(20,25)}$ $P2_{(40,25)}$ $P5_{(80,30)}$

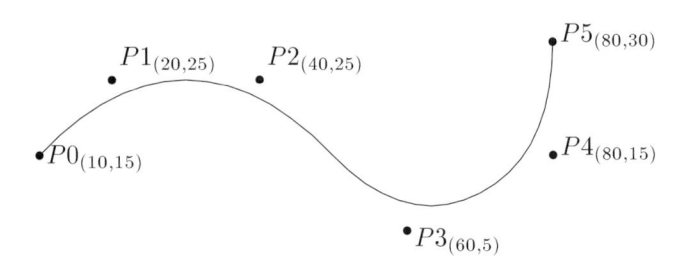

$P0_{(10,15)}$ $P4_{(80,15)}$

$P3_{(60,5)}$

B-spline curves do not always pass through the first and last control points.

Figure 8.3: A B-spline curve and its control points.

B-splines can be viewed as a collection of "Bézier-like" curves arranged so that the end of one flows smoothly into the next. The shape of each of these curve segments is generated by a sophisticated blending function which switches on and off the influence of various control points on the different elements of the curve.

Before looking at the recursive form of the B-spline blending function some fundamental concepts are illustrated in Figures 8.3–8.7.

Consider the curve shown in Figure 8.3. This quadratic (i.e. degree 2) B-spline approximates the position of the six control points shown. Although the curve appears continuous it is actually constructed from four separate segments shown in Figure 8.4.

Each segment is defined by a separate polynomial.

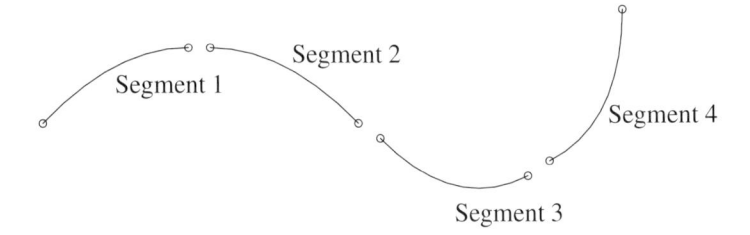

Segment 1, Segment 2, Segment 3, Segment 4

Figure 8.4: Segments of the B-spline curve.

The entire curve is referred to as a *piecewise polynomial*. This segmenting of the curve gives B-splines two important advantages over Bézier curves:

- The degree of the curve can be chosen independently of the number of control points.

- Local changes to the curve's shape are possible because individual control points have only local influences.

8.3 B-splines

> About geometric continuity
>
> The *smoothness* with which two curve segments meet is termed **continuity**. A distinction is made between *geometric* (termed G) and *parametric* (termed C) continuity. Different **degrees** of continuity are represented by the notation G^d and C^d. The physical meaning of the lower degrees of geometric continuity is easily illustrated:
>
> G^0: Continuity implies start and end points are common.
>
> $(x_0, y_0) = (x_1, y_1)$
>
> G^1: Continuity implies tangents have common direction.
>
> G^2: Continuity implies curvature is the same on both sides.

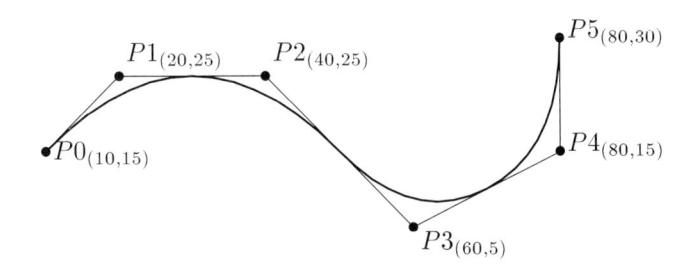

Figure 8.5: A B-spline curve and its control polygon.

Figure 8.5 shows the curve's control polygon. Like a Bézier curve, a B-spline curve lies within the convex hull of its control polygon. Although the curve shown is **tangent** to the sides of its control polygon, this behavior is peculiar to second-degree B-spline curves of the type shown and is not generally the case.

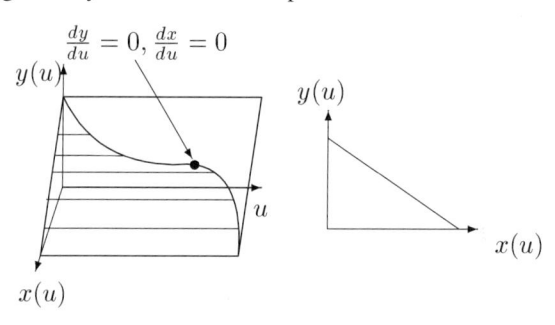

About parametric continuity

> In contrast to geometric continuity, parametric continuity (C^n) is defined in terms of derivatives of u. In general values of $\frac{d^n x}{du^n}$ are hard to visualize because the plot of a parametric curve does not show the independent variable u. Consequently C^n type continuity cannot be determined by inspection of displayed geometry. Consider the 2D parametric curve shown below:
>
> (a) Drawn in 3D (x, y, u) space   (b) Drawn in 2D (x, y) space
>
> The tangent vector to the curve can only be seen on the 3D plot (a); looking at the 2D representation it is impossible to deduce that the magnitude of $\frac{dx}{du}$ falls to zero around the midpoint.
>
> C^0: Continuity implies start and end points have common values of u.
>
> $$x_0(u) = x_1(u) \text{ and } y_0(u) = y_1(u)$$
>
> C^1: Continuity implies that the first derivative of position with respect to u is the same for both curves.
>
> $$x_0'(u) = x_1'(u) \text{ and } y_0'(u) = y_1'(u)$$
>
> C^2: Continuity implies that the second derivative of position with respect to u is the same for both curves.
>
> $$x_0''(u) = x_1''(u) \text{ and } y_0''(u) = y_1''(u)$$
>
> The segments of a degree d B-spline meet with C^{d-1} continuity.

Figure 8.6 shows the control points associated with each of the four segments. Notice three important things about these fragments of the control polygon:

1. The degree of the curve is one less than the number of control points for each segment. In this case we have a second-degree curve so each segment requires three control points. A cubic curve would require four control points per segment (see Figure 8.7). In other words:

$$n_s = n - d \tag{8.21}$$

where n_s = number of segments
d = degree of the curve
n = number of control points

8.3 B-splines

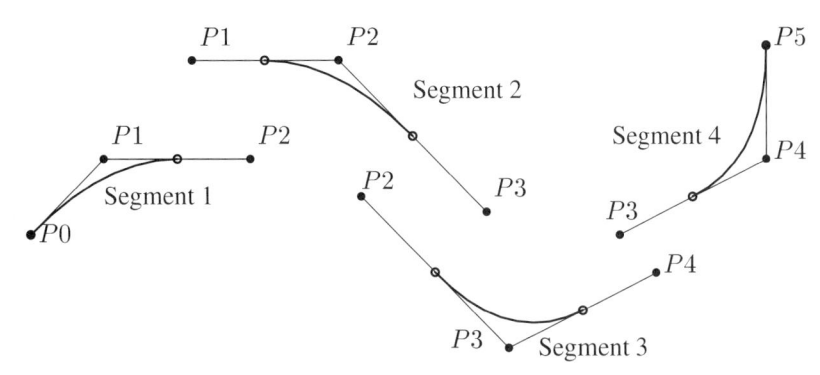

Figure 8.6: Segment's control polygons.

2. Adjacent segments have control points in common. The number of common points is determined by the degree of the curve and the type of continuity required between them. In this example the quadratic segments are united with C^1 continuity. Cubic curves, on the other hand, blend their pieces with C^2 continuity.

3. The segments interpolate (pass through) only the first and last points of the control polygon. Elsewhere the blending function prevents the curve from passing through the start and end points of the control polygon fragments. More generally a B-spline curve does not *have* to pass through *any* of its control points, but can easily be made to pass through the end ones if required.

Unlike Bézier curves (that are wholly specified by a set of control points) B-splines also require both the degree of the curve and the parametric values of each segment's start and end points (known as **knots**) before the blending function can calculate the final shape.

Having established that the degree of the curve affects the continuity with which segments meet, it is worth establishing the way in which control points and knots affect the shape before examining the blending function.

Control points

The arrangement of a B-spline's control points determines if it is **open** (i.e. the ends do not meet) or **closed** (i.e. the ends do meet).

A closed B-spline is created by repeating a number of the initial control points at the end of the sequence. The number of repeated points should be equal to the spline's degree. For example a **closed** cubic spline might have the following sequence of control points:

$$\{P_0, P_1, P_2, P_3, P_4, P_5, P_0, P_1, P_2\}$$

In both open and closed B-splines individual control points can be repeated a number of times. A control point repeated three times in a cubic spline causes the spline to pass through that point.

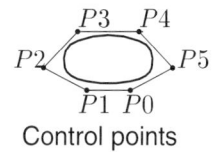

Control points for a closed spline

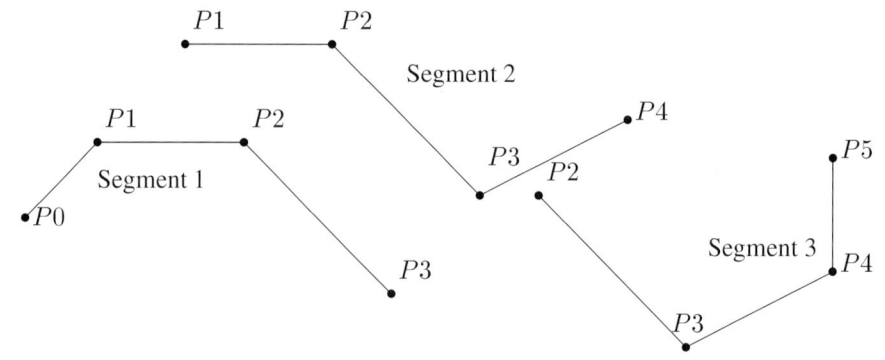

Figure 8.7: Control polygons for a cubic spline through the same points.

Knots

The start and end points of each segment are known as **knots** and are specified by values of the parametric variable u.

The set of knots for a given curve is called the **knot vector** and is required by the blending function.

The length of the **knot vector** is determined by:

1. The degree of the curve.
2. The number of control points.

The values in the knot vector must never decrease, but they can be repeated. The number of times the value of an individual knot is repeated is referred to as its **multiplicity**.

Multiple knot values reduce the continuity of the curve by one for each repeat of a particular value. Generally this is undesirable because it can create discontinuities in the curve. However, at the end points of a B-spline this behavior can be used to ensure that the resulting curve interpolates the first and last control points. For example, the following knot vector ensures that a third-degree B-spline passes through the first and last control points.

$$\underbrace{\{\ \overbrace{0, 0, 0}^{Multiplicity\ 3},\ \ 0.2, 0.4, 0.6, 0.8,\ \ \overbrace{1, 1, 1}^{Multiplicity\ 3}\ \}}_{Knot\ Vector}$$

The knot vector shown creates what is known as an **open knot vector**, because the knot spacing is uniform along the curve (except for the repeats at the ends to ensure it interpolates the first and last control points).

In contrast **non-uniform** knot vectors have unequally spaced knot values.

8.4 B-spline Blending Function

The behavior of the B-spline blending function is highly non-intuitive. Unlike Bézier curves there is no single set of blending functions because they are determined in part

8.4 B-spline Blending Function

by the knot vector. A B-spline curve is defined by a parametric function of the form:

$$\underline{p}(u) = \sum_{k=0}^{N} \underline{p}_k B_{(k,D)}(u) \quad (8.22)$$

where $N + 1 = n$, the number of control points,
$D - 1 = d$, the degree of the curve,
$B_{(k,D)}$ = point k's, order D, blending function,
u = a parameter value between the minimum and maximum knot values,
$\underline{p}_k$ = control point k

As with Bézier curves the blending function can be written in several different forms. The recursive form is:

$$B_{(k,D)}(u) = \frac{(u - t_k)}{(t_{k+D-1} - t_k)} B_{(k,D-1)}(u) + \frac{(t_{(k+D)} - u)}{(t_{(k+D)} - t_{(k+1)})} B_{(k+1,D-1)}(u) \quad (8.23)$$

where t_i is the u parameter value of knot i.

Consider how this function is evaluated for, say, the third control point ($k = 2$) on a quadratic ($D = 3$) curve:

$$B_{(k=2,D=3)}(u)$$

$$\underbrace{\frac{(u-t_2)}{t_4-t_2} B_{(2,2)}(u) \qquad \frac{(t_5-u)}{(t_5-t_3)} B_{(3,2)}(u)}_{+}$$

$$\underbrace{\frac{(u-t_2)}{t_3-t_2} B_{(2,1)}(u) \quad \frac{(t_4-u)}{t_4-t_3} B_{(3,1)}(u)}_{+} \quad \underbrace{\frac{(u-t_3)}{t_4-t_3} B_{(3,1)}(u) \quad \frac{(t_5-u)}{(t_5-t_4)} B_{(4,1)}(u)}_{+}$$

The recursion proceeds down to the level at which $d = 1$. For all values of k the blending function is implemented in such a way that:

$$B_{k,1} = 1 \quad \text{if} \quad t_k \leq u < t_{k+1}$$
$$B_{k,1} = 0 \quad \text{for all other values}$$

In other words if the value of u for which the blending function is being evaluated lies between knots t_i and t_{i+1} then it is **on** otherwise it is **off**.

Notice that although the blending function is being evaluated for the third control point ($k = 2$), the resulting expression involves terms for knots up to t_5.

From this we can deduce that the control point ($k = 2$) influences the shape of several other adjacent segments.

The expression can be evaluated further if the following knot vector[3] is specified:

$$\{t_0 = 0, t_1 = 0, t_2 = 1, t_3 = 2, t_4 = 3, t_5 = 4, t_6 = 4\} \tag{8.24}$$

This particular pattern of knots follows the form given on page 198 and allows the term $B_{(2,3)}$ to be evaluated as:

$$B_{(k=2,D=3)}(u)$$

$$\overbrace{\frac{u}{2} B_{(2,2)}(u) \qquad + \qquad \frac{(3-u)}{2} B_{(3,2)}(u)}$$

$$\overbrace{u B_{(2,1)}(u) \quad (2-u) B_{(3,1)}(u)} \qquad \overbrace{(u-1) B_{(3,1)}(u) \quad (3-u) B_{(4,1)}(u)}$$

Because the denominators of $B_{(k,d)}$ can evaluate to zero the convention $\frac{0}{0} = 0$ is used.

Thus with the stated knot pattern:

$$B_{(2,3)} = \frac{u^2}{2} B_{(2,1)}(u) + \frac{2u - u^2}{2} B_{(3,1)}(u) + \frac{4u - 3 - u^2}{2} B_{(3,1)}(u) + \frac{(3-u)^2}{2} B_{(4,1)}(u)$$

which can be written as:

$$B_{(2,3)} = \frac{u^2}{2} B_{(2,1)}(u) + \frac{1}{2}(-u^2 + 6u - 3) B_{(3,1)}(u) + \frac{1}{2}(3-u)^2 B_{(4,1)}(u)$$

In this way the blending function for each control point can be calculated. So for the curve shown in Figure 8.3 which has six control points:

$$B_{(0,3)} = (1-u)^2 B_{(2,1)}(u)$$
$$B_{(1,3)} = \tfrac{1}{2}(4 - 3u) B_{(2,1)}(u) + \tfrac{1}{2}(2-u)^2 B_{(3,1)}(u)$$
$$B_{(2,3)} = \tfrac{u^2}{2} B_{(2,1)}(u) + \tfrac{1}{2}(-u^2 + 6u - 3) B_{(3,1)}(u) + \tfrac{1}{2}(3-u)^2 B_{(4,1)}(u)$$
$$B_{(3,3)} = \tfrac{1}{2}(u-1)^2 B_{(3,1)}(u) + \tfrac{1}{2}(-2u^2 + 10u - 11) B_{(4,1)}(u) + \tfrac{1}{2}(4-u)^2 B_{(5,1)}(u)$$
$$B_{(4,3)} = \tfrac{1}{2}(u-2)^2 B_{(4,1)}(u) + \tfrac{1}{2}(-3u^2 + 20u - 32) B_{(5,1)}(u)$$
$$B_{(5,3)} = (u-3)^2 B_{(5,1)}(u)$$

---

[3] ACIS normalizes all curve and surface parameters so the equivalent knot vector in ACIS would be:

$$\{0, 0, 0.25, 0.5, 0.75, 1, 1\}$$

but since this makes the relationship between knots and control points harder to follow the discussion will assume a parameter range for each curve between 0 and $N - D + 2$ (i.e. $u \in [0, (N - D + 2)])$.

8.4 B-spline Blending Function

Finally the blending functions can be multiplied by the respective control points to give:

$$p(u) = p_0 B_{(0,3)} + p_1 B_{(1,3)} + p_2 B_{(2,3)} + p_3 B_{(3,3)} + p_4 B_{(4,3)} + p_5 B_{(5,3)}$$

Some insight into the working of this equation can be had by collecting together the terms active over individual segments:

All the $B_{(2,1)}$ terms : $\quad p_{Seg1}(u) = (1-u)^2 p_0 + \frac{1}{2}u(4-3u)p_1 + \frac{1}{2}u^2 p_2 \qquad t_1 \leq u < t_2$

All the $B_{(3,1)}$ terms : $\quad p_{Seg2}(u) = \frac{1}{2}(2-u)^2 p_1 + \frac{1}{2}(-2u^2 + 6u - 3)p_2 + \frac{1}{2}(u-1)^2 p_3 \qquad t_2 \leq u < t_3$

All the $B_{(4,1)}$ terms : $\quad p_{Seg3}(u) = \frac{1}{2}(3-u)^2 p_2 + \frac{1}{2}(-2u^2 + 10u - 11)p_3 + \frac{1}{2}(u-2)^2 p_4 \qquad t_3 \leq u < t_4$

All the $B_{(5,1)}$ terms : $\quad p_{Seg4}(u) = \frac{1}{2}(4-u)^2 p_3 + \frac{1}{2}(-3u^2 + 20u - 32)p_4 + (u-3)^2 p_5 \qquad t_4 \leq u < t_5$

Recall the knot vector:

$$\{t_0 = 0, t_1 = 0, t_2 = 1, t_3 = 2, t_4 = 3, t_5 = 4, t_6 = 4\}$$

u	$x(u)$	$y(u)$	$u_{normalized}$
0	10	15	0
0.7	24	24.1	0.175
2.5	60	8.75	0.625
3.7	79.1	21.9	0.925
4	80	30	1

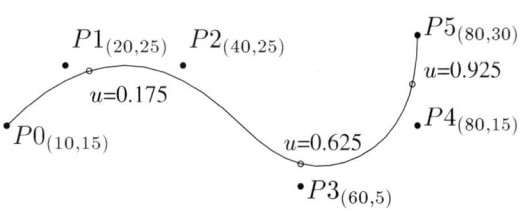

The longhand calculation of B-spline blending functions is not quick! A much faster way of exploring the effects of different parameter changes is to use ACIS. The program **nurbs.cxx** generates a fourth-degree curve through the six control points used in the worked example.

nurbs.cxx

```
// This program creates a spline curve and saves it to a file
#include <stdio.h>            // File I/O functions
#include <stdlib.h>           // defines exit
#include "constrct/kernapi/api/cstrapi.hxx" // Construction API's
#include "kernel/kernapi/api/api.hxx"       // Declares outcome class
#include "kernel/kernapi/api/kernapi.hxx"   // Declares kernel API's
#include "kernel/kerndata/top/alltop.hxx"   // Defines BODY class
#include "kernel/kerndata/lists/lists.hxx"  // Defines ENTITY_LIST class
#include "baseutil/vector/position.hxx"     //Defines point class
#include "kernel/kerndata/savres/fileinfo.hxx" // Declares fileinfo class
```

```
void save_ent(char*, ENTITY*);              // Function prototype; see save.cxx
void outcome_check(outcome res, char* string); // See program section.cxx

void main()
{
   api_start_modeller(0);
   api_initialize_constructors();
   api_checking(TRUE); // Turn API argument checking on
   int degree = 4;
   int rational = FALSE;  // Requires weighted control points
   int closed = FALSE;
   int periodic = FALSE;
   int num_ctrlpts = 6;
   position ctrlpts[6] = {position(10,15,0),position(20,25,0),
                   position(40,25,0), position(60,5,0),
                   position(80,15,0),  position(80,30,0)};
   double weights[1] = (double)NULL; // Only needed for rational splines
   double point_tol = resabs ;
   int num_knots = 9;
   double knots[9]={0,0,0,0,0.5,1,1,1,1};
   double knot_tol = resabs ;
   EDGE* spline = NULL;
   outcome res = api_mk_ed_int_ctrlpts (degree, rational,
              closed, periodic, num_ctrlpts, ctrlpts,
              weights, point_tol, num_knots, knots,
              knot_tol, spline);
   outcome_check(res, "Error in API call");
   save_ent("spline.sat",spline);
   api_terminate_constructors();
   api_stop_modeller();
}
```

The function **api_mk_ed_int_ctrlpts** takes the 12 arguments which can be summarized as follows:

Degree: This is an integer that defines the degree of the spline curve. The value also defines the continuity (i.e. smoothness) with which segments meet. In the program a spline of degree four is specified.

Control points: Defined in an array of position objects called ctrlpts, which is num_ctrlpts long. These points define the approximate location, orientation and general spread, or extent, of the resulting curve. The curve is *pulled* closer to repeated points and will interpolate any point repeated *degree* times. The variable point_tol is used to determine if two successive control points are coincident. In this example it is set to the system tolerance, resabs (see page 129).

Knot vector: An array of doubles num_knots long called knots. These points define where the segments of the curve meet. The values in the knot vector must

8.4 B-spline Blending Function

never decrease, but can be repeated. The continuity of the curve is reduced by one every time a knot is repeated. The variable **knot_tol** is used to determine if two successive knots are coincident. In this example it is set to the system tolerance, resabs (see page 129).

Control point weighting: Rational B-splines allow the curve to be guided closer to particular control points by *weighting* them to increase the amount of "pull" they appear to exert on the spline. In this example the variable **rational** has been declared false so no weights are needed and the array of doubles known as **weights** is declared NULL. For an example of rational B-splines see the program **spsol1.scm** later in this chapter.

Topology of the EDGE: The variables **closed** and **periodic** are used to (help) determine the topology of the EDGE created. If the curve is open, both a start and an end VERTEX are created for the edge. If the curve is closed, or periodic, a single VERTEX is created and used for the EDGE.

The program **nurbs.cxx** can be used to investigate how various parameters affect the shape of the curve. Firstly the **degree** of the curve can be varied. In each case the knot vector is arranged to cause the curve to interpolate the first and last control points and create spline segments of equal parametric range.

degree := 1

number of knots := 6

knot vector := {0,0.2,0.4,0.6,0.8,1}

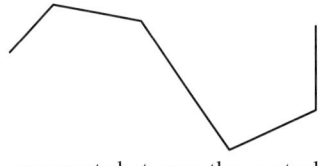

A first-degree curve appears as a series of straight line segments between the control points (i.e. the curve *is* the control polygon).

degree := 2

number of knots := 7

knot vector := {0,0,0.25,0.5,0.75,1,1}

A second-degree curve is tangent to the control polygon at mid-points of the control polygon's internal EDGEs.

degree := 3

number of knots := 8

knot vector := {0,0,0,0.33,0.66,1,1,1}

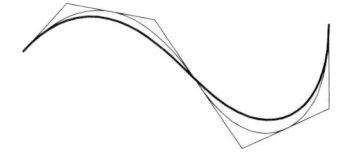

The segments of a cubic spline each join with C^2 continuity resulting in a smoother curve which follows the positions of the control points less closely. The degree 2 curve is also drawn here for comparison.

About choosing NURBS parameters

> The degree of the curve (d), the number of control points (n) and the number of knots (K_p) have the following relationships:
>
> 1. Number of knots given n and a choice for d:
>
> $$K_p = n + d - 1$$
>
> 2. Maximum degree that is possible:
>
> $$d_{max} = n - 1$$
>
> 3. Multiplicity of knots: The knot vector must start and end with d repeated knots if you want the spline to pass through the first and last control points.
>
> 4. Rest of the knots: The remaining knots can be set at any value between 0 and 1 but must form a non-decreasing sequence.

The effect of the knot vector can be demonstrated by keeping the control points and degree of the curve constant. The following four curves each have a different knot vector:

degree := 3

number of knots := 8

knot vector := {0,0,0,0.33,0.66,1,1,1}

The parameters create three segments of identical parametric range and produce a smooth even shape.

degree := 3

number of knots := 8

knot vector := {0,0,0,0.1,0.9,1,1,1}

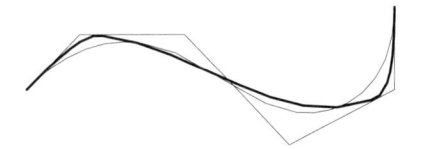

The knot vector creates a curve with heavier emphasis on the control points nearer the end by having a shorter parametric range there.

degree := 3

number of knots := 8

knot vector := {0,0,0,0.7,0.9,1,1,1}

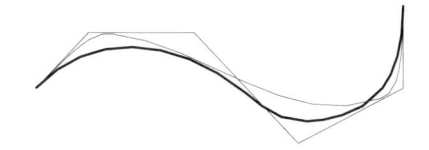

This curve has a long parametric range for the first segment, followed by two shorter ranges. The difference from the previous curve is small but noticeable.

degree := 3

number of knots := 8

knot vector :=
{0,0.2,0.3,0.7,0.8,0.85,0.95,1}

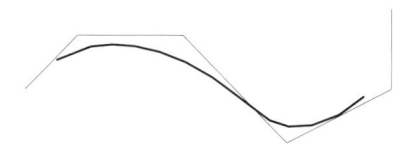

The knot vector contains no repeated knots at the end so the spline does not interpolate the first or last control points. The resulting curve starts and ends in space.

8.5 Rational Splines

In addition to the control points, knot vector and degree, there is a fourth set of parameters, known as weights, which arise in the extension of B-spline curves to rational B-spline curves (ACIS supports both rational and nonrational forms).

Mathematically a rational B-spline curve is defined as

$$p(u) = \frac{\sum_{k=0}^{n} w_k p_k B_{(k,D)}(u)}{\sum_{k=0}^{n} w_k B_{(k,D)}(u)} \qquad (8.25)$$

where p_k is the control point k,
$B_{(k,D)}$ is the degree D, B-Spline blending function for the point k,
w_k is the weighting factor for the point k, normally chosen to be positive.

Given that $1 = \sum_{k=0}^{n} B_{(k,D)}$ it is clear that when weighting factors of 1 are used the expression reverts to the standard B-spline equation.

However, when other values are used the amount of "pull" exerted by individual control points can be varied.

One well-known application of rational B-splines is in the generation of conic sections: ACIS's NURBS (Non-uniform Rational B-Spline) functions can be used to create a conic section by specifying three control points (P_0, P_1, P_2) and the knot vector {0,0,1,1} (i.e. the same as for a quadratic Bézier spline), with weights, as follows, assigned to each control point:

$$w_0 = 1$$
$$w_2 = 1$$

$$w_1 = \frac{r}{1-r}$$

where $0 \leq r < 1$.
This formulation allows four distinct type of conic to be generated:

$r > \frac{1}{2}$: a hyperbolic curve

$r = \frac{1}{2}$: a parabolic curve

$r < \frac{1}{2}$: an elliptical curve

$r = 0$: a straight-line curve

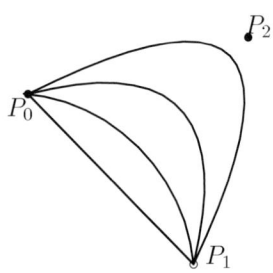

In the Scheme AIDE this NURBS construction appears as **edge:conic**; the program below uses this Scheme extension to generate an interesting hyperbolic surface.

The program **coniced.scm** creates a hyperbolic curve which is then swept around an axis to form a hyperboloid surface with negative Gaussian curvature.

coniced.scm

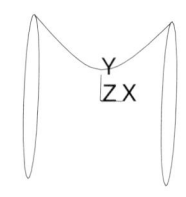

```
; This program makes two hyperboloids by sweeping an
; edge around the x-axis. The smaller is then subtracted
; from the larger and the resulting tube copied, rotated
; and intersected to form a cross.

(define hyperboloids
  (lambda() (let* (
    (se1 (edge:conic (position -80 50 0)
                     (position 80 50 0)
                     (position  0 5 0) 0.7)) ; r = 0.7
    (se2 (edge:conic (position -82 48 0)
                     (position 82 48 0)
                     (position  0 3 0) 0.7)) ; r = 0.7
    (outside  (wire-body (list se1)))
    (hypout (solid:revolve-wire outside (position 0 0 0)
                                (gvector 1 0 0) 360))
    (inside  (wire-body (list se2)))
    (hypin (solid:revolve-wire inside (position 0 0 0)
                               (gvector 1 0 0) 360))
    (hyp1 (solid:subtract hypout hypin))
    (hyp2 (entity:copy hyp1)))
  (begin
    (entity:transform hyp2
          (transform:rotation (position 0 0 0) (gvector 0 1 0) 90))
    (solid:unite hyp1 hyp2)
    (option:set "u_param_lines" 10)
    (option:set "v_param_lines" 8) hyp1))))  ;Return hyp1 solid
```

Section through
the resulting solid

The program unites two intersecting **hyperboloids of revolution**. The debugging information, shown below, confirms that the union operation has created an enclosed cavity, evidenced by the two SHELL records in the following summary of hyp1's data structure.

```
acis> (define h (hyperboloids))
acis> (entity:debug h 3)
     1 body record,           32 bytes
    11 attribute records,   2012 bytes
     1 lump record,           32 bytes
     2 shell records,         80 bytes
    20 face records,         880 bytes
    24 loop records,         768 bytes
    20 surface records,     3072 bytes
    56 coedge records,      2464 bytes
    28 edge records,        2016 bytes
    48 pcurve records,      4224 bytes
    16 vertex records,       384 bytes
    28 curve records,       2880 bytes
    16 point records,        768 bytes
Total storage 19612 bytes
"solid body"
```

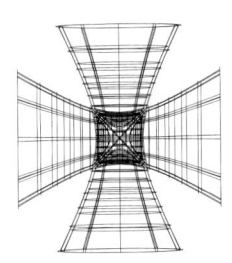

The location of this internal cavity is easily seen in the shaded view of the sectioned BODY, shown in the margin on page 206.

8.6 Creating Spline Surfaces

B-spline surfaces are created from rectangular arrays of control points, with separate knot vectors for parameters u and v. These run across the surface with the parameter lines for u cutting those for v and vice versa.

The position of a point on the surface can be calculated using an equation analogous to that of a B-spline curve:

$$p(u,v) = \sum_{i=0}^{m} \sum_{k=0}^{n} p_{(i,k)} B_{(i,du)}(u) B_{k,dv}(v) \qquad (8.26)$$

where $p_{(i,k)}$ is an $(m+1) \times (n+1)$ rectangular grid of control points
$B_{(i,d_u)}$ is the degree d_u blending function for each ith point going across in u
$B_{(j,d_v)}$ is the degree d_v blending function for each jth point going across in v

Notice that the surface can have different degrees in u and v directions. The program **spsol1.scm** demonstrates how a spline surface can be created from a grid of control points and specified blending functions in both u and v directions. The program is presented in two parts. First the data needed to define a spline surface is defined and stored in an **splsurf** data type. In the second part of the program the data is used to form a FACE which is intersected with a base plane to create a solid.

spsol1.scm

```
; This program makes a solid out of a spline and planar surfaces
; Create a spline-surface data-type to hold information needed to create it
(define Hump-Surface (splsurf))
(define Control-Points
  (list
    (position -36 36 5)(position 0 50 0)(position 36 36 5)
    (position -40 25 0)(position 0 25 -10)(position 40 25 0)
    (position -50 0 0)(position 0 0 50)(position 50 0 0)
    (position -40 -25 0)(position 0 -25 -10)(position 40 -25 0)
    (position -36 -36 5)(position 0 -50 0)(position 36 -36 5)
  ))
  ; Puts a list of 15 points (5 in "u", 3 in "v") into the Hump-Surface data-structure
(splsurf:set-ctrlpt-list Hump-Surface Control-Points 5 3)
  ; Specifies cubic polynomial curves, (3) which are open (1), rational (0) and
  ; have no singularities (0)
(splsurf:set-u-param Hump-Surface 3 1 0 0)
  ; Specifies quadratic polynomial curves, (2) which are open (1),
  ; rational (0) and have no singularities (0)
(splsurf:set-v-param Hump-Surface 2 1 0 0)
  ; Because "set-u-param" specified an open curve of degree 3,
  ; the knot list starts and ends with 3 equal entries to ensure
  ; interpolation of start and end points
(splsurf:set-u-knot-list Hump-Surface (list 0 0 0 0.5 1 1 1) 7)
  ; Because "set-v-param" specified an open curve of degree 2,
  ; the knot list starts and ends with 2 equal entries to ensure
  ; interpolation of start and end points
(splsurf:set-v-knot-list Hump-Surface (list 0 0 1 1) 4)
  ; Because we specified a rational type of spline, weights must be specified
(define Weights
  (list 1 1 1
        1 2 1
        1 8 1
        1 2 1
        1 1 1 ))
  ; Number of Weights must equal number of Control Points
(splsurf:set-weight-list Hump-Surface Weights)
```

This example starts by defining a grid (in list form) of the spline surface control points. The list is "inserted" into the **splsurf** data structure with the function **splsurf:set-ctrlpt-list** which also specifies the number of rows and columns making up the list.

8.6 Creating Spline Surfaces

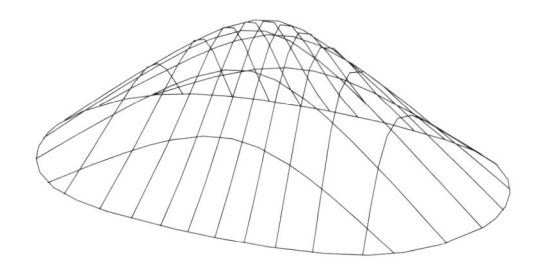

Figure 8.8: The hump solid.

The function **splsurf:set-u-param** is used to specify a set of cubic polynomial curves, (3) which are open (1), rational (0) and have no singularities (0). Similar parameters are specified using **splsurf:set-v-param** for the curves in the **v** direction.

One immediate consequence of this declaration is the need for a list of weights. Because the spline has been declared **rational** the relative influence of the control points must be defined.

Since both the degree and number of control points are now fixed the number of knot positions required can be determined.

Recall that for B-splines:

$$K_p = n + d - 1$$

In the **u** direction there are 5 control points and the degree is three, so 7 knot points will be required.

A similar calculation can be carried out in the **v** direction.

spsol1.scm (continued)

```
; Display six "v" lines and ten "u" lines
(option:set "v_param_lines" 6)
(option:set "u_param_lines" 10)

; Create a face with the specified spline surface
(define HumpFace
  (sheet:face (face:spline-ctrlpts Hump-Surface)))

; Create a base face
(define PlaneFace
  (sheet:face(face:reverse(face:plane
    (position -50 -50 10)
    100 100
    (gvector 0 0 1)))))
```

```
; Intersect the two infinite volumes above and below the faces
(define Hump (bool:intersect HumpFace PlaneFace))
; check it is a solid by working out its volume etc.
(solid:massprop Hump)
```

Having created a definition of a spline surface the **spsol1.scm** program uses it to first create a sheet FACE with **face:spline-ctrlpts** and then a BODY[4] with **sheet:face**. Lastly the sheet BODY is intersected with a BODY made from a planar FACE to form a 3D manifold solid.

8.7 Interpolation

Interpolation requires a system of equations to be solved (internally) by ACIS to generate some suitable control points so that the surface generated passes through the points to be interpolated.

The programs **slpgrid.scm** and **scherk.scm** demonstrate ACIS's spline interpolation functions.

slpgrid.scm

```
; This program creates a face with a surface that interpolates
; the given grid of points. Because points are repeated a
; closed spline surface is formed.
(define dip-face
  (lambda ()
   (let* ((spldata (splgrid))
     (grid (list
         ; Row 0
         ( position -40 -45 -40) ( position 30 -50 -40)
         ( position 20 -20 -40) ( position -10 -10 -40)
         ( position -20 20 -40) ( position -40 30 -40)
         ( position -60 -10 -40) ( position -40 -45 -40)
         ; Row 1
         ( position -40 -45 0) ( position 30 -50 0)
         ( position 20 -20 0) ( position -10 -10 0)
         ( position -20 20 0) ( position -40 30 0)
         ( position -60 -10 0) ( position -40 -45 0)
         ; Row 2
         ( position -40 -45 40) ( position 30 -50 40)
         ( position 20 -20 40) ( position -10 -10 40)
         ( position -20 20 40) ( position -40 30 40)
         ( position -60 -10 40) ( position -40 -45 40))))
     (begin
      (splgrid:set-point-list spldata grid 3 8)
```

---

[4]A sheet BODY whose only LUMP has just one FACE.

8.7 Interpolation

```
      (option:set "u_param_lines" 10)
      (option:set "v_param_lines" 8)
        ; The #t argument causes points to be interpolated rather than approximated (#f)
      (face:spline-grid spldata #t)))))
```

One of the really powerful uses of the Scheme AIDE's spline interpolation routines is the approximation of surfaces defined by an equation. The program **scherk.scm** demonstrates this by generating a number of points on Scherk's surface and then using them to create a spline FACE.

A minimal surface spans a given boundary with the smallest possible area. This type of surface can be generated experimentally by dipping a wireframe model of the boundary into a tub of soapy water. With luck a soap film will appear on the frame when it is removed in the shape of the wire's minimal surface. Scherk's minimal surface is unusual because of the very simple equation that defines it:

$$e^z \cos(y) = \cos(x)$$

About Scherk's surface

scherk.scm

```
; Creates a model of Scherk's Surface
(define scherk
  (lambda()
    (let* ((pt-array (list ))
        (scale (transform:scaling 25 25 25))
        (step 0.2) (xpts 0) (ypts 0)
        (spldata (splgrid)))
      (begin
      (do ((x-coord -1.5 (+ x-coord step)))
          ((> x-coord 1.5) )
        (set! xpts (+ xpts 1))
        (do ((y-coord -1.4 (+ y-coord step)))
            ((> y-coord 1.4) )
          (set! ypts (+ ypts 1))
          (let* ((diff (/ (cos x-coord) (cos y-coord)))
                 (z-coord 0.0))
            (if (not (zero? diff))
                (set! z-coord (log (abs diff))))
            (set! pt-array
                  (append pt-array
                          (list
```

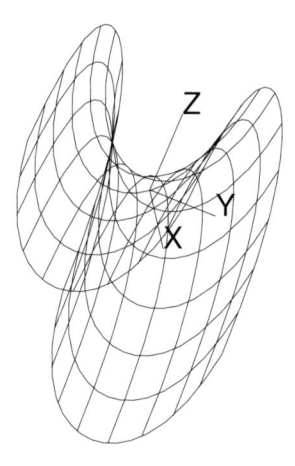

Figure 8.9: Scherk's surface : $e^z cos(y) = cos(x)$.

```
        (position:transform
          (position x-coord y-coord  z-coord) scale))))))))


; xpts = number of rows, ypts/xpts = number of columns
(splgrid:set-point-list spldata pt-array xpts  (/ ypts xpts) )

(option:set "u_param_lines" 10)
(option:set "v_param_lines" 8)

(face:spline-grid spldata #t))))
```

The program generates a grid of points which are stored in the list **pt-array**. The array is filled using two nested **do** loops. The outer one increments the x coordinate value while the inner varies the y axis location. At each step a value of z is calculated by the statements:

```
(let* ((diff (/ (cos x-coord) (cos y-coord)))
       (z-coord 0.0))
  (if (not (zero? diff))
    (set! z-coord (log (abs diff))))
```

The resulting z value is then used with the x and y coordinates to create a **position** which is added to the list.

The values used for step sizes and scaling were determined by trial and error.

8.8 Deformable Surfaces

Smooth surfaces are represented in ACIS by splines defined in terms of control points and knot vectors. Since neither of these parameters has any direct physical meaning, designing a curve, or surface, can be very difficult. The Deformable Surface Component (DS-Component) provides an alternative to control point or knot manipulation by providing a set of API's for interactive deformation of a spline surface.

The DS-Component works by allowing the user to specify a number of **loads** and **constraints** on a surface and then optimizing the shape for minimum energy. In this way it is possible to continually deform the geometry of a FACE while keeping certain points or curves on it in a fixed location. The energy-minimization algorithm is automatic and responds to interactively applied loads and constraints. The component supports the following general types of load:

- Point pressure: applies a force that always acts normal to the surface at a fixed point.

- Distributed pressure: applies a force that always acts normal to the surface over a subregion of a surface.

- Point spring: applies a force between two points that is proportional to the distance between them.

- Curve spring: applies a force proportional to the distance between two curves.

Constraints are used to ensure a set of points and/or a set of curves within the surface interpolate an exact set of positions in space. The component supports both point and curve constraints.

8.8.1 Deformable modeling theory

The shape of a deformable model is found by determining the minimum of an energy function. Energy based shape formulations were first introduced by Schweikert in 1966 who proposed a system based on a specific form of a differential equation. In 1974 Nielson noted that solving the equations proposed by Schweikert was equivalent to finding the minimum energy for a shape. This early work has since been refined and improved by a number of researchers. More recently, Jaar characterized the internal energy upon deformation as:

$$E(s) = \int_s \left\| G - G^0 \right\|_\alpha^2 + \left\| B - B^0 \right\|_\beta^2 ds \qquad (8.27)$$

G characterizes the stretching expression and B the bending expression. G^0 and B^0 represent the energetic state of the initial surface. Later Celnicker and Gossard generalized the energy equations used for the deformation of models so they could be expressed as:

$$E_{deformation} = \int_\theta \left(\alpha + \beta \right) d\theta \qquad (8.28)$$

α is a weighted *stretching* term and β is a weighted *bending* term. For a curve the equation becomes:

$$E_{curve} = \int_{curve} \left[\alpha(u)\left(\frac{\delta W}{\delta u}\right) + \beta(u)\left(\frac{\delta^2 W}{\delta u^2}\right)^2 - 2FW \right] du \qquad (8.29)$$

W is the parametric definition of the curve in terms of u, and F is a force vector applied at each point. The Deformable Surface Component is based on this general form of equation. Similar equations exist for the deformation of surfaces.

There are over eighty deformable surface commands and around twenty variables, such as the material's resistance to stretching, the gain of an applied constraint or the spring force. Each of these commands and variables interact with each other to perform deformations. The shapes will form a surface of minimal internal energy based on the above equation when the loads are applied.

Ripple.scm

Location and direction of applied spring loads

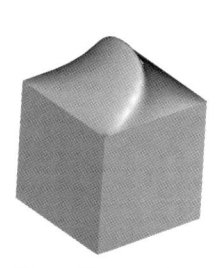

```
; Creates a block and adds a series of springs across the surface
(define blockdef
  (lambda()
    (let*(
      (block (solid:block (position -25 -25 -25) (position 25 25 25)))
      (cface (pick:face (ray (position 0 0 0) (gvector 1 0 0)))))
      (do
        ((param 0.1 (+ param 0.2)))
        ((> param 1) 'finish)
        (ds:add-spring cface 1 (par-pos 0.5 param)
                       (position 45 0 (+ (* param 50) - 20)) 1000)
        (ds:solve cface 1 1)
        (ds:commit cface))))))
```

Resulting surface

The program **Ripple.scm** demonstrates how a series of spring constraints can be used to *pull*, or deform, a planar FACE into a shallow "S" shaped hump. The program starts by locating a FACE cface to which the spring loads will be attached. A **do** loop is used to step the variable param through the values 0.1, 0.3, 0.5, 0.7 and 0.9. At each iteration of the loop a spring attribute is attached to the FACE at the u-v coordinates (0.5, param) by the Scheme Extension **ds:add-spring**. A position value is also specified in terms of param to define the end point of each spring (i.e. the direction in which it is pulling). Lastly each spring is assigned a **gain** (or strength) of 1000 units.

After each spring is added the extension **ds:solve** is called to assess the constraints on the FACE and calculate a new equilibrium shape. Every iteration of the loop ends with a call to **ds:commit** which replaces the geometry of the original FACE with that calculated by the **ds:solve** routine. In other words, it copies the deformable model's shape, and data, back to the model's data-structure.

8.9 Skinning and Lofting

Although skinning and lofting are probably the oldest methods of defining complex surfaces they are still the most commonly used in serious engineering design. Conceptually both methods are very similar and differ mainly in their boundary conditions:

- Skinning fits a FACE to (or through) a series of WIRE BODYs (i.e. sets of unconnected curves).

- Lofting create a new FACE which either:
    - Shares at least one EDGE with an existing FACE and also passes through a set of COEDGEs.
    - Starts from (or ends at) a COEDGE with a vector field and also passes through a set of COEDGEs.

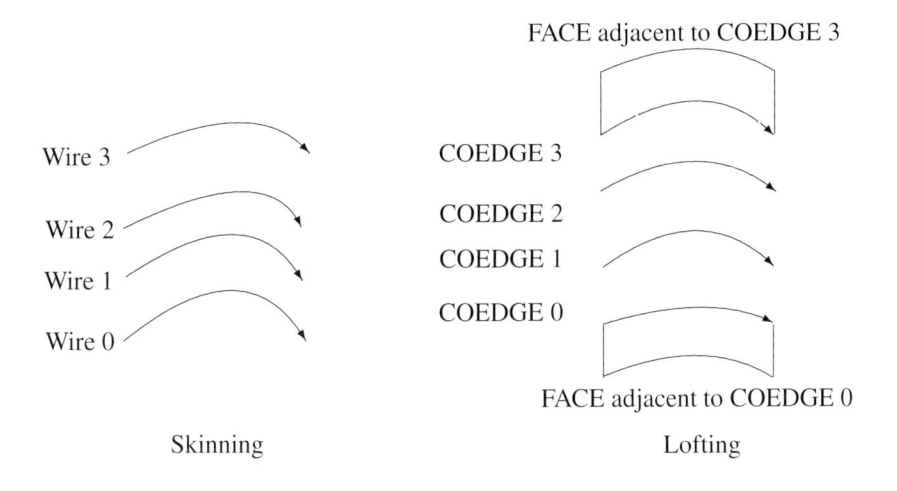

Figure 8.10: The Difference between Skining and Lofting.

The presence of initial adjacent FACEs allows the lofting routines to take account of tangent vectors associated with the initial COEDGE. So unlike skinned surfaces, the results of lofting operations can bulge out or even double-back, past their start and end COEDGEs.

As in many modeling operations "the devil is in the detail". For example, the sets of WIREs, or COEDGE, involved in these operations must be parameterized so their directions are aligned and the twist (between the start points of individual EDGEs) is minimized. Fortunately this sort of painful detail is hidden from the user behind the API interface.

The ACIS lofting operations require two basic categories of parameters. The first category defines the profiles to be lofted together with the control tangents that the curves of the loft surfaces must follow. The second category defines optional parameters to control what type of object is formed as a result of the loft operation.

About the origins of lofting

"Towards the end of 1939 the writer encountered the problem of aircraft contours and became convinced, as a result of subsequent investigations, that its many difficulties, - becoming more urgent as a result of the outbreak of war, would be largely resolved by a mathematical treatment of the subject. Following representations in 1943 to design office administration, it was finally decided that time would be allotted for the mathematical development of the front fuselage of the aircraft at that time "on the boards", and that future decisions on the subject would be based on a comparison with the lines as produced for the same aircraft by the method of full-scale layout. In spite of the fact that all calculations had to be performed with five-figure mathematical tables, the result of the comparison was so much in favor of the mathematical treatment, that two Friden electric calculating machines were obtained to facilitate the arithmetical work, and the decision was taken to develop the lines of the next aircraft by the same method."

"The Development of Curved Surfaces for Aero-design" by J.H. Shelley

The example **lofting.scm** demonstrates lofting between two profiles with differing numbers of EDGEs. The lofting function requires two separate lists of EDGE (i.e. one for either end of the the loft). Each of these EDGE lists (i.e. wire profiles) requires *take-off* vectors to define how the loft surface meets them (i.e. to control the tangency at the FACE boundary).

8.9 Skinning and Lofting

These vector fields are defined by a *law* (see Chapter 10) which evaluates to a vector given a position on the wire profile. In the following example the take-off vectors are the same at all points on the wire. Each EDGE list also has a take-off vector weighting factor; small values mean that the transition from the tangent direction to the lofted surface's natural shape happens abruptly. If, on the other hand a large value is used, the transition from the tangent to the lofted surface happens more gradually. Note that extremely high weight values could result in excessive whipping in the lofted surface, or a self-intersecting surface.

Once all the arguments have been defined the Scheme AIDE extension **sheet:loft-wires** then accomplishes the lofting.

lofting.scm

```
; This program lofts between 2 wire profiles to form a sheet body
; Create first wire-body
(define v1 (wire-body:points ; Top square
        (list (position 1 1 0) (position -1 1 0) (position -1 -1 0)
            (position 1 -1 0) (position 1 1 0) )))

; Create second wire-body
(define e0 (wire-body:points ; Bottom octagon
        (list (position 1 3 10) (position -1 3 10) (position -3 2 10)
            (position -3 -2 10) (position -1 -3 10) (position 1 -3 10)
            (position 3 -2 10) (position 3 2 10) (position 1 3 10) )))

; Get lists of the COEDGEs in each WIRE
(define coedge_list1 (entity:coedges e0))
(define coedge_list2 (entity:coedges v1))

; Define first take-off-factor vector
(define lawa (law "vec(0,0,-1)"))

; Define lawa's range of valid input variables (i.e. its domain)
(define dom0 (law "domain(law1,0,1)" lawa ))

; Define second take-off-factor vector and second law domain
(define lawb (law "vec(0,0,1)"))
(define dom1 (law "domain(law1,0,1)" lawb))

; Create a 'section' data-structure that holds the list of COEDGEs (defining the
; cross-sections of the loft) and their associated laws (which define the take-off vectors).
; A boolean argument is used to define the direction of the loft (i.e. #t = landing,
; #f = take-off) and, lastly, a weighting factor is specified (1 in both cases)
(define sec1 (section coedge_list1 (list dom0 dom0 dom0 dom0 dom0 dom0 dom0 dom0) #t 1))
(define sec2 (section coedge_list2 (list dom1 dom1 dom1 dom1) #f 1))

; Do the loft by supplying a list of "sections" and a number of
; optional boolean flag. In this case two (out of a possible 8 are specified)
```

```
; determine if 1) arc length (#t) or iso parameterization (#f) is used,
; and 2)whether twisting is permitted (#f) or not (#t)
(define loft1 (sheet:loft-wires (list sec1 sec2) #f #t))
(sheet:2d loft1)
```

Given the number of parameters involved in lofting operations it is not surprising that some very strange shapes can result.

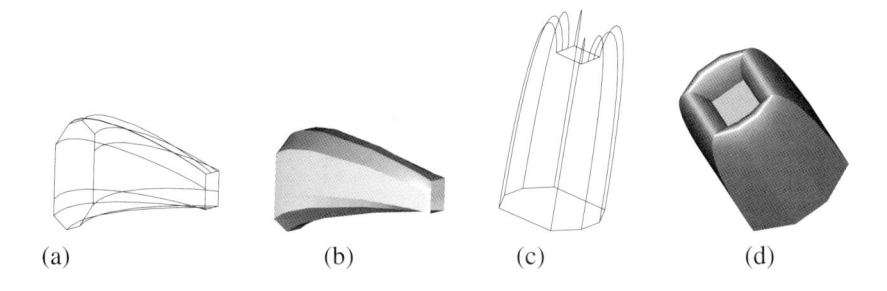

(a) (b) (c) (d)

Figure 8.11: The effects of lofting variables.

For example the object shown in Figure 8.11(a) and (b) were generated by the **lofting.scm** code but with a different take-off vector for the octagon profile (i.e. $vec(0, 1, -1)$) and which was added to the sector data-structure with a weighting factor of 5.

Likewise Figure 8.11(c) and (d) was generated with weighting factors of 5 and take-off vectors for both sections (octagon = $vec(0, 0, -10)$ and square = $vec(0, 0, 1)$) being declared as *landing* (i.e. #t). The call to **sheet:loft** was made with a series of option arguments (#t, #t, #t, #f) chosen to create the object shown.

8.10 Net Surfaces

Surfaces can also be interpolated through networks of bi-directional curves. This allows sheet BODYs to be interpolated through a number of wire profiles which define the surface's cross-section at various positions in space.

The curves in one direction are referred to as the **u curves**, while the ones in the other direction are known as the **v curves**. Not surprisingly there are some conditions: firstly there must be at least four wire BODYs (i.e. two u and two v), also the start points of the curves in the v direction must lie on the first curve in the u direction, and vice versa. Likewise the end points of the v curves must lie in the last curve in the u direction, and vice versa.

If all of the curves intersect, then the surface passes through the curves and their intersections. If any of the u curves of the network do not intersect all of v curves at some point, the intersection is interpolated. The maximum distance for the interpolation is governed by a tolerance argument (which defaults to resfit, see page 128).

8.10 Net Surfaces

(a) (b) (c)

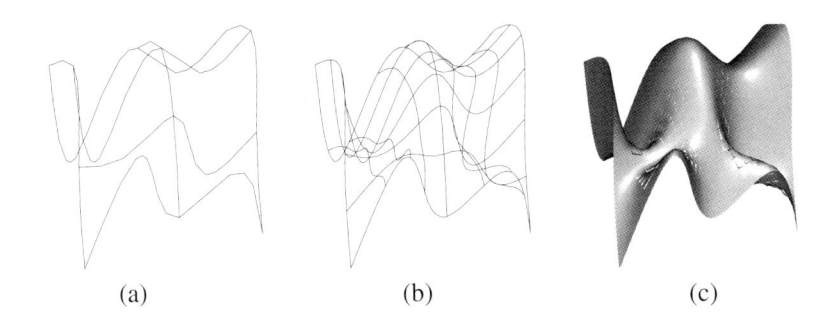

Figure 8.12: Net-surface definition.

netsurf.scm

```
; This program creates a net surface that interpolates
; a collection of bi-directional wires.

(define v1 (list (position 0 0 0) (position 5 10 0)
          (position 10 5 0) (position 15 15 0)
          (position 20 0 0)))

(define v2 (list (position 0 10 5) (position 5 5 5)
          (position 10 15 5) (position 15 10 5)
          (position 20 10 5)))

(define v3 (list (position 0 20 10)
          (position 5 15 10) (position 10 20 10)
          (position 15 5 10) (position 20 20 10)))

(define v4 (list (position 0 15 15)
          (position 5 10 15) (position 10 15 15)
          (position 15 0 15) (position 20 15 15)))

(define u1 (list (position 0 0 0) (position 0 10 5)
          (position 0 20 10) (position 0 15 15)))

(define u2 (list (position 10 5 0) (position 10 15 5)
          (position 10 20 10) (position 10 15 15)))

(define u3 (list (position 20 0 0) (position 20 10 5)
          (position 20 20 10) (position 20 15 15)))

; Create a series of spline curve wire-bodies
; in the u and v directions.
(define my_v1 (wire-body (edge:spline v1)))
(define my_v2 (wire-body (edge:spline v2)))
```

```
(define my_v3 (wire-body (edge:spline v3)))
(define my_v4 (wire-body (edge:spline v4)))

(define my_u1 (wire-body (edge:spline u1)))
(define my_u2 (wire-body (edge:spline u2)))
(define my_u3 (wire-body (edge:spline u3)))

; Create a net surface from the uv curves
(define net1 (sheet:net-wires
    (list my_v1 my_v2 my_v3 my_v4)
    (list my_u1 my_u2 my_u3) #t ))

; Make it double sided
(sheet:2d net1)
```

Figure 8.12(a) shows the four WIRE v-BODYs (i.e my_v1 to my_v2) and the three WIRE u-BODYs (i.e. my_u1 to my_u3). Figures 8.12(b) and (c) show wire frame (with parameter lines) and rendered views of the resulting surface.

8.11 Exercises

1. Generate a spline curve that interpolates a number of points lying on a catenary ($x = c\, cosh(\frac{y}{c})$) using the Scheme AIDE function **edge:spline**.

2. Alter the definition of the "hump" solid given in **spsol1.scm** to create a depression at the top of the protrusion (like a volcano).

Bulletin Boards and Rollback

9

Everybody makes mistakes! No matter how many times you've punched in sets of (x,y,z) coordinates it is still ludicrously easy to get things the wrong way up, back to front or slightly too far over. Because of this, users of 3D design systems frequently find themselves wishing to reverse, or undo, the last modification they made. Typically this arises after the result of a modeling operation, say a subtraction, is viewed and the user realizes that the objects were wrongly positioned in some way. Many types of software support recovery from this sort of user error by providing an **undo** button which **rolls back** the system to some previous state. This chapter gives a brief overview of the classes ACIS provides to allow system developers to easily create such logging (or rollback) mechanisms.

Essentially rollback (and rollforward) is done by carefully recording the effects of each modeling operation so they can be "undone" or "redone" at any time. This facility, which for the most part is quite invisible to both programmers and users alike, can support both low and high level rollback.

- At a **system level** it provides a mechanism for ensuring the system can elegantly recover from any problems encountered within an API routine. On detecting an error the API routine can simply roll back any changes made to the model since it was invoked and exit, thus leaving any models as they existed when the function was entered.

- At a **user interface level** it provides support for "what if" design allowing operators to *unmake* modifications to their models by simply pressing an **undo** button.

How is this done? The replay of changes to the model is supported by the following classes:

BULLETIN: A bulletin has a type recording the **creation**, **change** or **deletion** of an ENTITY. The type is not stored, but deduced from the presence or absence of **new** and **old** entity pointers.

BULLETIN_BOARD: Maintains a doubly-linked list of BULLETINs. A BULLETIN_BOARD detailing which ACIS ENTITYs have been created, changed or deleted is returned by *every* API function.

DELTA_STATE: Maintains a list of BULLETIN_BOARDs and so details the changes caused by applying a number of API functions. DELTA_STATEs can be arranged in linear lists or, in sophisticated applications, tree structures which allow design alternatives to be explored.

HISTORY_STREAM: Maintains a (possibly branched) list of DELTA_STATEs.

The programs in this chapter demonstrate the role played by each of these four classes.

9.1 Counting the BULLETINs

Fundamental to ACIS's rollback system is the **BULLETIN** class. This simple class contains two ENTITY pointers, known as **new** and **old**. The values assigned to these pointers determine the BULLETIN type in the following ways:

- Whenever an ENTITY is created a BULLETIN records the fact with its **new** pointer assigned to the newly constructed ENTITY and the value NULL assigned to its **old** pointer.

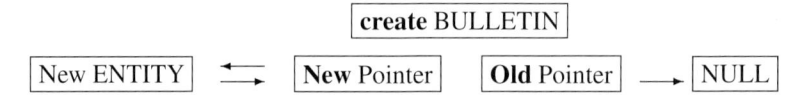

create BULLETIN
New ENTITY ⇌ New Pointer   Old Pointer → NULL

- Whenever an ENTITY is changed a BULLETIN is created with the **new** pointer assigned to the modified ENTITY and the **old** pointer referring to a copy of the ENTITY created before the *change* took place.

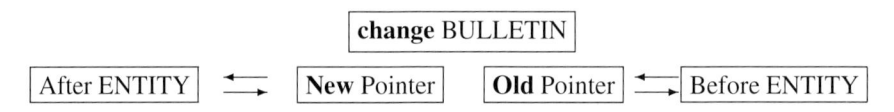

change BULLETIN
After ENTITY ⇌ New Pointer   Old Pointer ⇌ Before ENTITY

- Whenever an ENTITY is deleted a BULLETIN is created with the **old** pointer assigned to the (nominally) deleted ENTITY and **new** pointer assigned the value **NULL**. Note that the deleted ENTITY is not returned to the free store until the BULLETIN itself is deleted (see **api_delete_ds**), allowing it to be recovered when roll back is performed.

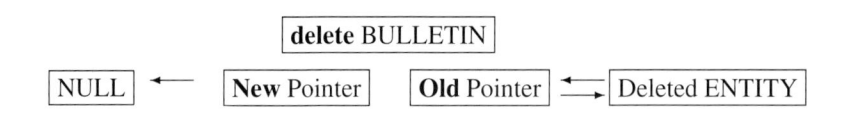

delete BULLETIN
NULL ← New Pointer   Old Pointer ⇌ Deleted ENTITY

9.1 Counting the BULLETINs

Notice that every ENTITY holds a pointer to the last BULLETIN that references it. This pointer (accessed through the ENTITY member function **rollback()**) provides one way of checking if a given ENTITY is deleted or not!

BULLETINs are chained together into lists maintained by the **BULLETIN_BOARD** class. Every API function returns an **outcome** object which contains a pointer to the BULLETIN_BOARD constructed during its execution and this can be accessed via the member function **bb()** of the outcome class.

More importantly, however, each BULLETIN_BOARD created is stored on a stack until such time as it is explicitly deleted (usually after being gathered into a DELTA_STATE).

The program **bb_size.cxx** demonstrates the existence of this stack by using the function **current_bb()** to return the BULLETIN_BOARD at the top of the stack. Once located the number of BULLETINs on the board are counted by the function **bb_size**.

bb_size.cxx

```
// This program counts the number of bulletins posted on the current
// bulletin board during the creation and intersection of two cylinders
#include <stdio.h>                              // Input / Output Functions
#include "boolean/kernapi/api/boolapi.hxx"      // Declares boolean API's
#include "constrct/kernapi/api/cstrapi.hxx"     // Construction API's
#include "kernel/kernapi/api/kernapi.hxx"       // Declares kernel API's
#include "kernel/kerndata/top/body.hxx"         // Topological Classes
#include "baseutil/vector/transf.hxx"           // Transform Class
#include "baseutil/vector/vector.hxx"           // Vector Class
#include "kernel/kerndata/bulletin/bulletin.hxx" // Bulletin Class
#include "kernel/kerndata/data/debug.hxx" "

int bb_size(FILE*); // Function Prototype

void main()
{
  api_start_modeller(0);
  api_initialize_booleans();
  printf("BB Length after api_start_modeller = %d \n",
                            bb_size(NULL));
  BODY *cyl1, *cyl2;
  api_make_frustum(100,20,20,20,cyl1);
  printf("BB Length after 1st call to api_make_frustum = %d BULLETINs\n",
                            bb_size(NULL));
  api_make_frustum(100,20,20,20,cyl2);
  printf("BB Length after 2nd call to api_make_frustum = %d BULLETINs\n",
                            bb_size(NULL));
  transf rotX = rotate_transf(pi/2,vector(1,0,0));
  api_apply_transf(cyl1,rotX);
  printf("BB Length after call to api_apply_transf = %d BULLETINs\n",
```

```
                                            bb_size(NULL));
        api_intersect(cyl1,cyl2);
        FILE *bfp = fopen("bb.dbg", "w");
        printf("BB Length after call to api_intersect = %d BULLETINs\n",
                                            bb_size(bfp));
        fclose(bfp);
        api_terminate_booleans();
        api_stop_modeller();
    }

    // Function to count the number of bulletins on a board
    int bb_size(FILE* bfp)
    {
        int count = 0;
        BULLETIN_BOARD *ourbb = current_bb();
        if (ourbb != NULL){
            BULLETIN *ourb = ourbb-> start_bulletin();
            while(ourb != ourbb-> end_bulletin())
                {
                    if(bfp != NULL) // Write Debug Info to File
                            ourb-> debug(bfp);
                    count++;
                     ourb = ourb-> next();
                }
                count++;  // Add one for the end bulletin
        }
        return count;
    }
```

Once this program is compiled and run the following text is printed on the screen:

```
Bulletin Board Size after api_start_modeller = 0 bulletins
Bulletin Board Size after 1st call to api_make_frustum = 25
Bulletin Board Size after 2nd call to api_make_frustum = 25
Bulletin Board Size after call to api_apply_transf = 2
Bulletin Board Size after call to api_intersect= 77 bulletins
```

Every ENTITY created, modified or deleted by the modeler is recorded by a BULLETIN. To some extent the number of BULLETINs can be checked by inspection. Table 9.1 shows that each cylinder (i.e. frustum) is composed of 25 ENTITYs and the BULLETIN_BOARD associated with these API calls shows 25 create BULLETINs. The list of 77 BULLETINs created by the call to **api_intersect** is composed of all three types of BULLETIN. The debug file, **bb.dbg** (a fragment of which is shown below), can be examined to determine exactly what ENTITYs have been created, changed or deleted.

```
44188: delete loop 55708 next 44204 previous 44172
44204: delete face 51648 next 44220 previous 44188
```

Table 9.1   ENTITY counts for **bb_size.cxx**.

ENTITY	Frustum	Intersection
BODY	1	1
LUMPS	1	1
SHELLS	1	1
FACES	3	4
LOOPS	4	4
EDGES	2	4
COEDGES	4	8
VERTICES	2	2
POINTS	2	2
SURFACES	3	2
CURVES	2	4
Total	25	33

```
44220: delete loop 55676 next 44236 previous 44204
44236: change shell 63868 to 63828 next 44252 previous 44220
44252: change lump 61792 to 61760 next 44268 previous 44236
44268: change body 59796 to 59700 next 44284 previous 44252
44284: change face 51912 to 51604 next 44300 previous 44268
44300: delete edge 45364 next 44316 previous 44284
44316: delete loop 55612 next 44332 previous 44300
```

The **bb.dbg** file gives a readable view of the BULLETIN_BOARD, the contents of which not only record changes but also give some idea of the *order* in which operations have been performed.

9.2 Creating DELTA_STATEs

The stack of BULLETIN_BOARDs created by a series of API calls can be collected into a list known as a **DELTA_STATE**. The ACIS Kernel provides several functions for manipulating the rollback system via DELTA_STATEs. The program **ds_size.cxx** shows how DELTA_STATEs can be created using **api_note_state** and their contents (i.e. a list of BULLETIN_BOARDs) examined. DELTA_STATEs themselves are recorded in collections known as **HISTORY_STREAMs** and the program ends by acquiring a pointer to the default stream and finding the length of the list of DELTA_STATEs it holds.

ds_size.cxx

```
// This program finds the number of bulletin boards in a series
// of delta states created during the running of the program
#include <stdio.h>                              // Defines file I/O functions
#include "boolean/kernapi/api/boolapi.hxx"      // Declares boolean API's
#include "constrct/kernapi/api/cstrapi.hxx"     // Construction API's
#include "kernel/kernapi/api/kernapi.hxx"       // Declares kernel API's
#include "kernel/kerndata/top/body.hxx"         // Defines the BODY class
#include "baseutil/vector/transf.hxx"           // Defines transform class
#include "kernel/kerndata/bulletin/bulletin.hxx" // Defines bulletin class

int delta_size(DELTA_STATE*);

void main()
{
    api_start_modeller(NULL);
    api_initialize_booleans();
    HISTORY_STREAM *hs = get_default_stream();
    BODY *cyl1, *cyl2;
    DELTA_STATE *d0 = NULL;
    api_note_state(d0);
    printf("\n Size 1 : Delta_State has %d BULLETIN_BOARDs in it \n",
                delta_size(d0));
    api_make_frustum(100,20,20,20,cyl1);
    printf("\n Size 2 : Delta_State has %d BULLETIN_BOARDs in it \n",
                delta_size(d0));
    api_make_frustum(100,20,20,20,cyl2);
    api_note_state(d0);
    printf("\n Size 3 : Delta_State has %d BULLETIN_BOARDs in it \n",
                delta_size(d0));

    api_delete_ds(d0);
    d0 = hs->get_current_ds()

    printf("\n Size 4 : Delta_State has %d BULLETIN_BOARDs in it \n",
                delta_size(d0));
    transf rotX = rotate_transf(pi/2,vector(1,0,0));
    api_apply_transf(cyl1,rotX);
    api_intersect(cyl1,cyl2);
    api_note_state(d0);
    printf("\n Size 5 : Delta_State has %d BULLETIN_BOARDs in it \n",
                delta_size(d0));

    DELTA_STATE_LIST dslist;
    hs->list_delta_states(dslist);

    printf("History Stream has %d DELTA_STATES in it\n",
                dslist.count());
```

9.2 Creating DELTA_STATEs

```
    if(hs-> can_roll_back())
      printf("Stream can roll back\n");
    if(hs-> can_roll_forward())
        printf("Stream can roll forward\n");

    api_terminate_booleans();
    api_stop_modeller();
}

int delta_size(DELTA_STATE *ds)
{
    int count = 0;
    BULLETIN_BOARD *bb = ds-> bb();
    while(bb != NULL)
        {
           count++;
           bb = bb-> next();
        }
    return count;
}
```

This program creates a series of DELTA_STATEs by calling **api_note_state**. This function gathers all the BULLETIN_BOARDs off the stack placed there since the last call to **note_state** and incorporates them in a list.

Although not done explicitly each DELTA_STATE is incorporated into a list maintained by a HISTORY_STREAM.

Once the program is compiled and run the following text is printed on the screen:

```
Size 1 : Delta_State has 0 bulletin boards in it
Size 2 : Delta_State has 0 bulletin boards in it
Size 3 : Delta_State has 2 bulletin boards in it
Size 4 : Delta_State has 0 bulletin boards in it
Size 5 : Delta_State has 2 bulletin boards in it
History Stream has 2 DELTA_STATES in it
Stream can roll back
```

The results demonstrate some important details of DELTA_STATE creation.

Size 1 The output from the program **bb_size.cxx** shows that although a BULLETIN_BOARD is returned by the function **api_start_modeller** it is empty. **api_note_state** disregards empty boards so the resulting DELTA_STATE is empty.

Size 2 The DELTA_STATE is still empty because although a BULLETIN_BOARD has been created by the call to **api_make_frustum** it has not yet been incorporated into a DELTA_STATE by a call to **api_note_state**.

Size 3 Shows that both the BULLETIN_BOARDs resulting from the creation of each cylinder have been incorporated into the DELTA_STATE.

Size 4 The function **api_delete_ds** returns both BULLETIN_BOARDs to free store so the memory used by them can be reallocated.

Size 5 After two more API calls the final call to **api_note_state** produces a DELTA_STATE with a list of two BULLETIN_BOARDs.

The default HISTORY_STREAM is seen to contain two DELTA_STATEs; one containing bulletins for the transform and intersection operations and the other being the current default to which any new bulletins will be added.

9.3 Rolling back the modeler

The information held in DELTA_STATEs can be replayed in reverse. By going through every BULLETIN held in a DELTA_STATE, and undoing every change, deletion and creation recorded, a model can be restored to its former state. Perhaps the most important thing to appreciate about this process is that *no new calculations are required*; the entire rollback is done by moving around pointers to existing data. The data itself is never moved, so consequently all DELTA_STATE and BULLETIN_BOARD manipulations are fast.

Because of this, each DELTA_STATE has an explicit direction. In the program **roll.cxx** two cylinders are intersected; if the state of the model prior to the intersection operation is referred to as **BEFORE** and subsequent to it as **AFTER** then the DELTA_STATE created by the call to **api_note_state** just after the Boolean is orientated to move the modeler from:

$$\text{AFTER} \Longrightarrow \text{BEFORE}$$

When the DELTA_STATE is used to roll the model back (undoing the intersection operation) its direction is reversed, becoming

$$\text{AFTER} \Longleftarrow \text{BEFORE}$$

roll.cxx

```
// This program use a delta state to roll back the modeler
// after the intersection of two cylinders
#include <stdio.h>                              // Defines I/O functions
#include "boolean/kernapi/api/boolapi.hxx"      // Declares boolean API's
#include "constrct/kernapi/api/cstrapi.hxx"     // Constructor API's
#include "kernel/kernapi/api/api.hxx"           // Declares outcome class
#include "kernel/kernapi/api/kernapi.hxx"       // Declares kernel API's
#include "kernel/kerndata/top/body.hxx"         // Declares BODY Class
#include "baseutil/vector/transf.hxx"           // Declares transform class
#include "kernel/kerndata/bulletin/bulletin.hxx" // Declares bulletin class
#include "kernel/kerndata/lists/lists.hxx"      // Declares ENTITY_LIST class
#include "kernel/kerndata/savres/fileinfo.hxx"  // Declares fileinfo class
```

9.3 Rolling back the modeler

```
int delta_size(DELTA_STATE*);      // See ds_size.cxx
void save_ent(char*, ENTITY*);     // See save.cxx, Chapter 4

void main()
{
    api_start_modeller(0);
    api_initialize_booleans();
    BODY *cyl1, *cyl2;
    api_make_frustum(100,20,20,20,cyl1);
    api_make_frustum(100,20,20,20,cyl2);
    transf rotX = rotate_transf(pi/2,vector(1,0,0));
    api_apply_transf(cyl1,rotX);

    DELTA_STATE *initial = NULL;
    DELTA_STATE *after_bool = NULL;
    DELTA_STATE *after_save = NULL;

    api_note_state(initial);

    outcome check = api_intersect(cyl1,cyl2);
    if(!check.ok())
        printf("error in Boolean \n");

    api_note_state(after_bool);
    printf("\n Delta_State after_bool contains %d Bulletin Board \n",
                    delta_size(after_bool));
    save_ent("roll0.sat",cyl2);

    api_note_state(after_save);

    outcome res1 = api_change_state(after_save);
    outcome res2 = api_change_state(after_bool);

    if(!res1.ok() || !res2.ok())
        printf("error in change states \n");

    save_ent("roll1.sat",cyl2);

    api_terminate_booleans();
    api_stop_modeller();
}
```

Once the program is compiled and run the following text is printed on the screen:

```
Delta State after_bool contains 1 Bulletin Board
```

Examination of the files produced by the program, shows that the intersection volume recorded in **roll0.sat** has been transformed back into a cylinder in **roll1.sat**. At first glance it can be a little hard to appreciate why **api_note_state** had to be called after the BODY has been saved to file, until one recalls that the **save_ent** function uses an API function.

Although the BULLETIN_BOARD returned by the operation is empty (i.e. nothing has been changed, deleted or created) it still increments an internal **state** counter, which means that the earlier Boolean cannot be undone until the save operation is unwound.

A better solution would be to roll back the modeler *within* the function so that the **save_ent** function leaves the model exactly as found. Although this could be done explicitly with calls to **api_note_state** and **api_change_state** it can also be done simply by including the macros **API_NOP_BEGIN** and **API_NOP_END** as shown in the following code fragment:

```
// A better save_ent routine
void save_ent( char *filename, ENTITY *ent)
{
    API_NOP_BEGIN
    FileInfo info;  // Create FileInfo Object
    info.set_product_id("HW-University");  // set info's data
    info.set_units(1.0);  // Millimeters

    // Sets header info to be written to sat file
    api_set_file_info(FileId | FileUnits, info);

    FILE *fp = fopen(filename, "w");
    if (fp != NULL) {
        ENTITY_LIST *savelist = new ENTITY_LIST;
        savelist->add(ent);
        api_save_entity_list(fp,TRUE,*savelist);
        delete savelist;
    } else
        printf("Unable to open file\n");
    fclose(fp);
    API_NOP_END
}
```

At the termination of the **save_ent** function the macro API_NOP_END will cause the modeler to roll back to its initial state.

9.4 Part history

The Scheme AIDE allows individual **PARTS** to have individual **HISTORY_STREAMs**, allowing a more selective rollback which leaves parts of the model unchanged.

The program **hist1.scm** demonstrates this facility.

hist1.scm

```
; This program shows how multiple parts can have separate histories.
; Create multiple parts and view windows for each of them
(define p2 (part:new))
(define p3 (part:new))
(view:new 0 0 400 400 p2)
(view:new 415 0 400 400 p3)
(env:set-active-part p2)    ; Create geometry in p2
(solid:block (position 10 20 30) (position 20 30 40))
; Create geometry in p3
(env:set-active-part p3)
(solid:sphere (position 0 0 0) 20)
; Roll back the block; the sphere is unchanged
(roll (history p2))
```

The Scheme extension **history** gets the HISTORY_STREAM associated with a given PART or ENTITY. Three types of history exist:

1. **PART histories** which contain rollback information for all ENTITYs in a PART, unless those ENTITYs have a history of their own.

2. **ENTITY histories** which can be attached to top-level ENTITYs via an attribute and contain history of all the subordinate ENTITYs. If an ENTITY with a history is deleted or rolled before its creation, its history still exists!

3. **Default history** which contains history for ENTITYs and PARTs without individual histories.

9.5 Saving and restoring of history

A history can be saved to file and restored so that rollback operations need not be constrained to a single session. The following Scheme AIDE commands can be used to demonstrate this:

rollsave.scm

```
; This program demonstrates state naming and save/restore histories.
; First we name states.
(roll:name-state "initial")
(define sph (solid:sphere (position 0 0 0) 5))
(roll:name-state "sphere")
(roll) ; rolls back to the initial state
(define cne (solid:cone (position 15 0 0) (position 35 0 0) 3 10 2))
```

(**roll:name-state** "cone")
(**roll**) ; *rolls back to the initial state*
(**roll:delete-previous-states**) ; *Should do nothing since there are no previous states*
(**define** cyl (**solid:cylinder** (**position** -15 0 0) (**position** -35 0 0) 6))
(**roll:name-state** "cylinder")
(**define** tor (**solid:torus** (**position** 10 -50 0) 15 8))
(**roll:name-state** "torus")
(**define** blk (**solid:block** (**position** 20 50 8) (**position** -20 20 0)))
(**roll:name-state** "block")
(**roll** -2) ; *rolls back 2 states*
(**define** pln (**sheet:2d** (**solid:wiggle** 120 120 0 "sym")))
(**roll:name-state** "wiggle")
(**part:save** "sr5_his.sat" #t #t) ; *Save as ASCII with history*
; *Clear everything out*
(**part:clear**)
(**roll:delete-all-states**)
; *Load it back in with its history*
(**part:load** "sr5_his.sat" #t #t)
; *Show the history restore worked by rolling to the saved and*
; *named states*
(**roll** "sphere")

The program leaves the model in the state labeled **sphere** from which it can be easily rolled forward again.

Laws and Graphs   10

Aerofoils, parabolic reflectors, mechanical cams and gear teeth are all examples of manufactured parts whose shapes have precise mathematical definitions. The **laws** system described in this chapter allows FACEs and EDGEs to be defined directly by design equations [1].

In other words the **law** class provides a general facility for the specification of curve and surface geometry, either directly in terms of **parametric** equations, or indirectly by means of offsets, sweeps and warps. In principle the ACIS data structure has always been open to the addition of user defined curves and surfaces. However, in practice prior to the introduction of laws new geometry was rarely added. By providing a high level language for writing mathematical expressions the laws system makes the introduction of custom geometry a mathematical exercise rather than a programming problem.

This chapter starts its overview of the law system by describing how mathematical expressions are constructed and evaluated. Subsequent sections demonstrate how laws can be used in offsetting, sweeping and space warping. The maths themes are continued at the end of the chapter when ACIS's graph representation and analysis utilities are introduced.

10.1 Law Expressions

Although it is possible to create a law expression directly from C++ classes it is, as always, easier to start with ACIS's Scheme extensions. Key to these is the **law** extension which allows mathematical expressions to be created from text strings. The syntax of these is undemanding. For example, a law defining the relationship $x^2 + \sqrt{x}$ can be coded in Scheme as:

(**define** sroot (**law** "x^2 + sqrt(x)"))

---

[1] At a low level this is facilitated by the parametric interface defined in the **surface** and **curve** classes from which the individual geometry classes (i.e. torus, sphere, ellipse etc.) are derived. Because all ACIS functions support the parametric interface it is possible to, say, blend or do Booleans, with any new type of curve or surface as long as they support this parametric interface.

Note that **law:eval** more commonly deals with a list of arguments.

Once a law is defined it can be evaluated for any value of x. This can be done interactively in the Scheme AIDE[2] using the **law:eval** function as follows:

acis>(**law:eval** sroot 1)
2
acis>(**law:eval** sroot 2.4)
7.30919333848297

Law expressions can be constructed from a rich mix of mathematical operators (summarized in Tables 10.1-10.5) and can incorporate any number of variables. For example the following expression $x^2 + y^3 + z^4 + w^5$ could be written in several different ways:

(**define** rl (**law** "x^2 + y^3 + z^4 + w^5"))
(**define** r2 (**law** "p1^2 + p2^3 + p3^4 + p4^5"))
(**define** r3 (**law** "b1^2 + b2^3 + a3^4 + a4^5"))

In general, a variable in a law expression is represented by a letter of the alphabet followed by an integer so a1, b5, h9 and x3 are all valid variable names.

The number part of a variable name determines the position of the argument (reading from left to right) in the argument list. For example:

$b1 = 82,$
$b2 = 22,$
$a3 = 133,$
$a4 = 4$

acis>(**define** rl (**law** "b1^2 + b2^3 + a3^4 + a4^5"))
acis>(**define** arg_list (**list** 82 22 133 4))
acis>(**law:eval** r1 arg_list)
312919117.0

The exceptions to this general rule are the letters (e, o, t, u, v, x, y, and z) which can each be used with, or without, an integer subscript.

When e is used without such a number attached it represents the base of the natural logarithms, but when followed by an integer (e.g. e6) it represents a normal variable.

The letter o represents the composition operator for helping to create complicated expressions (see page 246).

The remaining six letters provide a short-hand way of specifying 1,2,and 3D coordinates using common maths notation. So the letters t, u and x are equivalent to t1, u1 and x1. Likewise v, y and z are a shorthand form of v2, y2 and z3.

In addition to numerical arguments some law operators will accept ENTITYs as arguments. These are represented by the ENTITY name followed by a number. Thus, edge1, law3, trans2 and wire2 are all valid variable names in law expressions.

Once a law expression is defined ACIS provides functions for carrying out various mathematical operations. Some operators manipulate the expression symbolically. For example, the Scheme extension **law:derivative** creates a law object that is the derivative (in the calculus sense) of the given law with respect to a given variable.

For example, the law r2 could be differentiated with respect to p1 by typing:

---
[2] Scheme ACIS Interface Driver Extension

```
acis>(define r2 (law "p1^2 + p2^3 + p3^4 + p4^5"))
acis>(define dr1 (law:derivative r2 "p2"))
dr1
acis>(law:check dr1)
3*P2^2
```

If no second argument is given then the derivative is assumed to be with respect to x or a1. Note also the use of the **law:check** function to display the contents of the resulting law.

As every school child knows differentiation is easy but integration is quite another story. Consequently, to ensure a solution ACIS carries out integration numerically. For example:

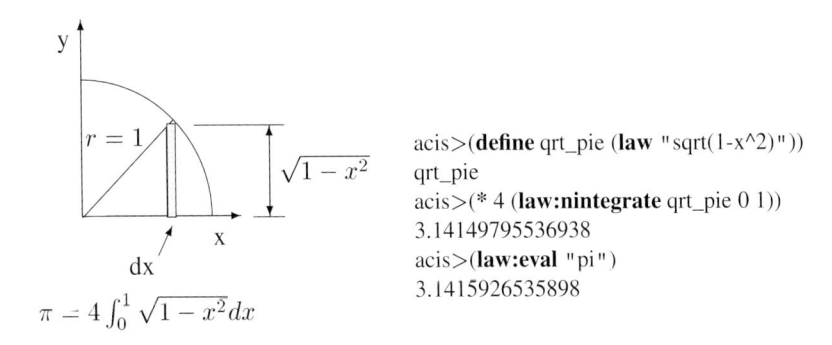

$$\pi = 4 \int_0^1 \sqrt{1-x^2}\,dx$$

```
acis>(define qrt_pie (law "sqrt(1-x^2)"))
qrt_pie
acis>(* 4 (law:nintegrate qrt_pie 0 1))
3.14149795536938
acis>(law:eval "pi")
3.1415926535898
```

In addition to the law to be integrated and the limits of the integration, the **law:nintegrate** function has an optional fourth argument that specifies the accuracy of the calculation. Consequently the accuracy of the previous approximation to π could be reduced by typing:

```
acis>(* 4 (law:nintegrate qrt_pie 0 1 0.1))
3.09076364904841
```

Although the writing of laws is very easy it should be noted that the law API's and Scheme extensions do not carry out any syntax checking. So the Scheme AIDE will appear to accept a law of the form "$x + x^2 - cos(x$" without any comment about the right-hand missing bracket. Later attempts to use this law, however, would fail. The solution when developing expressions is to use the **law:check** function to ensure the syntax has been correctly input.

10.2 Creating Laws in C++

Although the Scheme interface to the laws system allows expressions to be very quickly composed and tested it gives little insight into the C++ classes used to implement the system. Essentially each law is represented by a hierarchy of C++ objects.

The example **law1.cxx** illustrates this fact by using **api_str_to_law** to create a law data-structure from a string. The structure of the resulting tree is then explored by calling the different member functions of the classes involved. As well as providing access to the tree, each function knows the number of input arguments required to evaluate it and also the number of arguments it returns.

Table 10.1   Trigonometric laws.

Law Syntax	Math	Law Syntax	Math
cos(x)	$cos(x)$	cosh(x)	$cosh(x) = \frac{e^x + e^{-x}}{2}$
cot(x)	$cot(x) = \frac{cos(x)}{sin(x)}$	coth(x)	$coth(x) = \frac{e^x + e^{-x}}{e^x - e^{-x}}$
csc(x)	$csc(x) = \frac{1}{sin(x)}$	csch(x)	$csch(x) = \frac{2}{e^x - e^{-x}}$
sec(x)	$sec(x) = \frac{1}{cos(x)}$	sech(x)	$sech(x) = \frac{2}{e^x + e^{-x}}$
sin(x)	$sin(x)$	sinh(x)	$sinh(x) = \frac{e^x - e^{-x}}{2}$
tan(x)	$tan(x)$	tanh(x)	$tanh(x) = \frac{sinh(x)}{cosh(x)}$
arccos(x)	$cos^{-1}(x)$	arccosh(x)	$cosh^{-1}(x)$
arccot(x)	$cot^{-1}(x)$	arccoth(x)	$coth^{-1}(x)$
arccsc(x)	$csc^{-1}(x)$	arccsch(x)	$csch^{-1}(x)$
arcsec(x)	$sec^{-1}(x)$	arcsech(x)	$sech^{-1}(x)$
arcsin(x)	$sin^{-1}(x)$	arcsinh(x)	$sinh^{-1}(x)$
arctan(x)	$tan^{-1}(x)$	arctanh(x)	$tanh^{-1}(x)$

law1.cxx

```
// This program creates a simple law, explores its
// tree structure and then evaluates it.
#include <stdio.h> // Declares printf function
#include "kernel/kernapi/api/kernapi.hxx" // Declares kernel api functions
#include "kernel/kernutil/law/law.hxx" // Declares Law classes
#include "baseutil/vector/interval.hxx" // Used by Law classes

void main()
{
    api_start_modeller(0);
    api_initialize_kernel();
```

10.2 Creating Laws in C++

```cpp
    char *law_string = "x + y*cos(x)";

    law *function_law = NULL;

    api_str_to_law(law_string, &function_law);

// Explore the tree structure
    printf("Root Law --------------------------\n");
    printf("symbol = %s \n",function_law-> symbol());
    printf("class name = %s \n\n",function_law-> class_name());

    law *left_law = ((binary_law*)function_law)-> fleft();
    printf("Left Law --------\n");
    printf("String = %s \n",left_law-> string());
    printf("class name = %s \n\n",left_law-> class_name());

    law *right_law = ((binary_law*)function_law)-> fright();
    printf("Right Law -------\n");
    printf("symbol = %s \n",right_law-> symbol());
    printf("class name = %s \n\n",right_law-> class_name());

left_law = ((binary_law*)right_law)-> fleft();
    printf("Left Law --------\n");
    printf("String = %s \n",left_law-> string());
    printf("class name = %s \n\n",left_law-> class_name());

right_law = ((binary_law*)right_law)-> fright();
    printf("Right Law -------\n");
    printf("class name = %s \n\n",right_law-> class_name());

// Find number of arguments the law takes
 int in_size = function_law-> take_dim();
// Find number of arguments the law returns
 int out_size= function_law-> return_dim();

    printf("Input size %d, Output size %d\n\n", in_size, out_size);

    double *args = new double[in_size];
    double *answer= new double[out_size];


args[0] = 2.14;  // x
args[1] = 3.14;  // y

function_law-> evaluate(args,answer);

       printf("Law is %s, \n\nArguments x=%f, y=%f and result = %f\n\n",
              law_string, args[0], args[1], answer[0]);

    api_terminate_kernel();
```

```
        api_stop_modeller();
    }
```

When compiled and run **law1.cxx** produces the following output:

```
          Root Law  ---------------------------
          Symbol = +
          Class name = plus_law

          Left Law --------
          String = X
          Class name = identity_law

          Right Law -------
          Symbol = *
          Class name = times_law

          Left Law --------
          String = Y
          Class name = identity_law

          Right Law -------
          Class name = cos_law

          Input size 2, Output size 1

          Law is x + y*cos(x),
          Arguments x=2.14, y=3.140 and result = 0.447
```

Examination of the program shows that a law expression creates a tree data-structure where each node represents a mathematical operator, and each leaf a constant or variable.

After exploring the tree, **law1.cxx** uses the two functions **return_dim** and **take_dim** to establish the number of arguments required and returned by the expression. Arrays of the correct size are then created, filled with arguments, and finally the result calculated by the **evaluate** function.

All classes used in the creation of a law data-structure are derived from an abstract base class known as **law**. The seven classes derived from this root are:

- **constant_law**: represents constants (e.g. 34 or π).

- **identity_law**: represents variables (e.g. x or $a1$).

- **unary_law**: represents mathematical functions with *one* argument (e.g. $\cos\theta$).

10.2 Creating Laws in C++

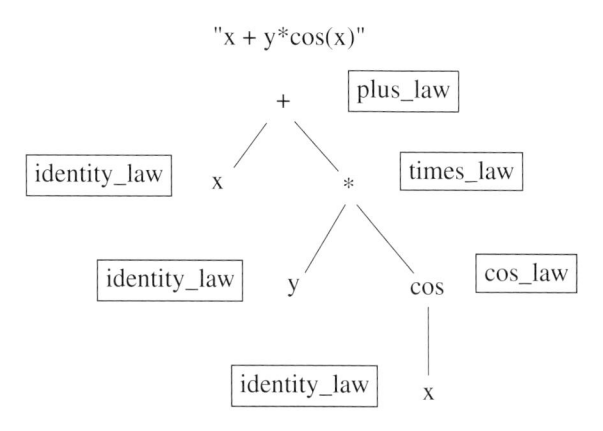

Figure 10.1: Law Tree Structure.

- **binary_law**: represents mathematical functions with *two* arguments (e.g. +).

- **multiple_law**: represents mathematical functions with *multiple* arguments (e.g. vector(x, y, z)).

- **law_data**: defines a base class used to support wrapper classes that allow ENTITY classes to appear as law functions. For example, an edge_law class takes a parameter value t as an argument and returns an (x, y, z) coordinate.

- **unary_data_law**: similar to unary_law but can accept as input an array of elements derived from the law_data class.

- **multiple_data_law**: similar to multiple_law but can accept as input an array of elements derived from the law_data class.

In addition to a normal ACIS **use_count** function (see page 124) each law must define or inherit the following functions:

- **derivative**: which returns a pointer to a law object representing the derivative of the expression.

- **evaluate**: which, given the correct arguments, will return a numerical result.

- **identification functions**: **isa**, **id** and **type** all return values which indicate the nature of the object.

- **return_dim**: returns the number of arguments *output* by the **evaluate** function.

- **take_dim**: returns the number of arguments required as *input* to the **evaluate** function.

Table 10.2   Law operation symbols.

Law Syntax	Description	Math/Example
law1 + law2	Addition operator	$x + y$
law1 - law2	Subtraction Operation	$x - y$
law1 * law2	Multiplication operator	$x \times y$
law1 / law2	Division operator	$x \div y$
law1^law2	Power operator	x^y
law1 o law2	Composition law	$f(g(x))$
abs(law1)	Absolute value	`"abs(1.09)"`
cross(law1, law2)	Cross product of two laws	$x \times y$
dot(law1, law2)	Dot product of two vectors	$x \cdot y$
e	Base of Natural Logs	2.71828182..
exp(law1)	Raises e to a given power	e^x
ln(law1)	Natural Log	$\ln(x)$
log(law1,law2)	Evaluates log of law1 to law2 (defaults to ten)	$\log_y(x)$
norm(law1)	Normalises any 1, 2 or 3D arguments	`"norm(vec(6,3,8))"`
pi	Mathematical Constant	π
rotate(law1, transf)	Function that transforms vectors	see trans
set(law1)	Function that returns 1 if law1 is positive and 0 if law1 is negative or zero (0)	sign(x)
size(law1)	Returns the square root of the sum of the squares of a given vector	`"size(vec(1,1))"`
sqrt(x)	Evaluates the square root of a given law	$\sqrt{x}$
step(law1, num1, law2, num2, ...)	Evaluates to define a function with disjoint intervals	
term(law1, item)	Returns a single item from a given multi-dimensional function	`"term(vec(1,2,3),2)"`
trans(law1, transf1)	Function that transforms positions	see rotate
vec(law1, law2, ...)	A vector of arbitrary dimensions	`"vec(1,7,12)"`

10.3 Planar Offsets

Obviously the creation and manipulation of symbolic equations is not an end in itself and the real application of the laws system is as a tool for defining, manipulating and analysing geometry.

Although laws expressions can be used to create geometry from scratch, perhaps it is easier to start by considering how existing shapes can be changed by offsetting with a law expression. Non-uniform offsets can be created by the optional law arguments that the offsetting function can take.

The scheme function **wire-body:offset** takes a planar wire body and offsets its profile by evaluating a user-defined law.

Figure 10.2 illustrates the results of offsetting circles of different diameter with the law $10 + 5 * sin(x)$. The picture graphically illustrates how the parameter x evaluates to the *arc length* of the wire. In otherwords the value of x runs from 0 to π on a unit circle and 0 to 4π when the diameter is 4.

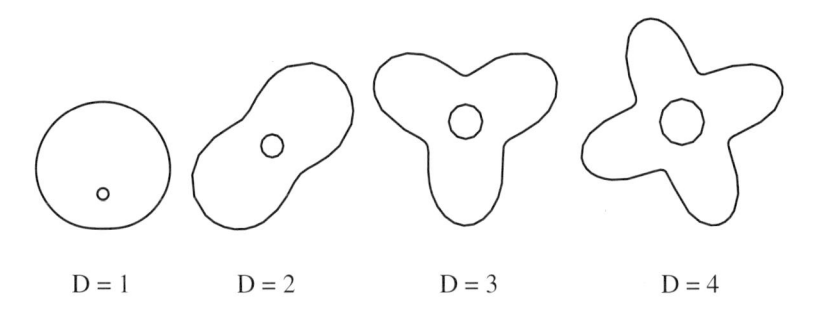

D = 1   D = 2   D = 3   D = 4

(**wire-body:offset** wire-circle-body "10 + 5*sin(x)")

$x =$ arc length

Figure 10.2: Offsetting Circles of different diameter.

Even if the wire is composed of a number of edges (as in the **u-offset.scm** example) the law evaluates to a continuous function (ie, the length) of the wire.

u-offset.scm

```
; This program creates a wire body
; and offsets it with a simple Law.

; Make the wire body
(define u-shape
 (wire-body(list
         (edge:linear (position 0 0 0) (position 20 0 0))
         (edge:circular-diameter (position 20 0 0) (position 20 20 0))
         (edge:linear (position 20 20 0) (position 0 20 0)))))
```

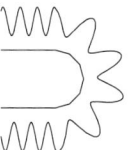

```
; Offset the wire
(define u-offset (wire-body:offset u-shape "10 + 5*sin(x)"))
```

Figure 10.2 illustrates how important it is for the programmer to anticipate exactly how the law will behave when applied to differing sizes of geometry.

10.3.1 Laws Offsetting: A practical example

Why would anyone want to generate such strangle offsets as those shown in Figure 10.2? One answer is found in mechanical engineering text books under the heading "Cam Design". These components are ideal candidates for demonstrating the facilities of the laws system because the design process requires that:

- They can have profiles defined in terms of polynomial equations.

- The *follower's* velocity and acceleration are found by differentiation of the polynomial equation[3].

- The maximum values of acceleration and velocity need to be determined to ensure that the *follower* does not lose contact with the cam at any point.

The example **cam-law.scm** demonstrates how the shape of a cam can be generated from its defining equation.

Preceding this are a few lines of Scheme which illustrate how Laws extensions can be used to determine when the maximum and minimum accelerations occur.

With reference to equation 10.2 on page 243, if we fix a value for R (the maximum displacement of the follower) at, say, 10 units, then the other design parameters of the cam can be easily computed as follows:

```
; cam displacement curve
  (define s (law " (10/3)*(20*x^3 -25*x^4 + 8*x^5)"))
; cam velocity curve
  (define v (law:derivative s))
; cam acceleration curve
  (define a (law:derivative v))

; Find when maximum acceleration occurs
(law:nmax a 0 1) ; 0.25
; Find maximum acceleration
(law:eval a (law:nmax a 0 1)) ; 45.83
; Find when minimum acceleration occurs
(law:nmin a 0 1) ; 1
```

---

[3]The follower is the part of the mechanism which is displaced (while remaining in contact) by the cam as it rotates.

10.3 Planar Offsets

> At the heart of many mechanical devices are small rotating components called cams whose function is to transmit motion to another element called a *follower*. The main advantage of a cam is that it can be easily designed to produce an exact linear output motion from a rotational input drive. In other words the designer can specify the displacement, velocity and acceleration of a cam's follower. One particular class of cams is created by polynomial displacement curves which have the general form:
>
> $$s = c_0 + c_1\theta + c_2\theta^2 + c_3\theta^3 + c_4\theta^4 + c_5\theta^5 \ldots + c_n\theta^n$$
>
> Differentiation of this expression (when $n = 5$) will produce equations for velocity and acceleration.
>
> $$v = \frac{ds}{d\theta} = c_1 + 2c_2\theta + 3c_3\theta^2 + 4c_4\theta^3 + 5c_5\theta^4$$
> $$a = \frac{ds^2}{d\theta^2} = 2c_2 + 6c_3\theta + 12c_4\theta^2 + 20c_5\theta^3$$
>
> where:
> s = deflection of follower,
> θ = a given rotation of the cam,
> v = velocity of the follower,
> a = acceleration of the follower.
>
> The constants (coefficients) $c_1, c_2, c_3, \ldots$ are determined by the boundary conditions which for a Dwell-Rise-Fall cam are:
>
> $$\theta = 0 : \quad s = 0, \quad v = 0, \quad a = 0$$
> $$\theta = \phi : \quad s = R, \quad v = 0, \quad \frac{da}{d\theta} = 0$$
>
> where:
> R = the maximum displacement of the follower.
> ϕ = cam rotation corresponding to R
>
> Application of the $\theta = 0$ conditions results in the disappearance of the coefficients c_0, c_1, c_2. The second set of boundary conditions at $\theta = \phi$ results in the following values:
>
> $$c_3 = \frac{20}{3}\frac{R}{\theta^3} \quad c_4 = -\frac{25}{3}\frac{R}{\theta^4} \quad c_5 = \frac{8}{3}\frac{R}{\theta^5} \quad (10.1)$$
>
> resulting in the displacement curve equation:
>
> $$s = \frac{R}{3}[20(\frac{\theta}{\phi})^3 - 25(\frac{\theta}{\phi})^4 + 8(\frac{\theta}{\phi})^5] \quad (10.2)$$

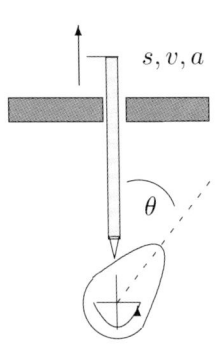

About Mechanical Cam Design

The code on page 242 applies **law:nmax** and **law:nmin** to the acceleration expression to determine the maximum and minimum values of the function for values of x between 0 and 1. The margin illustrates the behavior of the cam's acceleration and velocity expressions with plots of edges created by offsetting a horizontal line with the commands:

Displacement

Acceleration

(**define** baseline (**edge:linear** (**position** 0 0 0) (**position** 1 0 0)))
(**define** basewire (**wire-body** (**list** baseline)))
(**wire-body:offset** basewire a)).

To produce a symmetrical cam by offsetting a circle, a function is required that creates a rising profile for 180° which then falls back to its start point. The design equations only define half of the CAM so some adaptation is required to get the desired shape.

One simple way of switching between rising and falling profiles is to use a piecewise law which takes a series of arguments specifying alternate conditions and actions. The resulting law uses a different expression for its evaluation depending on its input. For example:

; define piecewise law
(**define** pl (**law** " piecewise(x<10,-1,x>10,1,0)"))
; Evaluate it
(**law:eval** pl 9) ; Evaluates to -1
(**law:eval** pl 11) ; Evaluates to 1
(**law:eval** pl 10) ; Evaluates to the default 0

A full list of the conditional expressions accepted by a **piecewise** law are given in Table 10.3.

Because the numerical arguments of the design (i.e. the base radius and the lift) must be embedded in the polynomial expression for the law as constants, the functions **number->string** and **string-append** are used to build up the law string. Two laws are constructed which differ only in the way the offset distance od is evaluated. In the rise-law $od = f(\frac{x}{\pi r})$ whereas in the fall-law $od = f(\frac{2\pi r - x}{\pi r})$. Thus the value of the offset will smoothly increase for half the circle and then fall for the rest. The piecewise law is then used to switch between these two expressions when $x = \pi r$.

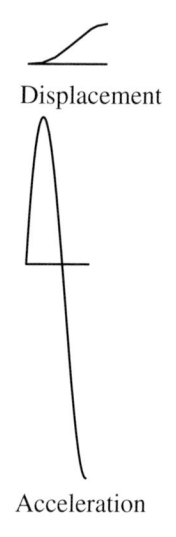

10.3 Planar Offsets

Table 10.3 Law relational symbols.

Law Syntax	Description
(law1) and (law2)	Logical and used with PIECEWISE operator
law1 = law2	Logical equals used with PIECEWISE operator
law1 > law2	Logical greater than used with PIECEWISE operator
law1 >= law2	Logical greater than or equal to used with PIECEWISE operator
law1 < law2	Logical less than used with PIECEWISE operator
law1 <= law2	Logical less than or equal to used with PIECEWISE operator
not(law1)	Logical not condition used with PIECEWISE operator
law1 != law2	Logical not equal condition used with PIECEWISE operator
max(law1,law2,...)	Evaluates the maximum of two or more input laws
min(law1,law2,...)	Evaluates the minimum of two or more input laws
law1 or law2	Logical or condition used with PIECEWISE operator
piecewise (cond1, law1, cond2, law2, default)	Evaluate differently based on conditional definition statements: IF cond1 THEN law1 ELSE IF cond2 THEN law2 ELSE default

cam-law.scm

```
; during offsetting x will vary between 0 and PI*D around the circle
(define cam-wheel
  (lambda (base-radius lift)
   (let*(
       (basecircle ( edge:circular ( position 0 0 0) base-radius))
       (basewire  ( wire-body (list basecircle)))
       (e (number->string (* base-radius PI 2)))
       (h (number->string (* base-radius PI )))
       (r (number->string lift))
       (p1 (string-append "(x/" h ")" ))
       (p2 (string-append "(("e"-x)/"h")" ))
       (rise-law (law(string-append "("r"/3)*(20*"p1"^3 - 25*"p1"^4 + 8*"p1"^5)")))
       (fall-law (law (string-append "("r"/3)*(20*"p2"^3 - 25*"p2"^4 + 8*"p2"^5)")))
```

```
                    (cam-default (law "sin(x)"))
                    (cam-cond1 (law (string-append "(x <" h")")))
                    (cam-cond2 (law (string-append "(x >= " h")")))
                    (cam_law (law  "piecewise(law1,law2,law3,law4,law5)"
                            cam-cond1 rise-law cam-cond2 fall-law cam-default)))
                  ( wire-body:offset basewire cam_law))))
(cam-wheel 8 4)
```

The program could also have been written using the "o" operator which allows the output of one law to be the input of another. The syntax is simply:

```
(law "law1 o law2")
```

which can be read as "law1 *of* law2". This allows us to compose two functions, (i.e. $(fog)(x) = f(g(x))$). So for example:

```
acis>(law:eval (law "a1^2 o a1^2") 2)
16
```

Using this operator the **cam-law.scm** program could, perhaps, have been more clearly written as:

```
; p1 evaluates to x for the rising profile function
(p1 (string-append "(x/" h ")" ))
; p2 evaluates to x for the falling profile function
(p2 (string-append "(("e"-x)/"h")" ))
; Define a general cam-law in terms of x
(cam-law (law (string-append "("r"/3)*(20*x^3 - 25*x^4 + 8*x^5)")))
; Equate x with p1 for rising profile function
(rise-law (law "law1 o law2" cam-law p1))
; Equate x with p2 for falling profile function
(fall-law (law "law1 o law2" cam-law p2))
```

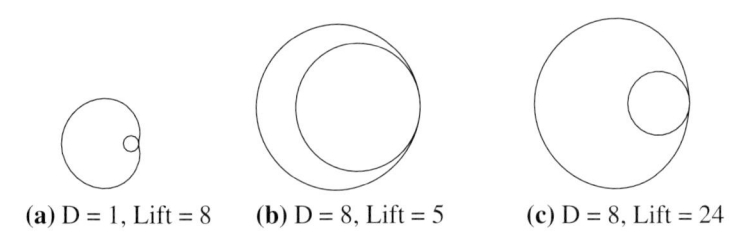

(a) D = 1, Lift = 8   (b) D = 8, Lift = 5   (c) D = 8, Lift = 24

Figure 10.3:   Cam profiles.

The resulting profiles shown in Figure 10.3 require further analysis before a design can be finalised. Even a casual glance suggests that the diameter of the inner base circle is a critial design parameter which, if badly chosen, can lead to problems known as *undercutting*, as can be seen in Figure 10.3 (a).

10.4 Helical Offsets

The preceding section was "economical with the truth" when it described the offset function. The most sophisticated types of offset have both an *offset* law and a *twist* law. The latter allows simple circles to be transformed into elegant helical shapes. If:

R = offset_law(t) (i.e. the size of offset (i.e. radius) at distance t from the start),
θ = twist_law(t) (i.e. the offset angle (i.e. direction) at distance t from the start).

then the relationship used to generate the resulting offset_curve from the original curve can be written:

$$Pt = pt + [(|T \times N|)R)]\cos(\theta) + [(|N|)R]\sin(\theta)$$

where:
Pt = Point on the offset curve at distance t along it,
pt = Point on the original curve at distance t along it,
T = the tangent to curve at distance t along it,
N = the normal to the plane of the wire body.

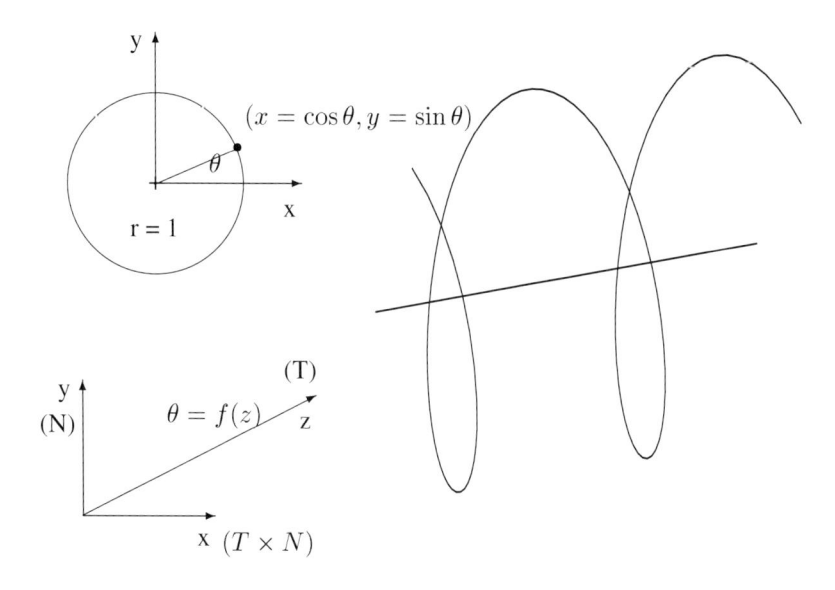

Figure 10.4: Parameters of a helical offset.

Although the functions **sine** and **cosine** were probably embedded in your mind during High School as ratios of certain lengths this is not the best mental model with which to understand their use in parametric equations. For the rest of this chapter you should think of these functions as the *x and y coordinates of points on a unit* *circle*. Figure 10.4 attempts to illustrate how an increasing value of θ simply causes a point on a unit circle to move around it.

Figure 10.4 envisages the offsetting of a linear edge lying along the z-axis. In this case N points up the y-axis and T points along the z-axis. Since the cross product results in a vector normal to the TN-plane, in this particular case the first term can be regarded as defining displacement along the x-axis. Likewise the second term represents the amount of y-axis offset. So in this case the equation could be written as:

$$x\text{-axis offset} = R\cos(\theta)$$
$$y\text{-axis offset} = R\sin(\theta)$$

Hopefully it should now be obvious from the above that the *twist_law* simply changes the proportions of the amount of horizontal and vertical offsets. In other words, the twist_law (which evaluates to a number input as radians to the cos and sin functions) determines how close the coils of an offset curve lie (i.e. it determines the pitch).

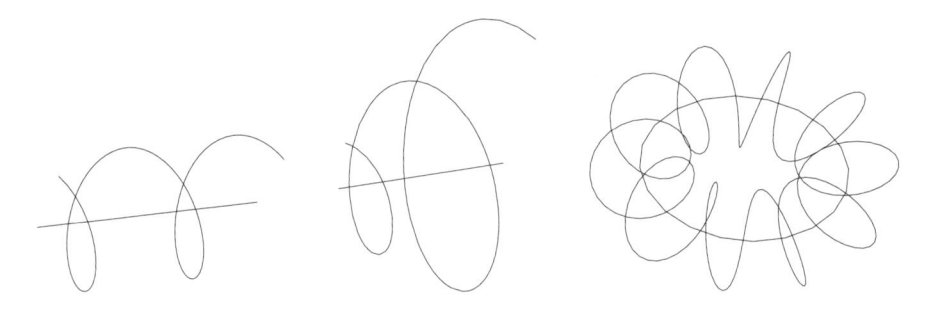

Figure 10.5:   Advanced offsetting.

If the offset_law is a constant, say 3, and the twist_law is simply the function **x** then the result is a helix that winds around the wire body used to generate it.

Note that the sin and cos functions will evaluate to both positive and negative value over a length of 0 to 6.28.

helical.scm

```
; This program uses offsetting to create three different types of helix

; Constant Helix
(define aWire0 (wire-body (edge:linear (position 0 0 -40)(position 0 0 40))))
; A helix with constant radius
(define aHelix0 (wire-body:offset aWire0 "20" "x/(2*PI)"))

;Diverging Spiral Helix
(define aWire (wire-body (edge:linear (position 0 0 -40)(position 0 0 40))))
; An expanding helix
```

10.5 Basic Sweeping

```
(define aHelix1 (wire-body:offset aWire "20+x/2" "x/(2*PI)"))

; Toroidal Round Helix
(define hoop (wire-body (edge:circular (position 0 0 0) (* 10 PI))))
(define aHelix2 (wire-body:offset hoop "15" "x/PI"))
```

Figure 10.5 shows three shapes generated by the offsets specified in the program **helical.scm**. The next example shows a different way in which helices can be created in ACIS.

10.5 Basic Sweeping

After the cylinder perhaps the most important shape in mechanical engineering is the helix. The first engineer to see useful applications in the fact a helix can transform itself into itself by rotating and moving along its axis must have been a remarkable person. Today helices are so common in engineering that the only place they get noticed is in geometric models! The program **helical.scm** showed how a number of different helices can be generated by offsetting an EDGE. However, much of the thought required to specify the correct *twist* law can be avoided by using the **api_edge_helix** function. In the example **helix-sw.cxx** a helical edge is created using this API and then used as a path along which a circular profile is swept.

helix-sw.cxx

```cpp
// This program creates a helix and then
// sweeps a circular disk along it, to create a spring
#include "constrct/kernapi/api/cstrapi.hxx"     //Declares api_make_planar_disk
#include "kernel/kerndata/top/alltop.hxx"       // Declares all topology classes
#include "baseutil/vector/position.hxx"         // Declares position classes
#include "baseutil/vector/unitvec.hxx"          // Declares unit_vector class
#include "baseutil/vector/vector.hxx"           // Declares vector class
#include "kernel/kerngeom/curve/curdef.hxx"     // Declares curve class
#include "kernel/kerndata/geom/curve.hxx"       // Declares CURVE class
#include "kernel/kerndata/geom/point.hxx"       // Declares POINT class
#include "sweep/sg_husk/sweep/swp_opts.hxx"     // Declares sweep_option class
#include "sweep/kernapi/api/sweepapi.hxx"       // Declares api_sweep_with_options
#include "offset/kernapi/api/ofstapi.hxx"       // Declares api_edge_helix
#include "kernel/kernutil/law/law.hxx"          //Declares law class
#include "kernel/kernutil/errorsys/errorsys.hxx" // Declare error functions
#include "kernel/kerndata/lists/lists.hxx"      // Declares ENTITY_LIST class
#include "kernel/kernapi/api/kernapi.hxx"       // Declares kernel api's
#include "kernel/kerndata/savres/fileinfo.hxx"  // Declares fileinfo class

// Function prototypes for save to file and error check functions
void save_ent(char*, ENTITY*); // see page 72
```

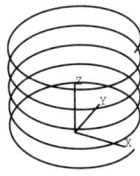

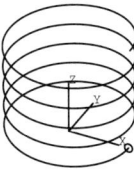

```
void check_result(outcome, char*); // see page 349

void main()
{
    api_start_modeller(0);
    api_initialize_sweeping(); // Initializes the sweep library

    position axis_start(0,0,0); // axis start position
    position axis_end(0,0,50);  // axis end position
    vector start_dir(1,0,0);    // from axis_start to helix start
    double radius = 30;         // major radius of helix
    double pitch= 10;           // distance between turns along axis
    logical handiness = FALSE;  // TRUE is right  FALSE is left

    EDGE *helix; // pointer to the helix created

    api_edge_helix (
                axis_start, //creates the helix
                axis_end,
                start_dir,
                radius,
                pitch,
                handiness,
                helix);

    VERTEX *v1= helix-> start(); // get the coordinates of
                                 // its start vertex
    APOINT *p1 = v1-> geometry();

    position origin = p1-> coords();

    const curve& cur = helix-> geometry()-> equation();
                    // curve direction at start
    unit_vector normal = cur.point_direction(origin);

    FACE *disk;

    api_make_planar_disk (
                origin,  //start of helix
                normal,  //along helix
                2,       //radius
                disk  );

    BODY *spring;
    sweep_options *opts = new sweep_options(); //use default values

    outcome res = api_sweep_with_options (
                disk, // profile to sweep
                helix, // path to sweep along
                opts, // sweep options
```

```
                         spring // swept body returned
                              );

    check_result(res,"Error in sweep");

    save_ent("spring.sat",(ENTITY*)spring);

    api_terminate_sweeping();
    api_stop_modeller();
}
```

After creating the helical EDGE the program uses the direct interface to obtain a tangent vector at its start point. This is done using the **point_direction** member function of the **curve** class as follows:

```
    // Get the geometry of the helix
        const curve& cur = helix->geometry()->equation();
                // curve direction at start
        unit_vector normal = cur.point_direction(origin);
```

Given a position and a direction the function **api_make_planar_disk** creates a circular FACE correctly located at the start of the helix.

The function, **api_sweep_with_options**, is overloaded so the number and type of its arguments can vary with its application, however, all variations require a **sweep_options** class to be given. This class is used to set the 17 different sweeping parameters (e.g. twist and draft angles). Fortunately, all these parameters have default values which are suitable for the creation of a "spring" so their explicit setting is not required in **helix-sw.cxx**.

Subsequent examples illustrate how this class can be used to create fancy sweeps and Table 10.4 on page 256 summarizes the member functions available to set sweeping parameters.

10.6 Complex Sweeping

The combination of law-based offsetting and sweeping enables the creation of a wide range of shapes. The **law_twist.cxx** example demonstrates this by creating a planar WIRE BODY which is offset with the trigonometric sin function and then swept along a vector while twisting through an angle of $180°$.

The program also illustrates two classic offsetting difficulties which programmers must anticipate.

The first twenty lines of code are concerned with the building of a wire BODY. Individual EDGEs are created for each arc and line in the profile before the function **create_wire_from_edge_list** is used to collect them into a single ENTITY (Figure 10.6(a)). The resulting shape has two sharp, convex, corners. Naïvely offsetting this profile with, say, the Law "$4 + sin(x/4)$" would result in the *broken* profile shown in Figure 10.6(b).

Gaps in profiles offset outwards are not unexpected and **api_offset_planar_wire** takes an integer argument (called gap_type in the example) which specifies how to fill them automatically.

Gaps can only be filled for constant offset (as opposed to Law offsets) so the program creates a constant Law using the command

api_str_to_law("4",&offset)

and sets gap_type to 0 (i.e. round corners) before carrying out the operation. The resulting profile has no gaps only smoothly rounded corners. However, attempts to apply directly a Laws offset of, say, "$5*sin(x/4)$" would still not produce the desired result.

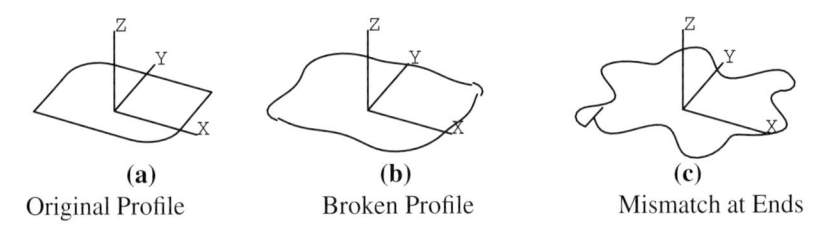

(a) Original Profile   (b) Broken Profile   (c) Mismatch at Ends

Figure 10.6: Offsetting problems.

Because the value of x evaluates to the arc-length of the profile the programmer must ensure that the first and last values of the offset distance are equivalent if the mismatch at the start and finish shown in Figure 10.6(c) is to be avoided. The program does this by scaling the laws offset to match the length, L, of the wire profile exactly. In other words instead of directly applying the offset law:

$$4 * cos(4 * x)$$

it applies the law:

$$4 * cos((4 * pi * x)/L)$$

Other applications might have other requirements, but in this case the approach solves the problem and results in the smooth, wavy, profile seen in the left-hand margin.

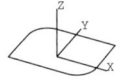

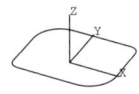

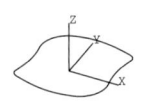

10.6 Complex Sweeping

law_twist.cxx

```cpp
// This program creates a planar wire, applies a constant
// offset to round the corners, applies a sin offset to
// make it wavy and then sweeps it along an axis with twist
#include <stdlib.h>                            // Declares float to string conversion
#include <stdio.h>                             // Declare output functions
#include <string.h>                            // Declares strcat function
#include "kernel/kernapi/api/kernapi.hxx"      // Declares kernel API's
#include "constrct/kernapi/api/cstrapi.hxx"    // Declares constructor API's
#include "kernel/kerndata/top/body.hxx"        // Declares Topological Class BODY
#include "constrct/geomhusk/wire_utl.hxx"      // Declares api_wire_len
#include "offset/kernapi/api/ofstapi.hxx"      // Declares offsetting API
#include "kernel/kerndata/top/alltop.hxx"      // Declares all topology classes
#include "kernel/kerndata/lists/lists.hxx"     // Declares ENTITY_LIST class
#include "kernel/kernutil/law/law.hxx"         // Declares list classes
#include "baseutil/vector/position.hxx"        // Declares position class
#include "baseutil/vector/vector.hxx"          // Declares vector class
#include "baseutil/vector/unitvec.hxx"         // Declare unit vector class
#include "kernel/kerndata/savres/savres.hxx"   // Declares save/restore api
#include "kernel/geomhusk/geom_utl.hxx"        // Declares degrees_to_radians
#include "sweep/sg_husk/sweep/swp_opts.hxx"    // Declares sweep_options class
#include "sweep/kernapi/api/sweepapi.hxx"      // Declare sweeping api
#include "kernel/kerndata/savres/fileinfo.hxx" // Declares fileinfo class

// Function prototypes for save to file and error check functions
void save_ent(char*, ENTITY*); // see page 72
void check_result(outcome, char*); // see page 349

void main()
{
   api_start_modeller(0);
   api_initialize_sweeping();

   // Create a planar wire body
   EDGE *line1, *arc2, *line3, *line4, *arc5, *line6 = NULL;

   api_curve_line(position(-20,-10,0), position(10,-10,0), line1);
   double start_angle = degrees_to_radians(270);
   double end_angle = degrees_to_radians(360);

   api_curve_arc(position(10,0,0),10,start_angle,end_angle, arc2);
   api_curve_line(position(20,0,0), position(20,20,0), line3);
   api_curve_line(position(20,20,0), position(-10,20,0), line4);
   start_angle = degrees_to_radians(90);
   end_angle = degrees_to_radians(180);
   api_curve_arc(position(-10,10,0),10,start_angle,end_angle,arc5);
   api_curve_line(position(-20,10,0),position(-20,-10,0), line6);
   // Declare varables
```

```
BODY *wire_body;
ENTITY_LIST elist;
elist.add(line1); elist.add(arc2); elist.add(line3);
elist.add(line4); elist.add(arc5); elist.add(line6);

create_wire_from_edge_list(elist, wire_body);

BODY *offset_wire;
unit_vector normal(0.0, 0.0, 1.0);
law *offset;

api_str_to_law("4", &offset); // Constant offset
int gap_type = 0; // round corners
logical trim = TRUE;

api_offset_planar_wire(wire_body, offset, NULL, normal,
                       offset_wire, gap_type, trim);

double length = 0.0;
api_wire_len (offset_wire, length); // 156.548668

char law_str[50]; // Define law string
strcpy(law_str,"4*cos((4*pi*x)/");
char len[20]; // Convert 'double' length to 'string' len
_gcvt(length, 12, len );
strcat(law_str,len); // Add len to the end of law_str
strcat(law_str,")"); // Add the closing bracket
printf("string = %s\n",law_str);

BODY *offset_wire2;
law *offset2;
outcome r2 = api_str_to_law(law_str, &offset2);
check_result(r2,"error str to law");

r2 = api_offset_planar_wire(
            offset_wire, // input
            offset2,     // distance
            NULL,        // twist
            normal,      // normal to plane of wire
            offset_wire2, // output
            gap_type,
            trim);
check_result(r2,"error in second offset");

// Make a path to sweep along
position array_pts[2];
array_pts[0].set_x(0.0);
array_pts[0].set_y(0.0);
array_pts[0].set_z(0.0);
array_pts[1].set_x(0.0);
```

10.6 Complex Sweeping

```
        array_pts[1].set_y(0.0);
        array_pts[1].set_z(50.0);

        BODY *path = NULL;
        r2 = api_make_wire(NULL,2,array_pts,path);
        check_result(r2,"error in make wire");

        BODY *twist;
        sweep_options *opts = new sweep_options();
        opts-> set_twist_angle(M_PI);
        opts-> set_solid(TRUE);

        // Profile may be lost with modification of BODY while
        // sweeping, so we want to keep the owner of the profile.
        ENTITY *new_body;
        r2 = api_get_owner(offset_wire2, new_body);
        check_result(r2, "error in get owner");
        r2 = api_sweep_with_options(
                    (ENTITY*) offset_wire2, // Profile to sweep
                    (ENTITY*) path,         // Path to sweep along
                    opts,                   // Sweep options
                    (BODY *&) new_body);    // Swept body returned
        check_result(r2, "error in sweep");

        twist = (BODY *&)new_body;
        save_ent("twistsweep.sat", twist);
        delete opts;
        offset-> remove(); //  Delete law classes
        offset2-> remove();

        api_terminate_sweeping();
        api_stop_modeller();
}
```

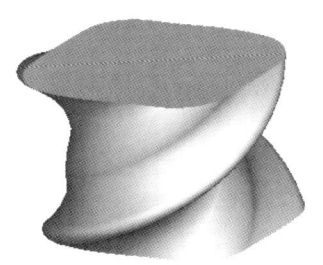

Figure 10.7:   Result of law_twist.cxx.

Table 10.4 Sweep options.

Member Function	Argument	Description
set_draft_angle	double	Sets the angle (in degrees) by which the swept profile is expanded (+) or contracted (-) while sweeping.
set_draft_law	law*	Sets a pointer to a law class which evaluates to a draft angle.
set_end_draft_dist	double	Sets the offset distance (in the plane of the profile) at the end of sweep (see also set_start_draft_dist).
set_gap_type	int	Sets fill method for gaps created by draft offsetting natural (2), round (0), and extend (1).
set_miter	miter_type	Specifies how mitering is performed. Enumeration values are default_miter (default), old_miter, new_miter, crimp, and bend.
set_param_rail	logical	If TRUE (default), the parameterization is taken into consideration; however, when path is nonplanar the arc-length is used (FALSE).
set_rail_law	law*	Law which specifies the orientation of a profile as it is swept. A non-planar path defaults to minimum rotation, otherwise default is a constant vector equal to the planar normal.
set_rigid	logical	Specifies whether cross-sections are perpendicular to the sweep path (FALSE, the default) or parallel to one another (TRUE).
set_scale_law	law*	Vector law which specifies the amount of scale in xyz directions. The x-axis is parallel to the rail; z is parallel to the tangent of the path ($y = z \times x$).
set_solid	logical	If TRUE the result of sweeping is a solid, otherwise (FALSE) a sheet-body.
set_start_draft_dist	double	Sets an offset distance in the plane of the profile for the start of the sweep (see also set_end_draft_dist).
set_steps	int	Specifies a number of linear segments which a sweep around an axis should be broken into.
set_sweep_angle	double	Set the angle (in radians) to sweep around an axis (default 2π).
set_to_face	surface*	The SURFACE to be used for clipping.
set_twist_angle	double	Sets how much a profile twists as it is swept along a path.
set_twist_law	law*	Evaluates to the angle of twist (in radians) as a function of x (between $x = 0$ and $x =$ length of the wire).
set_which_side	logical	Establishes which side of a profile should be swept.

Having created a suitably exotic profile the final task is to define the *type* of sweep required. This is achieved by creating a **sweep_options** object, setting its data members to appropriate values and then supplying it to **api_sweep_with_options** as an argument.

A glance at Table 10.4 on page 256 (which summarizes the sweep_options) gives an idea of the mind boggling range of sweeps ACIS allows[4].

The program **law_twist** chooses to sweep along a vector while twisting the profile smoothly through π radian. After saving the resulting shape the program calls the **remove** methods of the two law classes that have been created. This does not directly delete the law object but instead (like an ENTITY) reduces its *use_count* by one. Only when the use_count is reduced to zero is the object deleted.

10.7 Creating EDGEs with Laws

The function **api_edge_law** provides a convient way of creating EDGEs whose geometry is defined by a parametric equation expressed as a law. Parametric equations express the coordinates of the shape being modelled in terms of the *geometric dimension* of the object. So curves (or EDGEs) which are one-dimensional are defined by equations with only one variable, commonly called t. To map between the one-dimensional world of the parameter and the three-dimensional world in which the parametric curve is *embedded* three functions are needed to calculate the coordinates (x_c, y_c, z_c) of every point on the curve. These can be written:

$$\begin{aligned} x_c &= f_x(t) \\ y_c &= f_y(t) \\ z_c &= f_z(t) \end{aligned}$$

In the previous example a helix was created behind the closed doors of **api_edge_helix**. However, parametric equations for the helix can be found in many textbooks. The next example creates a helical EDGE from a law string.

The general equation of a spiral (with a constant radius r) can be adapted easily to create a helical spiral with a radius which smoothly increases from R:

$$\begin{aligned} f_x &= r\cos(t) &\rightarrow& \quad (R + c_1 t)\cos(t) \\ f_y &= r\sin(t) &\rightarrow& \quad (R + c_1 t)\cos(t) \\ f_z &= c_2 t &\rightarrow& \quad c_2 t \end{aligned}$$

where :

t varies between zero and the length of the spiral,

c_1 is a constant which determines the final radius of the spiral,

c_2 is a constant which determines the pitch of the spiral.

---

[4] Watch the small print when experimenting with sweeps! For example the options: "draft_angle", "draft_law", and the combination "start" and "end_draft_dist" are mutually exclusive. In other words when one is defined, the others are cleared. Likewise "twist_angle" and "twist_law".

Both "start" and "end_draft_dist" need to be stated even when one is set to zero. When "start_draft_dist" = 0, the result is like "draft_angle" only that an offset distance is used instead of an angle.

The program **asweep.cxx** uses the following parametric equations to define a helical spiral:

$$\begin{aligned} x_c &= (10 + 15t)\cos(t) \\ y_c &= (10 + 15t)\sin(t) \\ z_c &= 20t \end{aligned}$$

(i.e. with $c_1 = 15$ and $c_2 = 20$) to create an EDGE along which a square, planar, FACE is swept. To illustrate the flexiblity of the system **sweep_options** are selected which cause the profile to rotate as it is swept.

asweep.cxx

```
// This program sweeps a square along a spiral using a
// twist_path_law for orientation.
#include<stdio.h>                             // Declares printf function
#include "kernel/acis.hxx"                    // Declares system-wide information
#include "kernel/kernapi/api/kernapi.hxx"     // Declares kernel API's
#include "constrct/kernapi/api/cstrapi.hxx"   // Declares constructor API's
#include "kernel/kerndata/top/alltop.hxx"     // Declares all topology classes
#include "kernel/kerndata/lists/lists.hxx"    // Declares ENTITY_LIST class
#include "kernel/kernutil/law/law.hxx"        // Declares law classes
#include "baseutil/vector/position.hxx"       // Declares position class
#include "baseutil/vector/vector.hxx"         // Declares vector class
#include "baseutil/vector/unitvec.hxx"        // Declares unit_vector class
#include "kernel/kerndata/savres/savres.hxx"  // Declares save/restore api
#include "sweep/sg_husk/sweep/swp_opts.hxx"   // Declares sweep_options class
#include "sweep/kernapi/api/sweepapi.hxx"     // Declares sweep api's
#include "kernel/kerngeom/curve/curdef.hxx"   // Declares curve class
#include "kernel/kerndata/geom/curve.hxx"     // Declares CURVE class
#include "kernel/kerndata/geom/point.hxx"     // Declares APOINT class
#include "kernel/kerndata/savres/fileinfo.hxx"// Declares fileinfo class

// Function prototypes for save to file and error check functions
void save_ent(char*, ENTITY*); // see page 72
void check_result(outcome, char*); // see page 349

void main()
{
    api_start_modeller(0);
    api_initialize_sweeping(); // Initializes the sweeping library

    FACE *basef;
    api_make_plface (
            position(-10,-10,0), // origin of the plane
            position(10,10,0),   // left point on the plane
            position(10,-10,0),  // right point on the plane
            basef); // created plface returned
```

10.7 Creating EDGEs with Laws

```
law *spiral;
api_str_to_law(
            "vec((10+15*t)*cos(t),(10+15*t)*sin(t),20t)",
            &spiral);

EDGE *path;
api_edge_law (
            spiral, // defining law
            2, // start value of t
            25, // end value of t
            path); // created edge

// Get the position of one of the EDGE's ends
VERTEX *v1= path-> start();
APOINT *p1 = v1-> geometry();
position origin = p1-> coords();

const curve& cur = path-> geometry()-> equation();

//Find the tangent direction at the end of the EDGE
unit_vector z_dir = cur.point_direction(origin);
vector x_dir = 10*(cur.point_curvature(origin));
vector y_dir = 10*(x_dir * z_dir);

transf move = coordinate_transf(origin,
            normalise(x_dir),
                normalise(y_dir));

// Transform the planar FACE so it's normal is tangent to the EDGE
outcome res = api_transform_entity(basef,move);
check_result(res, "error in apply transform");

// Create array of laws to hold the min_rotation_law's arguments
law *inputs[4];

// Create a law vector, normal to the curve's start-point
law *array[3];
array[0] = new constant_law(x_dir.x());
array[1] = new constant_law(x_dir.y());
array[2] = new constant_law(x_dir.z());
law *inital_o = new vector_law(array, 3);

inputs[0] = spiral; // sweep path
inputs[1] = inital_o; // starting orientation
inputs[2] = new constant_law(10);
inputs[3] = new constant_law(20);

// Create a vector field law for minimal rotation about the curve.
min_rotation_law *rlaw = new min_rotation_law(inputs,4);
```

```
        law *tw = new identity_law(0);
        law *four = new constant_law(4);

        law *twr = new times_law(tw,four);

        twist_path_law *twp = new twist_path_law (
                        rlaw, // input vector field
                        spiral, // curve law
                        twr // twist law  );

        BODY *twist;
        sweep_options *opts = ACIS_NEW sweep_options;

        opts−> set_rail_law(twp);

        res = api_sweep_with_options (
                        basef, // profile to sweep
                        path, // path to sweep along
                        opts, // sweep options
                        twist // swept body returned );

        check_result(res, "error in sweep");

        save_ent("fsweep.sat",twist);

        api_terminate_sweeping();
        api_stop_modeller();
}
```

Figure 10.8: Twisting sweep along a helix.

Figure 10.8 illustrates how **asweep.cxx** generates the shape. First the helical EDGE and planar FACE are created. Initally created at the origin, the planar face is translated to the bottom of the spiral using the function **coordinate_transf**. This function creates a transformation which moves the origin to the given position, and aligns the x and y axes to two given unit vectors: The two orthogonal unit vectors required are obtained from the helical EDGE's tangent and curvature vectors at its start point.

In order to maintain the desired relative orientation of the profile during the sweep a **min_rotation_law** is used. The **min_rotation_law** constructor takes an array of four sub_laws which define:

1. A law, (spiral), defining the sweep path,
2. A law, (inital_o), defining a constant starting vector.
3. A law defining the lower bound of the domain.
4. A law defining the upper bound of the domain.

Having set the options the sweep is performed as in the previous example. The resulting shape is a fascinatingly complex shape, reminiscent of razor wire.

The laws system allows FACEs to be created whose geometry is specified by an *equation*. The type of equation required is a parametric equation specifying a mapping from a two-dimensional world (i.e. (u,v)) to a three-dimensional space (i.e. (x,y,z)). The challenge for the user of this facility is to identify appropriate parametric equations. Often this is a trivial task, but sometimes it can be, quite literally, impossible. Consider the humble unit sphere.
Implicit :
$$f(x,y,z) = x^2 + y^2 + z^2 - 1$$
Parametric :
$$f(u,v) = (\cos(u)\cos(v), \sin(u), \cos(u)\sin(v))$$
This is an example of a shape that can be described *exactly* by both parametric and implicit equations.
However, conversion from implicit, (x,y,z), to parametric, (u,v), forms is not always possible. As a rule any quadratic polynomial, implicit equation will be paramatizable, but expressions containing terms raised to 3rd, 4th or higher powers will not necessarily have polynomial parametric forms.

About Laws for Surface Equations

10.8 Creating FACEs with Laws

The laws facility allows users to create FACEs with arbitrary parametric equations. This is an extremely powerful facility but great care must be taken in its use to ensure that the resulting FACEs do not self-intersect.

Table 10.5 Miscellaneous Law symbols.

Law Syntax	Description
cur(edge1)	Allows standard EDGE ENTITYs to be used in laws expressions. The resulting law returns a position on the EDGE at a given parameter value.
surf(face1)	Allows standard FACE ENTITYs to be used in laws expressions. The resulting law returns a position on the FACE at the given u,v parameters.
wire(wire1)	Allows standard WIRE BODYs to be used in laws expressions. The resulting law returns a position on the WIRE at a given parameter value.
dcur(edge1, num)	Allows standard ACIS EDGE entities to be used in laws expressions. Evaluates parametric derivatives at a given parameter value, **num** specifies the order of the derivative (i.e. **num** = 0 returns the same result as **cur**, **num** = 1 returns the tangent, etc.).
dsurf(face1, nu, nv)	Evaluates the parametric derivative of a surface. Argument **nu** is the order of the derivatives with respect to u, and **nv** is the order of the derivative with respect to v.
dwire(wire1, num)	Evaluates a scaled parametric derivative of a wire. Argument **num** specifies the order of the derivative.
surfnorm(face1)	Allows standard ACIS FACE entities to be used in laws expressions. Evaluates the vector surface normal of its argument at given u,v parameters.
surfperp(face1, law1)	Returns the u,v position on the given surface that is closest to the position given by law1.
surfvec(face1, para_law1, vec_law2)	Returns a vector on face1 at para_law1 that is tangent to vec_law2.
twist(vector_field, path_law, twist_law)	Function that returns a twisted vector field about a given path.

10.8 Creating FACEs with Laws

One well known family of parametric surfaces[5] is the *superquadrics*. These are a collection of objects whose form can be controlled easily using a small number of parameters.

These shapes are created by observing that the trigonometric forms of the quadric surfaces can be extended with arbitrary exponents. In this way the equation of an ellipsoid, for example, can be rewritten as:

$$
\begin{array}{llll}
x = & a_1 \cos(u) \cos(v) & \rightarrow & a_1 \cos(u)^{e_1} \cos(v)^{e_2} \quad -\pi/2 \leq u \leq \pi/2 \\
y = & a_2 \cos(u) \sin(v) & \rightarrow & a_2 \cos(u)^{e_1} \sin(v)^{e_2} \quad -\pi \leq v < \pi \\
z = & a_3 \sin(u) & \rightarrow & a_3 \sin(u)^{e_1}
\end{array}
$$

where a_1, a_2, a_3 represent the radius of the principal axes (i.e. $a_1 = a_2 = a_3 = 1$ specifies a unit sphere). If $e_1 = 1$ and $e_2 = 1$ we get an ellipsoid, and if we choose other values we get super-ellipsoids.

Similarly a torus can be made *super*:

$$
\begin{array}{llll}
x = & (R + r\cos(u)) \cos(v) & \rightarrow & (R + r\cos(u)^{e_1}) \cos(v)^{e_2} \quad 0 \leq u < 2\pi \\
y = & (R + r\cos(u)) \sin(v) & \rightarrow & (R + r\cos(u)^{e_1}) \sin(v)^{e_2} \quad 0 \leq v < 2\pi \\
z = & r \sin(u) & \rightarrow & r \sin(u)^{e_1}
\end{array}
$$

At first glance these relationships look like they can be turned easily into laws expressions. In fact they can, but there are a couple of caveats.

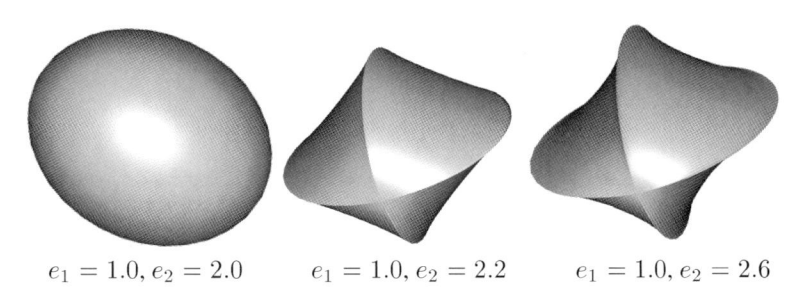

Figure 10.9: Super ellipsoids.

$e_1 = 1.0, e_2 = 2.0 \qquad e_1 = 1.0, e_2 = 2.2 \qquad e_1 = 1.0, e_2 = 2.6$

$e_1 = 1.0, e_2 = 1.0$

$e_1 = 1.0, e_2 = 2.2$

$e_1 = 1.8, e_2 = 2.2$
Super Tori

First, taking the formula for superquadrics literally introduces terms which could require the evaluation of fractional powers of negative numbers, such as $\cos(-1)^{2.2}$, and so should involve complex numbers. Consequently the example **suq.scm** modifies the behaviour of the standard trigonometric functions with the **piecewise** operator (introduced in the **camlaw.scm** example, on page 246) in the following way:

(**define** mysin_u (**law** "piecewise(sin(u) > 0,sin(u)^1,-(abs(sin(u))^1))"))

The following program **suq.scm** demonstrate how this construct can be used to generate super ellipsoids and tori FACEs.

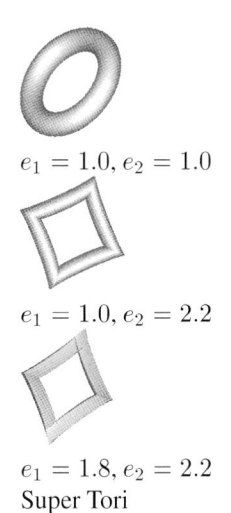

---
[5] See paper by Alan Barr in 1981, IEEE CG&A Vol1

Kernel *suq.scm*

```
; This program demonstrates the use of the face:law function by
; creating a super ellipsoid face and then a super toroid.

; define piecewise trig functions to avoid fractional powers
; of negative numbers.

; e1 = 1
(define mysin_u (law "piecewise(sin(u) > 0,sin(u)^1,-(abs(sin(u))^1))"))
(define mycos_u (law "piecewise(cos(u) > 0,cos(u)^1,-(abs(cos(u))^1))"))

; e2 = 2.2
(define mysin_v (law "piecewise(sin(v) > 0,sin(v)^2.2,-(abs(sin(v))^2.2))"))
(define mycos_v (law "piecewise(cos(v) >0,cos(v)^2.2,-(abs(cos(v))^2.2))"))

; Parametric equation of super ellipsoid
(define sue (law "vec(10*law1*law3, 8*law1*law4, 8*law2)" mycos_u mysin_u mycos_v mysin_v))

; Define parametric limits
(define hPI (/ PI 2))       ; Half PI
(define mhPI (* -1 hPI)) ; Minus half PI
(define mPI (* -1 PI))    ; Minus PI

; Make the super ellipsoid face
(define supe (face:law sue  mhPI hPI  mPI PI ))

;----------------------------------------
; Normal torus use e1 = 1, e2 = 1
; Super "square" torus use e1 = 1, e2 = 2.2

(define stor (law "vec( (20 + 5*law1)*law3,
                        (20 + 5*law1)*law4,
                        5*law2)" mycos_u mysin_u mycos_v mysin_v))

(define tPI (* 2 PI))
; Make the super torus face
(define torf (face:law stor  0.01 tPI  0.01 tPI ))
```

Once created a laws FACE can be treated like any other ACIS entity. Consequently normal Booleans (unite, subtract and intersect) can be used to incorporate a law's surface into other shapes.

The following example creates a funnel[6]-shaped surface using the implicit equation $0 = \frac{1}{2}\ln(x^2 + y^2) - z$. This has the following parametric form:

---

[6]Based on http://www.astro.virginia.edu/~eww6n/math/Funnel.html

10.8 Creating FACEs with Laws

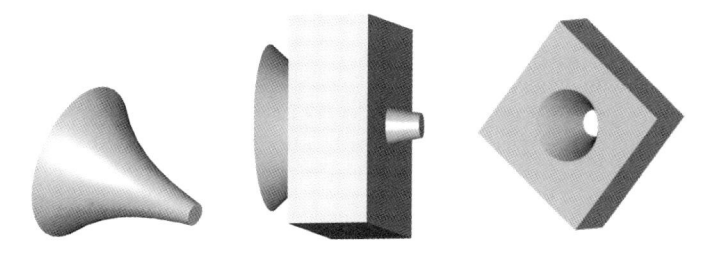

$$x = r\cos(u) \quad 0 \le u < 2\pi$$
$$y = r\sin(u) \quad r_1 \le r < r_2$$
$$z = \ln(r)$$

Figure 10.10: Creating a funnel shaped hole.

Having created the surface it is then reversed (to ensure a positive volume), transformed into a sheet body (i.e. a body consisting of a single face) and then covered to form a solid.

Kernel *funnel.scm*

```
; This example makes a funnel-shaped FACE which is
; used to form a hole on a BODY by subtracting it from a block

(define fun (face:law (law "vec(p1*cos(p2), p1*sin(p2), 20*ln(p1))") 3 20 0 (* 2 PI)))
(face:reverse fun) ; To ensure a positive volume
(define funbod (sheet:face fun))
(define funsolid (sheet:cover funbod))

(define base (solid:block (position -25 -25 30) (position 25 25 50)))
(define outlet (bool:subtract base funsolid))
```

About Surfaces of Revolution and Ruled Surfaces

A surface created by a curve when it revolves about a fixed axis (lying in the same plane as the curve) is called a *surface of revolution*. The parametric equation for such surfaces is easily found.

If we assume that the axis of rotation is the z-axis and that the curve lies in the y-z plane. Then we can define the curve by two equations:

$$y = g(u)$$

$$z = h(u)$$

If the curve rotates through an angle v then the resulting surface is defined parametrically by:

$$x = g(u)\cos(v)$$

$$y = g(u)\sin(v)$$

$$z = h(u)$$

The funnel on page 265 has this form. Using these relationships a laws definition of a catenoid (the surface of revolution created by rotating a catenary) could be derived in the following way:

$g(u)$ = distance from the z-axis = $a \cosh(\frac{u}{a})$
$h(u)$ = distance from the y-axis = u

Resulting in:

$$(x, y, z) = (a\cosh(\frac{u}{a})\cos(v), a\cosh(\frac{u}{a})\sin(v), u)$$

Similar relationships exist for *ruled surfaces* which are formed by sweeping a line along a curve.

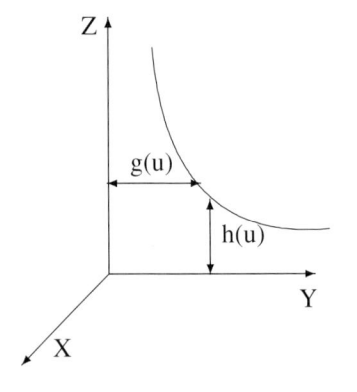

10.9 Space Warping and Scaling

As well as opening up the possibility of modelling strange and exotic geometry the laws system also supports the *deformation* of shapes. Without doubt the simplest form of deformation is non-uniform (or un-equal) scaling where every point's (x, y, z) coordinates are transformed by a different amount parallel to the coordinate axes.

Adopting the convention[7] that points in the undeformed solid have coordinates denoted by a lower case x,y,z; while points in the deformed shape are denoted in upper case by X,Y,Z, then a differential orthogonal scaling operation can be written as:

$$X = a_1 x$$
$$Y = a_2 y$$
$$Z = a_3 z$$

where a_1, a_2 and a_3 are constants.

After a moments reflection, it comes as no surprise that the ratio between a shape's undeformed (v_o) and deformed (v_d) volumes can be stated as:

$$a_1 a_2 a_3 = \frac{v_d}{v_o}$$

More generally, a_1, a_2 and a_3 can be functions defined by laws so X,Y,Z can depend on x,y,z in more complex ways. In C++ differentially scaled models can be created through the application of a scale transform (using the function **scale_transf** in a manner similar to the example on page 80) or by application of a law which explicitly defines the mapping of the coordinates (i.e. (x, y, z)) of one ENTITY to new, scaled, locations (i.e. (X, Y, Z)). The following example **head.scm** demonstrates how both law and unequal orthogonal scaling can be used to simulate the ageing of a human head.

head.scm

; *This program deforms a wire profile with a non-uniform scaling*
; *and a law which results in a shear transform.*

; *Define Head profile to transform*
(**define** poslist
  (**list** (**position** 17 -22 0) (**position** 16.5 -16 0)(**position** 17 -12 0) (**position** 21 -4 0)
    (**position** 23.5 1 0)(**position** 24 6 0)(**position** 23 11 0)(**position** 21 16 0)
    (**position** 17 21 0)(**position** 12 24.5 0)(**position** 7 26 0)(**position** 0 26.5 0)
    (**position** -6 26 0)(**position** -12 25 0)(**position** -16 23 0)(**position** -19 21 0)
    (**position** -20 16 0)(**position** -19.5 11 0)(**position** -19.5 6.5 0)(**position** -21 4 0)

---

[7]Used in *Global and Local Deformations of Solid Primitives* A.H.Barr, Comp Graphics, Vol 18, No 3, July 84.

(**position** -23 2 0)(**position** -25 0.5 0)(**position** -26.3 -1.5 0)(**position** -25.5 -3 0)
(**position** -24 -4 0)(**position** -24.5 -6 0)(**position** -25.5 -9.5 0)(**position** -24 -11 0)
(**position** -25.2 -11.5 0)(**position** -25 -13 0)(**position** -24 -14.2 0)(**position** -26 -18 0)
(**position** -23.5 -22.5 0)(**position** -17 -22.5 0)(**position** -8 -20 0)(**position** -8 -31 0)
(**position** 17 -22 0)))

(**define** head (**edge:spline** poslist))
(**define** head_bod (**wire-body** head))

; Create shear law transform, k = -0.1
(**define** age_law (**law** "vec((x+y*tan(-0.1)),y,z)"))
; Apply it
(**law:warp** head_bod age_law)
; Differentially scale the resulting head, X=0.9x
(**entity:transform** head_bod (**transform:scaling** 0.9 1 1))

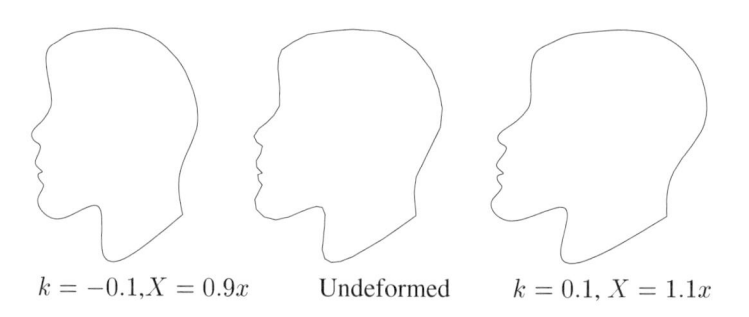

$k = -0.1, X = 0.9x$ Undeformed $k = 0.1,\ X = 1.1x$

Figure 10.11:   Non-uniform scaling of head profiles.

The face of a baby with its exaggerated cranium and diminutive face is changed radially by growth. However, because the face grows more rapidly than the cranium not only the size of the head but also its proportions change. The **head.scm** example simulates some of the effects of growth using an x-axis strain implemented as a non-uniform scaling and a shear transform created by a warping law.

The x-axis deformation is caused by the growth of soft integuments which strain the bone tissue in the horizontal direction. The slope of a face has been known to be an important identifying characteristic since the 1940's. Interestingly, as a person grows from infancy to adulthood the profile of the face changes its angle obliquely towards being more prognathic (i.e. a shear transform). The example takes the profile of a ten year old boy and then (for effect) exaggerates both the aging effects (younger on the left, old on the right)[8].

The shear transform uses the following deformation:

$$X = x + y * \tan \theta$$

---

[8]Aging Faces as Visual-Elastic Events, Pittenger & Shaw, Experimental Psychology, 1975, Vol 1:4.

10.9 Space Warping and Scaling

$$Y = y$$
$$Z = z$$

10.9.1 A Tapering Law

Tapering along the z axis is not really much different from scaling since it incrementally changes the x and y components while leaving z unchanged. Consequently a tapering law could be written simply as:

$$r = f(z)$$
$$X = rx$$
$$Y = ry$$
$$Z = z$$

The above equations are used to create a warping law in the example **taper.scm**.

taper.scm

```
; This program creates a global taper along the z-axis

(define c1 (solid:cylinder (position 0 0 40) (position 0 0 -40) 10))

(define p1 (solid:prism 40 20 50 9))

(solid:unite c1 p1)

(law:warp c1 (law "vec(0.05*z*x,0.05*z*y,z)"))
```

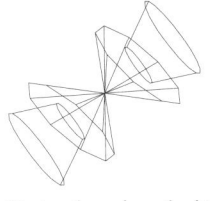

Tapering warps have practical applications in the design of moulded components where a small taper (or draft angle) is used to ensure the parts can be removed easily from the die.

Note the singularity at z = 0 !

10.9.2 A Twisting Law

The global twisting of an object about the z axis can be simulated by rotating the x and y coordinates as a function of height.

The warping law falls neatly out of the standard transform for rotation about the z-axis:

$$\begin{bmatrix} X \\ Y \\ Z \end{bmatrix} = \begin{bmatrix} \cos\theta & \sin\theta & 0 \\ -\sin\theta & \cos\theta & 0 \\ 0 & 0 & 1 \end{bmatrix} \begin{bmatrix} x \\ y \\ z \end{bmatrix}$$

Hence the warping equations are:

$$\theta = f(z)$$
$$X = x\cos(\theta) - y\sin(\theta)$$
$$Y = x\sin(\theta) + y\cos(\theta)$$
$$Z = z$$

twist.scm

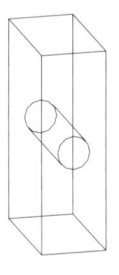

```
; Global twist about the z axis
; First make a block with a hole in it
(define s-beam (solid:block (position -10 -20 -30) (position 10 20 30)))
(define hole (solid:cylinder (position 0 -40 0) (position 0 40 0) 5))
(solid:subtract s-beam hole)

(solid:massprop s-beam) ; 44858.4

(define twist_z_law (law "vec(x*cos(z*(pi/60))-y*sin(z*(pi/60)),
                              x*sin(z*(pi/60))+y*cos(z*(pi/60)),z)"))

(law:warp s-beam twist_z_law)

(solid:massprop s-beam)  ; 44858.4
```

In the example $\theta = f(z) = z\frac{\pi}{60}$ so the object will twist through π radians every 60 modelling units. Note that as the theory predicts (see page 271) the volume of the solid is unchanged by the transform. Twisting about the z-axis is of course a special case but the same approach could be used to define many other general twisting warps. For example, knowing that the transform for rotation by θ radians about the unit vector (c_x, c_y, c_z) is:

$$\begin{bmatrix} c_x^2 + (1-c_x^2)c\theta & c_x c_y(1-c\theta) - c_z s\theta & c_x c_z(1-c\theta) - c_y s\theta \\ c_x c_y(1-c\theta) + c_z s\theta & c_y^2 + (1-c_y^2)c\theta & c_y c_z(1-c\theta) - c_x s\theta \\ c_x c_z(1-c\theta) - c_y s\theta & c_y c_z(1-c\theta) + c_x s\theta & c_z^2 + (1-c_z^2)c\theta \end{bmatrix}$$

Where $c\theta = \cos\theta$ and $s\theta = \sin\theta$. A warp which twists about any axis could be produced.

10.9 Space Warping and Scaling

> The way in which a deforming transformation locally affects a volume can be determined by the Jacobian matrix J of a transform function F. The Jacobian matrix defines the ratio by which (x, y, z) coordinates are changed by a deforming transform. So if a warping function is thought of as a transform between the original coordinates (x, y, z) and the final ones (X, Y, Z) the process can be described as :
>
> $$X = F_x(x)$$
> $$Y = F_y(y)$$
> $$Z = F_z(z)$$
>
> And the Jacobian can be written as:
>
> $$J = \begin{bmatrix} \frac{dF_x}{dx} & \frac{dF_y}{dx} & \frac{dF_z}{dx} \\ \frac{dF_x}{dy} & \frac{dF_y}{dy} & \frac{dF_z}{dy} \\ \frac{dF_x}{dz} & \frac{dF_y}{dz} & \frac{dF_z}{dz} \end{bmatrix}$$
>
> The determinant of the Jacobian determines the volume ratio between an infinitesemal deformed region and the corresponding undeformed region. For example, consider the twisting deformation:
>
> $$\theta = f(z)$$
> $$X = x\cos(\theta) - y\sin(\theta)$$
> $$Y = x\sin(\theta) + y\cos(\theta)$$
> $$Z = z$$
>
> Then the Jacobian for the function can be written as:
>
> $$J = \begin{bmatrix} \cos\theta & -\sin\theta & -x\sin\theta f'(z) - y\cos\theta f'(z) \\ \sin\theta & \cos\theta & x\cos\theta f'(z) - y\sin\theta f'(z) \\ 0 & 0 & 1 \end{bmatrix}$$
>
> And hence, everywhere:
>
> $$\det J = \cos^2\theta + \sin^2\theta = 1$$
>
> Thus such a twisting transform preserves the volume of the original solid.

About Warping and Volume

10.10 Representation and Analysis of Graphs

Mathematical graph theory provides ways of representing and reasoning about relationships between all manner of things. Composed only of vertices and edges[9] graphs model arbitrary networks of connections simply by defining which edges connect which vertices. This simple structure has proved to be astonishingly flexible and applications for graph theory have been found in such diverse situations as the layout of floor plans to the analysis of electrical circuits. In each case graphs have been able to analyze the essential structure of a set of relationships. Nowhere is this seen more clearly than in geometric modelling where the great value of graphs arises from their ability to represent the connectivity between geometric entities.

Although the ACIS libraries provide classes (i.e. **generic_graph**, **gvertex** and **gedge**) for creating and analyzing arbitrary graphs, API routines are provided to generate VERTEX-EDGE, FACE-EDGE and CELL-adjacency graphs automatically. The following sections discuss what each of these graphs represents before presenting two programs that demonstrate how graphs can be created and analyzed in both C++ and the Scheme AIDE.

10.10.1 VERTEX-EDGE Graphs

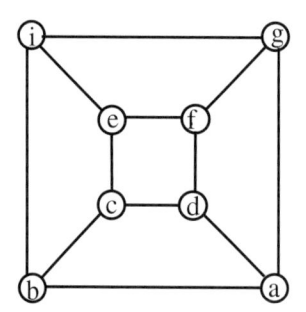

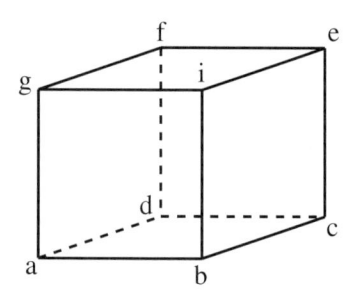

Figure 10.12: VERTEX-EDGE graph of a cube.

The function **api_create_graph_from_edges** generates a VERTEX-EDGE graph from a list of EDGEs. In a VERTEX-EDGE graph there is a one to one mapping between the VERTEXs of the model and the vertices of the graph (Figure 10.12). Likewise the EDGEs of the model and the edges of the graph.

Notice that in Figure 10.12 each vertex is connected to three others and the resulting graph is planar, in other words it can be drawn on a plane without any of its edges crossing. Intriguingly VERTEX-EDGE graphs associated with objects con-

---

[9]Note that the vertices and edges of graphs are written in lowercase while the uppercase VERTEXs and EDGEs refer to ENTITYs in ACIS models.

10.10 Representation and Analysis of Graphs

taining holes are always non-planar. In other words they can never be drawn on a flat surface without their edges crossing at some point.

10.10.2 FACE-EDGE Graphs

The function **api_create_graph_from_faces** supports the generation of FACE-EDGE graphs, in which every vertex represents a FACE, and each edge an EDGE, in the model. Thus for the FACE-EDGE graph of the cube, shown in Figure 10.13, each FACE is connected by four edges to other vertices in the graph.

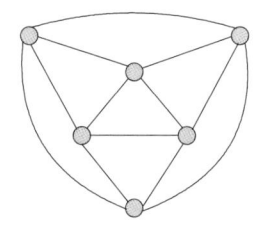

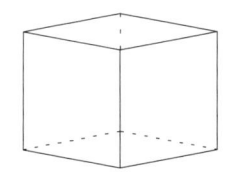

Figure 10.13: FACE-EDGE graph of a cube.

Interestingly planar polyhedra FACE-EDGE graphs are *duals* of their associated VERTEX-EDGE graphs. This relationship is clearly seen in drawings of VERTEX-EDGE graphs if one notes that each FACE on a BODY can be associated with a **region**, or area, bounded by the graph's edges[10]. Because regions are adjacent at the edges of the graph it is not too surprising that these two graphs are closely related.

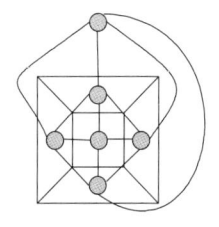

Planar VERTEX-EDGE and FACE-EDGE graphs are duals

10.10.3 CELL-Adjacency Graphs

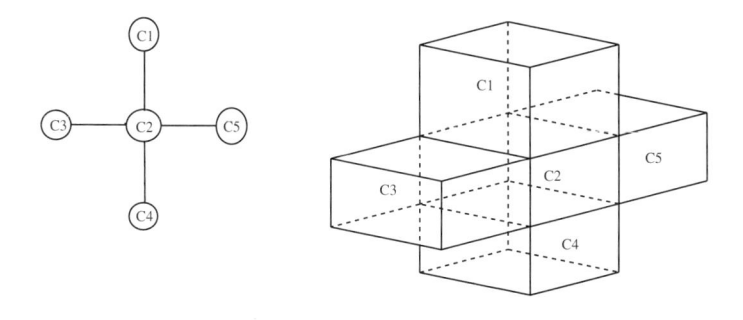

Figure 10.14: CELL-adjacency graph of a nonmanifold BODY.

---

[10]To ensure a one-to-one correspondence between faces on the polyhedron and regions in the graph, one has to associate the initial face with the infinite exterior region.

The nonmanifold cellular structure described in Chapter 1 (page 14) can also be described by a graph. In the CELL-Adjacency graph (CAG), generated by the function **api_create_graph_from_cells**, each vertex represents a CELL and each edge records the fact that a common boundary ENTITY is shared by two CELLs. Because CELLs can be either 2D or 3D, adjacency relationships between them can be formed by either EDGEs or FACEs. Figure 10.14 illustrates the CAG generated by the nonmanifold union of two blocks.

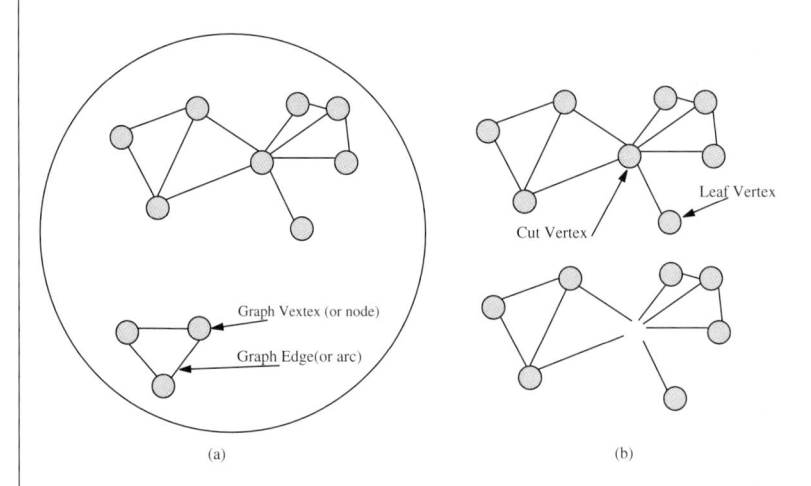

About Graph Theory

> Graphs have a rich taxonomy of classifications which describe their connectivity. The terms which appear most frequently in the generic_graph class can be defined as followed:
>
> (figure showing graph examples labeled (a) and (b), with annotations: Graph Vertex (or node), Graph Edge (or arc), Cut Vertex, Leaf Vertex)
>
> - A connected graph is a graph in which a path (or chain) of edges and vertices exist between every pair of vertices in the graph. A connected graph is said to have only one component. A disconnected graph is one that is not connected and so splits into several distinct components.
> - A vertex whose removal will cause a graph to become disconnected is called a cut-vertex (because removing it would cause the graph to fall apart).
> - A leaf-vertex is connected to only one edge.
> - A cycle is a closed sequence of edges.
> - A tree is a graph with no cycles.

The program **cutvertex.cxx** demonstrates how the functionality of the **generic_graph** class can be used to support an elegant form of feature recognition invented by some Italian mathematicians in the late 1980s. They noted that

10.10 Representation and Analysis of Graphs

DP-features (i.e. depressions or protrusions) which emanated from a single FACE, were signalled by the presence of a cut-vertex in the object's FACE-EDGE graph[11]. Consequently removal of such cut-vertices in a BODY's FACE-EDGE graph causes the boundaries of singularly connected DP-Features to separate from the FACEs on which they are located. Although this method allows the sub-graphs associated with DP-features to be easily identified further tests are required to determine if a particular network of FACEs and EDGEs represents the boundary of a protrusion or a depression.

The program **cutvertex.cxx** uses this method to locate the FACEs and EDGEs associated with a square depression on a simple block. The program starts by making a test object and then calls **api_create_graph_from_faces** to generate a **generic_graph** object that represents the BODY's FACE-EDGE graph. The generic_graph's member functions **cut_vertices** and **get_entities_from_vertex** are then used to locate the FACE, cf, which connects the depression to the BODY.

Removal of cf with **api_unhook_face** causes the FACE-EDGE graph associated with the BODY to disconnect into two components.

cutvertex.cxx

```
// This program finds the FACEs of the Depression/Protrusion feature
// associated with the cut-vertex in a FACE-EDGE Graph
#include <iostream.h>
#include "constrct/kernapi/api/cstrapi.hxx"   // Declares constructor APIs
#include "kernel/kernapi/api/kernapi.hxx"     // Declares kernel APIs
#include "kernel/kerndata/top/alltop.hxx"     // Declares topology classes
#include "kernel/kerndata/lists/lists.hxx"    // Declares ENTITY_LIST class
#include "boolean/kernapi/api/boolapi.hxx"    // Declares boolean api
#include "base/baseutil/vector/position.hxx"  // Declares position class
#include "base/baseutil/vector/vector.hxx"    // Declares vector class
#include "base/baseutil/vector/transf.hxx"    // Declares the transform class
#include "sbool/kernapi/api/sboolapi.hxx"     // Declares graph APIs
#include "kernel/kernutil/law/generic_graph.hxx" //Declares graph class

void main()
{
    api_start_modeller(0);
    api_initialize_booleans();

    // Make test component (block with square hole)
    BODY *block, *hole;
    api_make_cuboid(70,70,70,block);
    api_make_cuboid(20,10,50,hole);
    transf moveZ = translate_transf(vector(0,0,30));
    api_apply_transf(hole,moveZ);
```

---

[11] Leila De Floriani, *Feature Extraction from boundary models of three dimensions*, IEEE Transactions on Pattern Analysis and Machine Intelligence, 1989.

```
            api_subtract(hole, block);

            ENTITY_LIST faces;  // Get list block's FACEs
            api_get_faces(block,faces);

            // FACEs of the model represent nodes, or vertices (ie, gvertexs),
            // of the graph. EDGEs of the model represent the arcs of the graph
            generic_graph *fag1, *fag2;
            api_create_graph_from_faces(faces,fag1);

            // Output characteristics of graph
            cout << "Components = " << fag1-> components() << endl;
            cout << "Edges = " << fag1-> number_of_edges() << endl;
            cout << "Vertices = " << fag1-> number_of_vertices() << endl;

            // Create a new graph structure containing all the cut vertices in "fag1"
            generic_graph *cut_v = fag1-> cut_vertices();
            cout << "Number of Cut Vertices = "
                 << cut_v-> number_of_vertices() << endl;

            ENTITY_LIST cutList;

            // Fill "cutLists" with all entities associated
            // with all gvertexes of the graph "cut_v"
            cut_v-> get_entities_from_vertex(cutList);

            FACE *cf = (FACE*)cutList[0];
            BODY *cut_face;

            //Remove the cut vertex FACE from the block BODY
            api_unhook_face(cf,cut_face);

            faces.clear(); // Clear ENTITY_LIST
            api_get_faces(block,faces);

            // Make second FACE-EDGE Graph
            api_create_graph_from_faces(faces,fag);

            cout << "After Unhook Components = " << fag2-> components() << endl;
            cout << "Edges = " << fag2-> number_of_edges() << endl;
            cout << "Vertices = " << fag2-> number_of_vertices() << endl;

            api_terminate_booleans();
            api_stop_modeller();
        }
```

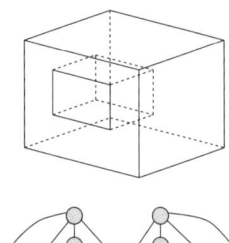

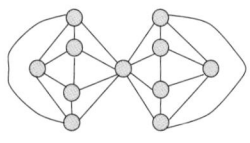

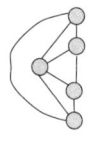

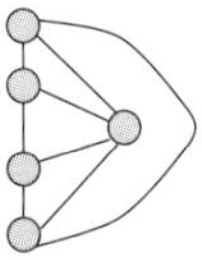

10.10 Representation and Analysis of Graphs

When executed the program prints out the following:

```
Components = 1
Edges = 24
Vertices = 11
Number of Cut Vertices = 1
After Unhook Components = 2
Edges = 16
Vertices = 10
```

An alternative, and less destructive, way of removing the cut-vertex would be to subtract **cut_v** from **fag1**. Although this operation would cause **fag1** to disconnect the change would be limited to the generic_graph's data-structure and so leave the model unchanged.

Graphs can also be generated from the cellular representation of nonmanifold topology. The next example, **cell_graphs.scm**, illustrates this facility.

cell_graph.scm

```
; This program creates a mixed dimension BODY, adds
; cellular topology data to the model and then generates
; a graph of the cells adjacency relationships.
; Lastly sub-graphs of 2D and 3D cells are extracted and
; examined.

; Make test part
(define b1 (solid:block (position -40 0 0) (position 40 10 10)))
(define b2 (solid:block (position 30 -10 -10) (position 45 20 20)))
(define face1 (face:planar-disk (position -10 0 0) (gvector 1 0 0) 20))
(define sheet1 (sheet:face face1))
(define sheet1 (sheet:2d sheet1))
(define face2 (face:planar-disk (position -20 15 0) (gvector 0 1 0) 30))
(define sheet2 (sheet:face face2))
(define sheet2 (sheet:2d sheet2))
(define e (bool:nonreg-unite b1 b2 sheet1 sheet2))

; Attach cellular data
(cell:attach e)

; Generate Cell Adjacency Graph (CAG)
 (define g (graph (entity:cells e)))

; Extract subgraphs of 2D CAG
(define g_2d (graph:subgraph_2dcell g))

; Extract subgraph of 3D CAG
(define g_3d (graph:subgraph_3dcell g))
```

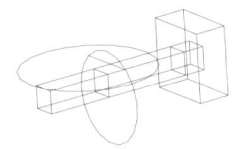

```
; Change display to see graph contents
(entity:set-highlight (graph:entities g_2d) #t)
(env:set-highlight-color 4)
(entity:set-highlight (graph:entities g_3d) #t)
```

The structure of the two subgraphs can be examined by means of their external representation (see page 52) in which edges between vertices are indicated by a dash (note that entity numbers vary from session to session).

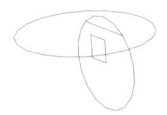

```
acis> g_2d
#[graph "(entity 58 1)-(entity 59 1) (entity 58 1)-(entity 60 1) (entity 59 1)-(entity 60 1)"]
```

Three 2D CELLs, one of which contains a square hole.

The graph of 2D CELL adjacencies forms a cycle which arises as a consequence of their common EDGE.

```
acis>g_3d
#[graph "(entity 54 1)-(entity 55 1) (entity 54 1)-(entity 57 1) (entity 56 1)-(entity 57 1)"]
```

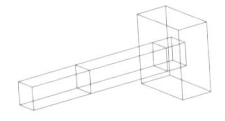

The graph of 3D CELL adjacencies forms a linear graph (i.e. no branches or cycles). Notice that although the CELL embedded in the large cuboid shares three FACEs with its neighbors only one edge appears in the graph. In other word multiple edges, that define the same adjacency relationship, are removed from the CAG.

Four 3D Cells

10.11 Exercises

1. The Klein bottle[12] is formed by taking a cylinder and twisting one end round so that it passes through its own wall. The resulting shape is a self intersecting surface which has no inside. It has a parametric equation of the form:

$$0 \leq u \leq 2\pi$$

$$0 \leq v \leq 2\pi$$

$$r = 4(1 - \frac{\cos(u)}{2})$$

---
[12]Created by Felix Klein in 1882.

10.11 Exercises

$$x = \begin{cases} 0 \leq u < \pi & 6\cos(u)(1+\sin(u)) + r\cos(u)\cos(v) \\ \pi < u \leq 2\pi & 6\cos(u)(1+\sin(u)) + r\cos(v+\pi) \end{cases}$$

$$y = \begin{cases} 0 \leq u < \pi & 16\sin(u) + r\sin(u)\cos(v) \\ \pi < u \leq 2\pi & 16\sin(u) \end{cases}$$

$$z = r\sin(v)$$

(a) Using the scheme extension **face:law** create a model of a Klein Bottle.

(b) Rotate the model around and explain the curious rendering artifact visible from some angles.

2. The following equations represent a constant radius bend around the x-axis. The object bends at a rate of k radians per model unit.

$$X = x$$

$$Y = -\sin(ky)\left(z - \frac{1}{k}\right)$$

$$Z = \cos(ky)\left(z - \frac{1}{k}\right)$$

Using the scheme extension **law:warp** create a law which bends a cylinder ($r = 10$, $l = 80$) around the x-axis at a rate of $k = 0.01$ radians per model unit.

Model Modification 11

"Your design is great, but could you ..."

- Change these holes to take M12 bolts instead of M14?
- Round off these edges to make it easier to assemble?
- Change the taper so we can press it from this new material?

Every engineer knows that the 'devil is in the detail' and a long, iterative, process of design modification is usually required before everyone, from manufacture to marketing, is happy with a new design. One inevitable consequence of this process is that 3D models frequently require small changes to their shape.

This chapter explores some of the operations available within ACIS to support the editing and modifications of 3D models. Although the functionality is spread across several different components (i.e. Local Operations, Blending, Shelling and Healing) the common use of surface-surface intersection operations provide an underlying theme to which the text repeatedly returns. For example:

- Local Operations (such as *tweak* and *move*) **intersect** a FACE's underlying surface (i.e. its geometry) and with the surfaces of adjacent FACEs to define its EDGEs.

- Shelling **intersects** the offset surfaces of a BODY to determine the EDGEs or a shrunken, of inflated, copy of the original BODY. A Boolean subtraction operation between the offset and original BODYs can then be used to create a thin-walled solid.

- Blending translates FACEs and **intersects** adjacent geometry to create the *spine* curve central to blend construction.

- Healing uses geometry repair functions which, in certain cases, **intersect** FACEs to define the underlying curve of their shared (i.e. adjacent) EDGEs.

The chapter starts by presenting a C++ program that, in a naïve way, carries out a tweak in a manner similar to the Local Operations (LOP) component. The more robust functionality of the LOP, Shelling and Blending components are then demonstrated using the Scheme interface. The chapter ends with a brief look at how both the LOP and Blending components supply concepts that are exploited by the Healing component. The chapter also discusses the numerous options associated with these components and tabulates both their names and default values.

> There are no known references to the word *tweak* (to pitch and twist sharply, pull with sharp jerk) prior to its appearance in Shakespeare's Hamlet

11.1 Simple Tweaking Procedure

One of the key observations underlying many of the model modification API's is that if a FACE has been slightly changed (i.e. transformed) in some way, the EDGEs which bound it can be recalculated by:

1. Intersecting the new (i.e. transformed) surface with the surface of **each** adjacent FACE: remember that the surface is the underlying, unbounded geometry on which the FACE sits (Figure 11.1(a)) - even if the existing FACE does not extend far enough to perform the intersection, the surface (probably) does.

   This process will define the curves underlying the EDGEs which bound the "new" geometry of the FACE.

2. Intersecting each new curve with its neighbors will determine the bounding VERTEXs of the updated EDGEs.

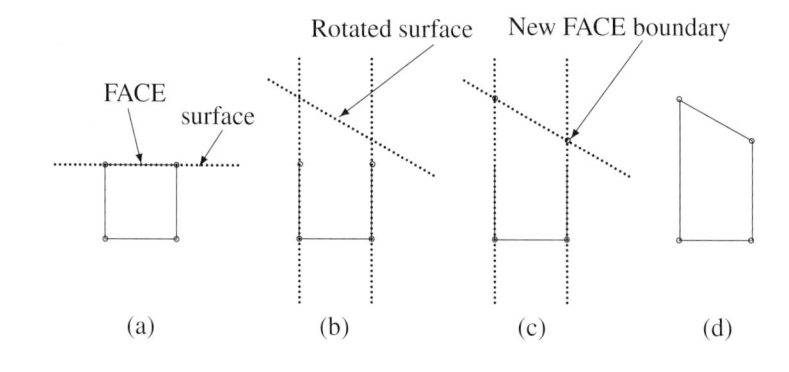

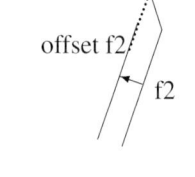

> Offset surfaces frequently intersect when faces do not.

Figure 11.1: A simple tweak.

SURFACES are used in preference to FACEs because there are many situations in which the original FACEs will no longer intersect after offsetting. Consider, for example, a move operation applied to one of the FACEs which shares an obtuse EDGE: the margin picture illustrates the problem.

This process of movement and re-intersection is illustrated in Figure 11.1 by the $45°$ rotation of a cube's FACE. Transforming the FACE's underlying surface

11.1 Simple Tweaking Procedure

is easily done but the original EDGEs and VERTEXs no longer bound its surface (Figure 11.1(b)). To re-calculate the boundary, the rotated surface is intersected (or tweaked[1]) with the surfaces of its adjacent FACEs (Figure 11.1(c)). The new geometry calculated from the intersections can be used to update the original data-structure to produce an object which is topologically identical to the original, but has different geometry (Figure 11.1(d)).

Implementation of this approach to tweaking is helped by the separation of topology from geometry in the ACIS data-structure. This enables a FACE's underlying geometry to be changed while the record of which EDGEs bound it, and so which FACEs are adjacent to it, is maintained until their own geometry can be updated.

Of course, in general, it is not really as simple as Figure 11.1 suggests as re-calculation of a FACE's EDGEs must ensure that:

- All the newly calculated geometry is correctly *orientated* relative to the existing data-structure.

- Multiple solutions to intersection calculations are correctly handled.

- Degenerate cases (e.g. where FACEs or EDGEs collapse to zero or LUMPs divide) are correctly handled.

Before looking at some examples of these difficulties, the program **diy_tweak.cxx** demonstrates how the Figure 11.1 tweak can be implemented using only non-component APIs and the direct interface. In order to limit the size of the program it is assumed that surface-surface, and curve-curve, intersections return only a single curve, or point, respectively.

diy_tweak.cxx

```
// This program creates a block and then rotates the surface of one face
// by 25 degrees. The surface is intersected with the surfaces of the
// adjacent faces to recalculate the edges. Finally the edge curves
// are intersected to find the new vertex coordinates
#include <stdio.h>                              // Declares File I/O functions
#include "constrct/kernapi/api/cstrapi.hxx"     // Constructor API's
#include "kernel/kernapi/api/kernapi.hxx"       // Kernel API's
#include "kernel/kerndata/top/alltop.hxx"       // TOPOLOGY classes
#include "kernel/kerndata/geom/point.hxx"       // Point Class
#include "kernel/kerndata/geom/allsurf.hxx"     // SURFACE classes
#include "kernel/kerndata/geom/allcurve.hxx"    // GEOMETRY classes
#include "kernel/kerndata/lists/lists.hxx"      // ENTITY_LIST class
#include "kernel/kerngeom/surface/allsfdef.hxx" // non-ENTITY surface classes
#include "kernel/kerngeom/curve/allcudef.hxx"   // non-ENTITY curve classes
#include "baseutil/vector/position.hxx"         // position class
```

---

[1] Elsewhere in solid modeling literature the word *tweaked* is associated purely with the transformation of FACEs. In ACIS, however, it is synonymous with transformation and intersection operations.

```cpp
#include "baseutil/vector/unitvec.hxx"      // unit_vector class
#include "baseutil/vector/vector.hxx"       // vector class
#include "baseutil/vector/transf.hxx"       // transform class
#include "kernel/kernint/intsfsf/sfsfint.hxx"   // surface/surface intersect classes
#include "kernel/kernint/intcucu/intcucu.hxx"   // curve-curve intersection record
#include "intersct/kernint/intsfsf/intsfsf.hxx"  // surface/surface intersectors
#include "intersct/kernint/intsfsf/gensfsf.hxx"  // int_surf_surf class
#include "intersct/kerndata/makeint/intsect.hxx" // face-face intersection record
#include "intersct/kernapi/api/intrapi.hxx"  // Intersector API
#include "kernel/kerndata/savres/fileinfo.hxx"  // Declares fileinfo class

void save_ent(char*, ENTITY_LIST&); // see page 72
logical same_sense(EDGE*); // Test if an EDGE and its curve have the same sense

void main()
{
        api_start_modeller(0);
        api_initialize_constructors();

        BODY* block;
        api_make_cuboid(10,5,20,block);

        // Get top FACE to tweak
        FACE *ff;
        api_find_face(block,unit_vector(0,0,1), ff);

        PLANE *fs = (PLANE*)ff->geometry();

        // Rotate its surface (assume sense not changed)
        *fs *= rotate_transf(M_PI/4,vector(1,0,0));

        // get the lower case surface
        const surface& sfs=fs->equation();

        // Update the curves of all EDGEs adjacent to the FACE
        // by intersecting the surfaces

        for(LOOP *lp = ff->loop(); lp != NULL; lp = lp->next())
        { // In this case only one loop exists
          COEDGE *ce = lp->start();
          do {
            // cast to loop because the owner could be a wire
            LOOP *adjloop = (LOOP*)ce->partner()->owner();

            const surface& adjfs= adjloop->face()->geometry()->equation();

            surf_surf_int *ints = int_surf_surf(sfs,NULL,
                                        *( const transf * )NULL,
                                        adjfs,NULL,
                                        *( const transf * )NULL);
```

11.1 Simple Tweaking Procedure

```
            // Assuming only one straight intersection curve
            STRAIGHT *newc = new STRAIGHT((const straight&)(*(ints->cur));
            ce-> edge()-> set_geometry((CURVE*)newc);
            ce = ce-> next();
        } while (ce != lp-> start());
    } //end for each loop on ff

    // Intersecting EDGE curves to update VERTEXs
    for(lp = ff-> loop(); lp != NULL; lp = lp-> next()) {
        COEDGE *ce = lp-> start();
        do {
            curve_curve_int *results = NULL;

            api_intersect_curves(ce-> edge(),ce-> next()-> edge(),FALSE,results);

            APOINT *apt = new APOINT(results-> int_point);

            // update VERTEX coordinates
            if (ce-> sense() == FORWARD)
                ce-> edge()-> end()-> set_geometry(apt);
            else
                ce-> edge()-> start()-> set_geometry(apt);

            // update EDGE sense
            if (same_sense(ce-> edge())) // flip sense
                ce-> edge()-> set_sense(FORWARD);
            else
                ce-> edge()-> set_sense(REVERSED);

            ce = ce-> next();
        } while (ce != lp-> start());
    } //end for each loop on ff

    ENTITY_LIST slist;
    slist.add(block);
    save_ent("tweak.sat", slist);
    api_terminate_constructors();
    api_stop_modeller();
}

// check that EDGE is heading in the same direction as its curve
logical same_sense(EDGE *e)
{
```

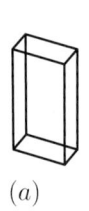

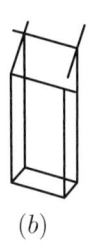

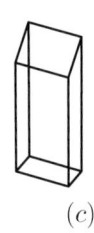

```
        position p1 = e−> start()−> geometry()−> coords();
        position p2 = e−> end()−> geometry()−> coords();
        unit_vector uv0 = normalise(p2 - p1);
        unit_vector uv1 = e−> geometry()−> equation().point_direction(p1);

        if (uv0 % uv1 >= 0)  // Scalar product of the two vectors
                    return TRUE;
        else
                    return FALSE;
}
```

(a)　　　　　　　　　(b)　　　　　　　　　(c)

Figure 11.2:　Tweaking the top of a block.

The **diy-tweak** code starts by obtaining a pointer (**ff**) to a planar FACE whose normal is aligned with the x-axis. A pointer to the FACE's geometry (**fs**) is extracted from the data-structure and the infix operator *= is used to apply a rotation transform to it. Once moved, the lower case (i.e. non-ENTITY) geometry class is obtained (**sfs**) and the process of recalculating the EDGEs and VERTEXs started. The routines **int_surf_surf** and **api_intersect_curves** are use to calculate the new geometry of these ENTITYs.

Lastly the ENTITY member function **set_geometry** is used to update the geometry of both EDGEs and VERTEXs. Although it is possible to directly insert new points, or curves, into a model's data-structure via the direct interface, such an approach would not update the dependent geometry of other ENTITYs. For example the resulting object would have EDGEs with the correct geometry but incorrect boundaries (e.g. incorrect start and end parameters, as in Figure 11.2(b)).

In contrast, the **set_geometry** method used to update the VERTEX coordinates ensures that the parameters of all EDGEs that use it are correctly updated. Similar problems arise if the **set_sense** function is by-passed using the direct interface.

The **diy-tweak.cxx** example is a special case because the procedure as presented will only work for small changes to convex polyhedra. The most obvious limitation is the assumption that each intersection operation will result in a single ENTITY. In practice however this is the exception rather than the rule. For example, consider a system that performs a 2D tweak on the straight line segment of the truncated-circle shown in Figure 11.3. After translating the line a small distance to the left, in order

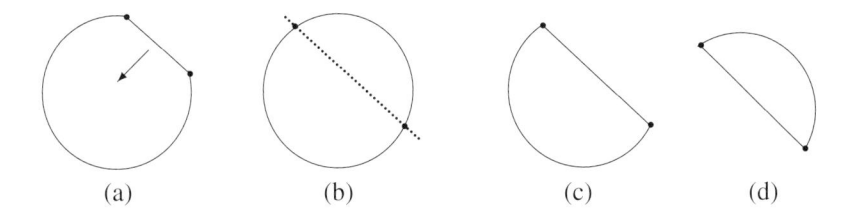

Figure 11.3: Multiple Solutions, or chiralities, in a 2D tweak.

to modify the surrounding geometry the intersection of the new position of the line with the circle needs to be found.

The intersection operation returns two points which can be used equally well to bound a large or a small arc of the circle. Because these and many other types of multiple solution can be viewed as arising from a choice of *turning* left, or right, (i.e. at the end of the straight line segment) they are referred to as **chiralities**. The problem is not insurmountable and various geometric tests can be used to automatically determine the correct solution (for instance one could test if the new arc contains points that also lie on the original arc). However these tests all add complexity to the code and the infinite variety of 3D geometry continues to present challenges to the writers of tweaking procedures. For instance, the reader might like to consider how the correct intersection curves could be determined for the object shown in the margin (assume the planar disk on top of the cylinder is moved upwards). In this case both of the adjacent cylindrical FACEs produce the same circular curve when intersected with the tweaked FACE.

Fortunately many of these difficulties are hidden from the user by the default behaviour of the LOP component. For example the component will attempt to ensure that EDGEs before and after tweaking retain the same vexity[2]. This, and other, heuristics resolve many ambiguities (including the one in the margin) but do not always produce the answer users want. In these cases the default behaviour can be changed, or over-ridden, by the user.

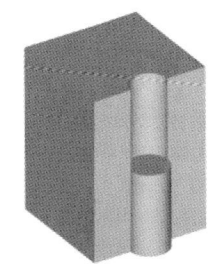

Intersecting the top of the cylinder with its adjacent geometry produces two identical curves

11.2 Local Operations

Having established the feasibility of defining a modified surface's new boundary (by intersecting it with the surrounding geometry) it is possible to envisage how the functions of the LOP component are built on this principle. Let us start by reviewing the four generic types of local operation:

1. **Tweak**: This is the key function of the LOP component which calculates the geometry of each bounding EDGE and VERTEX by re-intersection of the ad-

---

[2]Convex or concave.

jacent geometry. In addition to handling a number of special cases, the orientation of the new EDGEs must be carefully considered in order to maintain the validity of the tweaked solid. The tweak API requires the user to supply the new surface as an argument. The other local operations call tweak internally after generating a new surface and so are generally much more useful than the 'stand-alone' tweak functions.

2. **Move**: This function allows a transform to be applied to one, or more, FACEs. After calculating each FACE's new (i.e. transformed) surface the routine calls *tweak* to determine the resulting EDGEs.

3. **Offset**: Unlike the 'move' function which applies a general transform this function first offsets a FACE's surface and then calls *tweak* to determine the new boundaries (i.e. EDGEs).

4. **Taper**: This function is a specialized form of *move* which allows each line on a ruled surface to be rotated about a different axis. Hence a circular cylinder can be tapered to a cone by specifying a draft angle and a point on its axis. There are three distinct types of taper:

    - Plane taper : Rotates each FACE about its intersection with a given taper plane.
    - Edge taper : Rotates each FACE about a given EDGE.

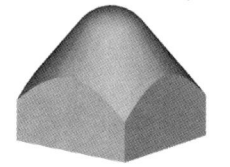

5. **Shadow taper**: Adds new FACEs, or modifies existing FACES, to remove regions where a given FACE is considered to be in the shadow. The shadow is specified by the user specifying the draft direction and angle. Once again *tweak* is used to calculate the EDGEs bounding the new FACEs.

The following sections illustrate several of these operations and some of the classic special, or degenerate, cases that are supported by the LOP component.

11.2.1 Creation and Deletion of ENTITYs

The program **move_sep.scm** illustrates two interesting behaviors which occur when a spherical FACE is subjected to a series of move operations that progressively separate it from an embedded cube. This process causes the circular EDGE which arises from the intersection of sphere and plane to reduce in diameter and ultimately collapse to a point before the LOOP disappears altogether.

This process results in the creation of a new LUMP ENTITY when the spherical FACE ultimately separates from the cube (Figure 11.4(d)).

11.2 Local Operations

move_sep.scm

```
; A block with a spherical face is created, located (with a
; mouse pick) and the local operation move-faces used to transform its position
(define ball (solid:sphere (position 0 0 20) 10))
(define block (solid:block (position -10 -10 0) (position 10 10 20)))
(define object (solid:unite ball block))   ; Create test part
; Select spherical face, with left mouse click in view window
(define hemi (pick:face (read-event)))

(lop:move-faces hemi (gvector 0 0 2))  ; Move face 2 units along z axis
; Repeated calls to the move operation will eventually separate the sphere
```

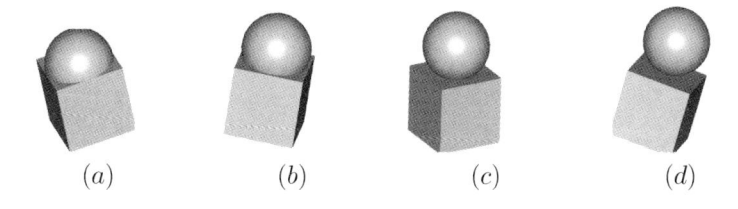

(a)   (b)   (c)   (d)

Figure 11.4:   The move FACE LOP.

This behaviour is easily demonstrated because the analytical surfaces underlying FACEs have mathematical definitions that are either infinite in extent (e.g. a plane) or closed (e.g a sphere) so their extension is easily calculated. Spline surfaces, however, are only defined over limited areas (i.e. in the region of the control points). Consequently, all local operations, apart from tweak, extend geometry as necessary in the following way:

- B-Spline curves are extended in the directions of their end tangents.

- Intersection curves are extended by further intersecting the surfaces (extending the surfaces if necessary).

- B-Spline surfaces are extended by ruled surfaces in the direction of their EDGE tangents.

Some restrictions apply to this process. For example, it is not possible to extend surfaces or curves through singularities.

When the FACEs being moved, tapered or offset by LOP functions are tangential to their neighbours (a situation is usually created by blends), surface extensions alone will not fill any gaps created (see Figure 11.10). However the component ensures that during local operations blend FACE tangency is preserved if the blend FACE has been identified by the addition of a blend attribute (added automatically if the option "add_bl_atts" was **on** when the blend was created).

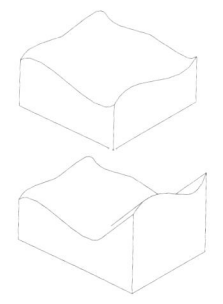

Spline surface extension to support a 'move' LOP

11.2.2 Self-intersections

Up to this point the examples of local operations have all moved FACEs cleanly away from the BODY into conveniently open space. However the LOP component has to anticipate situations where the movement of a FACE produces a self intersecting (ie. non-manifold) BODY.

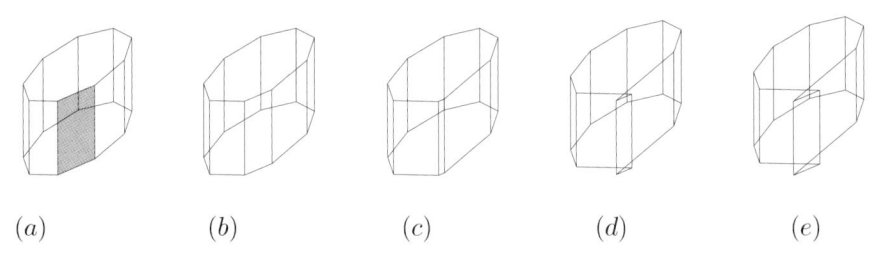

(a)   (b)   (c)   (d)   (e)

Figure 11.5:   Local Operation offsetting problem.

By default the user is forewarned about changes to a BODYs topology and any self-intersections are fixed by means of a Boolean operation which creates a valid manifold. Consequently the problem is largely invisible to LOP component users unless the default behaviour is changed. The program **lop_off.scm** does this using the **option:set** command.

Indeed the behaviour of all the local operations can be altered dramatically by changing the default settings of the options listed in Table 11.1 on page 291.

lop_off.scm

```
; This program offsets one face of a prism. Because the options "lop_repair_self_int"
; and "lop_check_invert" are switched off a large offset can result in a non-manifold
; body.

; Create test object
(define tp (solid:prism 40 20 50 9))

; Select side face
(define aface (list-ref (entity:faces tp) 2))

; Switch off checking for non-valid faces
(option:set "lop_check_invert" #f)

; Switch off automatic repair of self intersections
(option:set "lop_repair_self_int" #f)

; Offset face by 2 units along its face normal
(lop:offset-faces (list aface) 2)
```

11.2 Local Operations

Figure 11.5 illustrates the results of repeatedly offsetting the shaded FACE. If the option `lop_check_invert` is switched on then the program will generate an error message every time an invalid tweak is attempted (i.e. moving from Figure 11.5(c)-(d)) and will leave the BODY unchanged. However if `lop_check_invert` is off and `lop_repair_self_int` is on an invalid BODY is generated and then repaired by deleting the offset FACE and reintersecting its adjacent FACES.

Table 11.1 Local Operation options.

Option Name	Description
lop_check_invert	Controls the checking of invalid (i.e. inverted or inside-out) FACEs.
lop_fail_on_no_part_inv_sf	Controls whether or not a local operation fails when a FACE surface can only be partially offset.
lop_ff_int	Sets additional validity checking of the BODY (i.e. improper FACE/FACE intersections, SHELL and LUMP containment).
lop_ff_error_prevent_roll	Controls whether or not errors in FACE/FACE checks cause a roll back.
lop_merge_vertex	Sets merging of redundant VERTEXs.
lop_prefer_same_convexity_sol	Controls whether to pick a solution that most closely matches the original EDGE convexities, or that has the least common EDGE convexities.
lop_repair_self_int	Controls whether or not self-intersections are repaired on the bodies that result from local operations and shelling.
lop_sort_on_convexity	Controls whether sorting of solutions proceeds first on convexity and then distance, or first on distance and then on convexity.
lop_prefer_nearest_sol	Controls whether the nearest, or farthest, solution is preferred when picking between multiple solutions in the tweak algorithm (see below).
lop_use_euclidean_dist_score	Controls whether Euclidean or parametric distance is used when scoring different solutions.

11.2.3 Multiple Solutions

The intersections carried out by the tweak function may result in several possible solutions. However, once again, the default settings of the LOP component largely shield the user from the problems this causes. For example, if possible the component will automatically choose a solution that leaves the BODY locally valid and **with the same EDGE convexity** as the original. In the next two examples this default behaviour is switched off to allow this behaviour of the tweak function to become visible.

Unlike the previous programs the next example does not explicitly move any surfaces, but instead it creates new boundaries (ie. EDGEs) for an existing surface by re-intersecting it with adjacent geometry and choosing different solutions to those used initially.

tweak_pos.scm

```
; This program carries out a classic tweak by re-intersecting a spherical
; face on a block so that a second solution is adopted.
; Subsequent to this several planar faces are also tweaked.

; Make a block
(define block (solid:block (position -10 -10 -10) (position 10 10 10)))

; Make a sphere with the radius of the block's diagonal length
(define ball (solid:sphere (position 0 0 10) (sqrt 200)))

; Make the test object
(define bod (solid:subtract block ball)))

; Get top face of the ball
(define b (pick:face (ray (position 0 0 -9) (gvector 0 0 1))))

; Set option to false to ensure a different convexity solution
; is generated.
(option:set "lop_prefer_same_convexity_sol" #f)

(lop:tweak-faces b b #f) ;Change convex to concave
; The boolean argument indicates if the sense
; (i.e. orientation) of the tweak FACE should be reversed

; Now get different solutions for each planar side face
;Tweak one of the planar side FACEs

(define s1 (pick:face (ray (position 0 0 -8) (gvector 0 1 0))))
(lop:tweak-faces s1 s1 #t)

;Do the same to the rest
(define s2 (pick:face (ray (position 0 0 -8) (gvector 0 -1 0))))
```

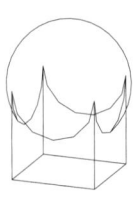

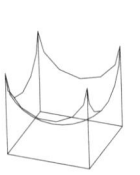

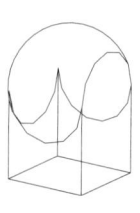

11.2 Local Operations

(**lop:tweak-faces** s2 s2   #t)

(**define** s3 (**pick:face** (**ray** (**position** 0 0 -8) (**gvector** 1 0 0))))
(**lop:tweak-faces** s3 s3   #t)

(**define** s4 (**pick:face** (**ray** (**position** 0 0 -8) (**gvector** -1 0 0))))
(**lop:tweak-faces** s4 s4   #t)

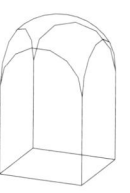

As the geometry of the objects involved in the tweak become more complex the notion of what constitutes a "correct" solution becomes increasingly hard to define. Consequently facilities exist within the LOP component to generate a list of possible solutions from which the user can choose. The next example, **tweak_list.scm**, illustrates this behaviour by tweaking the planar FACE at the end of an open trefoil to create four curves that result from its intersection with the surface of the knot.

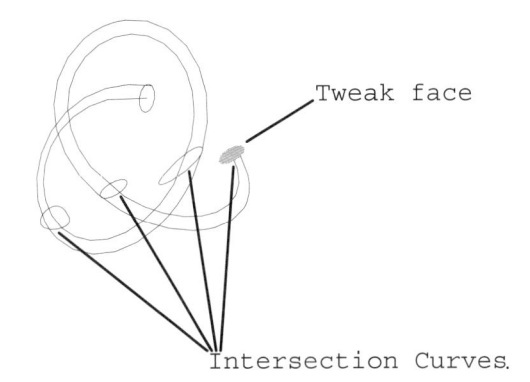

Figure 11.6:   Tweaking choices.

Once again this behaviour is only visible if the default behaviour of the component is changed so that the closest solution is not automatically selected.

tweak_list.scm

```
; This program uses the laws system to create an open
; trefoil knot. One of the planar end faces is then
; tweaked. A list of possible solutions is generated
; and one chosen before finishing the tweak.

(option:set "sil" #f) ;Make display easier for the laws surface

; Switch off LOP component defaults
(option:set "lop_prefer_same_convexity_sol" #f)
(option:set "lop_prefer_nearest_sol" #f)
```

```
(option:set "lop_sort_on_convexity" #f)

; Define trefoil knot law
(define sknot-x
  (law "4.1*cos(t)-1.8*sin(t)-8.3*cos(2*t)-8.3sin(2*t)-1.1*cos(3*t)+2.7*sin(3*t)"))
(define sknot-y
  (law "3.6*cos(t)+2.7*sin(t)-11.3*cos(2*t)+3.0*sin(2*t)+1.1*cos(3*t)-2.7*sin(3*t)"))
(define sknot-z
  (law "4.5*sin(t)-3.0*cos(2*t) + 11.3*sin(2*t)-1.1*cos(3*t)+2.7*sin(3*t)"))
(define tknot (law "vec(law1,law2,law3)" sknot-x sknot-y sknot-z))

; 0 to 2PI creates a closed knot, but for this example
; we want an open one so use parameter range 0 to 1.8PI.
(define tknotcurve (edge:law tknot  0 (* 1.8 PI)))

; Get position and tangent vector at t=0 on knot curve
(define start_pt (curve:eval-pos tknotcurve 0))
(define start_tan (curve:eval-tan tknotcurve 0))

; Create planar face
(define pdisk (face:planar-disk start_pt start_tan 2.5))

; Sweep plane along the laws curve to create knot shape
(define trefoil (sweep:law pdisk tknotcurve))

; Get a list of faces on the knot
(define face_list (entity:faces trefoil))

;Get one of the planar end faces
(define tface (list-ref face_list 1))

;Check that it is the right one
(face:planar? tface)

; First step in tweaking.
; Attaches attributes to edges involved in tweaking the given face.
; Returns a list of those edges with more than one possible solution.

(define edge (car (lop:tweak-faces-init tface tface #f)))

; Display the possible EDGE solutions (view:dl window only)

(define solution_list (lop:tweak-query-edge-solutions edge))

; Choose one solution from the list
(define solgeom (list-ref solution_list 2))

; Choose an edge solution for a subsequent tweak to use.
(lop:tweak-pick-edge-solution edge solgeom)
```

11.3 Offsetting and Shelling

; Finish the tweak using the chosen solution
(**lop:tweak-faces** tface tface #t)

The program starts by switching off the defaults:

- lop_prefer_same_convexity_sol
- lop_prefer_nearest_sol
- lop_sort_on_convexity

The Laws system is then used to create an EDGE with the parametric equation for a trefoil knot[3].

The output of the laws expression is a single EDGE, so a planar disk is then swept along it to form a solid BODY. When one of the knot's planar FACEs is tweaked with the laws surface four intersection curves are generated (Figure 11.6) and the user is able to select which of these curves the system should use to complete the operations. The program for doing this has four distinct steps:

1. **lop:tweak-faces-init**: One of the end FACEs, tface, is extracted from the model and used as the first and second[4] arguments for the function **lop:tweak-faces-init**. **lop:tweak-faces-init** attaches an attribute to the edge.

2. **lop:tweak-query-edge-solutions**: Displays all the potential solutions in a **view:dl** window (Figure 11.6(b)) and returns references to them in a list called solution_list.

3. **lop:tweak-pick-edge-solution**: The program selects solution number 2 from the list.

4. **lop:tweak-faces**: Completes the tweak by creating a new "end" for the knot at the location of intersection curve number 2 (see margin).

11.3 Offsetting and Shelling

The manufacturing process of blow moulding inflates a tube of hot plastic inside a cavity until it takes up the shape of the surrounding die. The resulting object is a thin-walled container used to hold anything from shampoo to soft-drinks. So big is the business of designing and manufacturing these objects that it has a solid modelling operation all of its own to address its unique requirements.

Because no one over the age of three ever looks inside a shampoo bottle, all the design effort goes into the outside of the bottle. Consequently the containers are

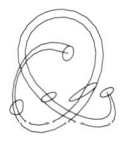

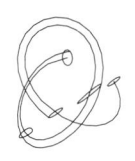

---

[3]From Aaron Trautwein's PhD Thesis "Harmonic Knots" (University of Iowa, 1995) which contains parametric equations for all prime knots through eight crossings.
[4]The function's second argument also facilitates the substitution of new geometry into an existing FACE. So, for example, a planar FACE could be changed into a spherical one.

designed on CAD systems as solids that are only hollowed out when the appearance has been finalized and detailed analysis (e.g. cooling rates and stiffnesses etc.) of the mould cavity begins. The process of automatically hollowing out the solid is called shelling and is built, largely, on the offset function of the LOP component.

By moving every FACE on a BODY in the direction of its surface normal an object can be uniformly enlarged (positive offset) or reduced (negative offset).

Once offset inwards the resulting BODY can be subtracted from the original to create the thin-walled solid required to represent blow moulded products. The principle challenge in implementing offsetting routines arises from the need to handle topology changes similar to those seen in other local operations.

The program **lop_offset.scm** demonstrates how the LOP component handles the self-intersections which frequently occur during the offsetting of BODYs containing concave features (e.g. slots and holes).

lop_offset.scm

```
; This program demonstrates how the offsetting of bodies requires the
; removal of faces. First a slotted block with a hole is made, this is
; then offset outward causing the "features" to be removed. A final
: negative offset is used to return the body to its original size.

; Make the test object
(define block (solid:block (position -25 -15 -10) (position 25 15 10)))
(define slot (solid:block (position -5 -18 -6) (position 5 18 12)))
(define hole (solid:cylinder (position 12 7 0) (position 12 7 12) 4))
(define bod (solid:subtract block hole slot)))

; Set Local Ops options
(option:set "lop_repair_self_int" #t)
(option:set "lop_merge_vertex" #t)

; Offset all FACEs outwards
(lop:offset-body bod 3)

; Offset all FACEs outwards
(lop:offset-body bod 2)

; Offset all FACEs inwards 5 units
(lop:offset-body bod -5)
```

Once again the demonstration of this behaviour requires a particular setting of the LOP component options. In this case the operation would fail to collapse the slot unless **lop_merge_vertex** and **lop_repair_self_int** are both turned **on**.

Note that both positive and negative offsets are used first to swell the object and then to return it to its original dimensions.

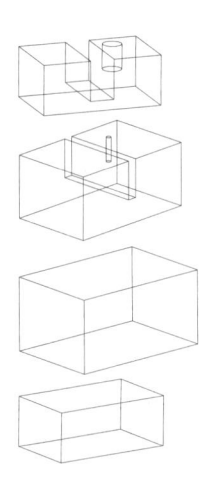

11.4 Blending

The **Shelling** component uses offset BODYs to create thin-walled, hollow solids (see Figure 11.7).

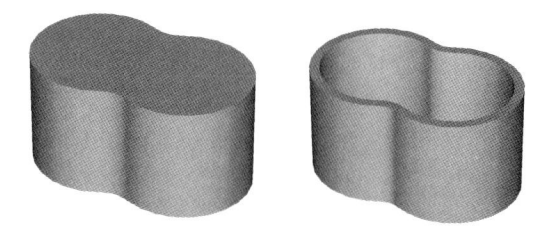

Figure 11.7: Shelling applied to a solid BODY.

11.4 Blending

Blends are small FACEs that create smooth transitions between larger FACEs. This role dictates that they must be added to a shape after the larger surfaces have been defined (i.e. once there are areas to provide transitions between).

Since the interface between FACEs is synonymous with EDGEs it is not surprising that blending is frequently seen as removal of sharp corners (i.e. EDGEs). However as the later examples in this section will demonstrate, many non-EDGE blends are also possible. The section starts by outlining the basic approach used to create EDGE blends and emphasizes the many challenges presented by the approach. Several Scheme programs are then presented which illustrate the facilities ACIS provides for creating blends.

11.4.1 Blending Terminology

The key concept underlying the approach adopted by ACIS to blending is that blend FACEs are created as separate ENTITYs (2D sheets) which are incorporated into the BODY by means of a Boolean operation. Although this approach demands the code used to construct the blend is complex it has two important advantages:

1. A blend FACE can be treated exactly like any other FACE. Consequently blends on blends, or Booleans on blends, are possible.

2. Local changes to the boundary of a blend FACE are relatively easy to implement. In other words the boundaries of blends can be made to "flow" around obstructions (such as holes) close to the EDGE being blended (see margin).

Pivotal to the creation of blend sheets (almost irrespective of their complexity) are the concepts of **spring** and **spine** curves. Consider the blend illustrated in Figure 11.8 where a single EDGE on a cube is blended with a constant radius.

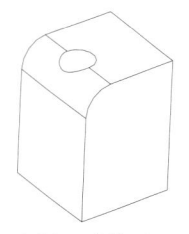

A blend that "overflows" a hole

Spine curves represent the blend's skeleton and provides valuable reference points when the ends of the blend are being constructed. Again they are easily generated by intersection of the surfaces of the lateral FACEs after they have **both** been offset inwards by the radius of the blend.

The spine curve is used to determine the type of surface the blend will require: a straight spine curve implies a cylindrical surface, whereas a circular spine indicates that a toroidial surface is required.

The two spring curves define the geometry of the EDGEs between the blend and the lateral FACEs on either side of it. By definition the EDGEs formed by spring curves are smooth (i.e. the blend FACE meets the lateral FACE tangentially along a spring curve).

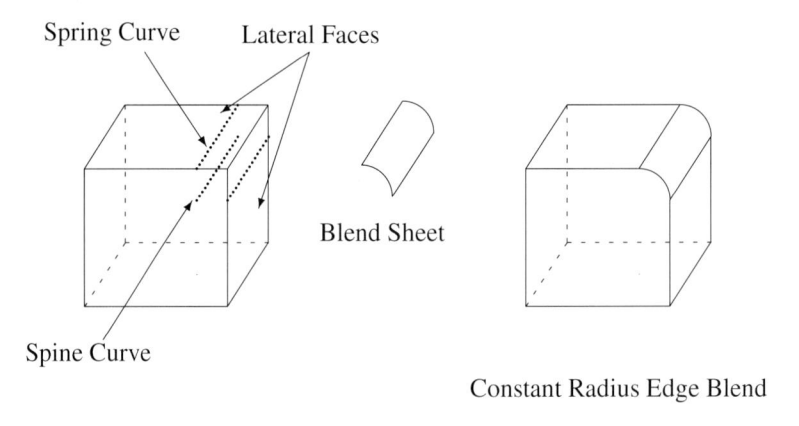

Figure 11.8: Blend terminology.

Internally spring curves are generated by projection of the spine curve onto the lateral FACEs. In simple cases the end points of the spring EDGEs can be found by intersecting the spring curves with the terminal FACEs at either end of the blend. These intersection points, together with the capping curves, allow the creation of the four EDGEs needed to bound the FACE that defines the blend.

After the blend sheet FACE has been created it is added to the BODY using a standard Boolean operation.

11.4 Blending

Table 11.2 Blending options.

Option Name	Description	Default
abl_remote_ints	Sets the use of the global Boolean for ENTITY-ENTITY blending.	F
abl_require_on_support	Sets whether both sides of a blend must be on their supports.	T
blend_make_simple	Sets whether or not blends are simplified, if possible.	T
bl_cap_box_factor	Sets the size of the box used by the capping algorithm.	2
bl_envelope_surf	Controls the type of blend surface used when a variable-radius blend is created.	F
bl_preview_approx_sf	Sets whether blends are previewed using a spline approximation.	T
bl_preview_tightness	Sets how tight the blend preview is to the true blend.	8
bl_remote_ints	Sets whether or not global interference checking is performed.	F
cap_preference	Determines whether capping tries to use as many or as few capping FACEs as possible.	"old"
vbl_check	Sets whether or not further checks are done on VERTEX blend surfaces created by blending.	F
vbl_quick_check	Sets whether or not quick checks are done on VERTEX blends.	F
vtx_blnd_simple	Sets whether or not the new algorithm for VERTEX blend boundaries is used.	F
blend_mix_convexity	Controls how blending handles mixed convexty EDGEs.	T
force_capping	Determines whether capping will fail, or remove large parts of the blend in some special cases.	F
v_blend_nsurf	Enables or disables construction of general VERTEX blend surfaces.	T
v_blend_rb	Sets whether or not a VERTEX blend surface can be a rolling ball surface.	T

Many of the problems arising with tweaking reappear in blending when surfaces are intersected. For example, although the spring and spine curves are uniquely defined by a few parameters, determining exactly where a blend should start and finish is frequently problematic.

Many of the problems do not have a single solution. The "correct" behaviour is frequently determined by the wishes of the users rather than a mathematical principle.

Consequently the blending component supports a long list of ACIS options which can be used to customize the behaviour of the software (see Table 11.2).

11.5 Variable Radius Edge Blends

By changing the geometry of the spring curves and the blend surface a range of variable radius EDGE blends can be created. The example **edgeblends.scm** demonstrates several of the options available.

edgeblends.scm

```
; This program takes a loop of four EDGEs on a block and
; creates a different type of blend on each.

; Create a solid block.
(define block (solid:block (position -20 -20 -20)   (position 20 20 20)))

; Get a list of all the LOOPs of EDGEs on the block
(define looplist (entity:loops block))

; Get the list of EDGEs in first loop in the list
(define edgelist (entity:edges (car looplist)))

; Define a name for each EDGE in the list
(define e0  (list-ref edgelist 0))
(define e1  (list-ref edgelist 1))
(define e2  (list-ref edgelist 2))
(define e3  (list-ref edgelist 3))

; Create a constant radius blend attribute
(define rad1 (abl:const-rad 5))

; Attach the blend attribute rad1 to EDGE e0
(abl:edge-blend e0 rad1)

; Create and attach a variable radius blend attribute to EDGE e1.
; The blend has a radius of 5 at the start and 3 at the end
(blend:var-rad-on-edge e1 5 3)

; Create a blend attribute on EDGE e2 whose radius is
```

Blends with variable radius (left) and constant radius (right)

Blends with elliptical cross-section (left) and tapered radius (right)

11.5 Variable Radius Edge Blends

```
; specified by a series of positions along the EDGE.

; Get the list of VERTEXs on e2
(define vert (entity:vertices e2))

; Call the first one in the list start
(define start (vertex:position (car vert)))

; Call the second one in the list end
(define end (vertex:position (car (cdr vert))))

; Create variable radius blend defined by position-radius pairs

(define rad2  (abl:pos-rad
                    (list start   ;Position list
                         (position:interpolate start end 0.2)
                         (position:interpolate start end 0.4)
                         (position:interpolate start end 0.7)
                         end)
                    (list 3 2 7 3 6) ;Radius list
                    e2)) ; EDGE on which blend lies
;Attach the blend attribute on the EDGE e2
(abl:edge-blend e2 rad2)

; Create a blend attribute with an elliptical cross section whose
; reference FACE is on the left (#f for right) relative to the EDGEs
; start and end.
; Length of the major and minor axis and the angle turned through are
; also given as arguments (the option also exists to specify different
; major, minor axis lengths at the end of the blend).
(define rad3 (abl:ell-rad #t 5 6 60))

; Attach the blend attribute on the EDGE e3
(abl:edge-blend e3 rad3)

; Get a list of all EDGEs and/or VERTEXs that make-up the
; blend network by following EDGEs marked as blended from e2.
(define blnet (blend:get-network e2))

; Create the blends
(blend:network blnet)
```

Each of the four EDGE blends is created in the same way. First, a blend attribute is created which holds the parameters of the blend. This attribute is then attached (or added) to an EDGE.

The function **blend:get-network** returns a list which represents a network of ENTITYs whose blends will interact. To be part of a blend network, an EDGE or

VERTEX must be part of a solid BODY, have an attached blend attribute and one of the following:

- Meets an EDGE smoothly (i.e. with tangent continuity) at either a blended or an unblended VERTEX.

- Meets an EDGE or several EDGEs at a blended VERTEX.

- Meets no more than one other blended EDGE at an unblended VERTEX, (this is known as a blend pair).

This network is contained in a list returned by **blend:get-network** (including the ENTITY given as a seed). The list is empty if the given ENTITY has no blend attribute.

The actual formation of the blends occurs on the last line of the program when the network is passed to the function **blend:network**.

11.6 Vertex Blends

The procedure for blending individual EDGEs is not too hard to envisage. However the challenges become a magnitude greater when smooth transitions between blended EDGEs are also required. In the previous example the EDGE blends are simply intersected to create a "mitre".

However designers frequently require blends to run smoothly into each other. In ACIS this sort of transition is created by the specification of VERTEX blends.

v_blend.scm

;*This program blends three EDGEs and a VERTEX on prism*

; *Create the test part*
(**define** tp (**solid:prism** 40 20 50 9))

; *Get a list of its EDGEs*
(**define** elist (**entity:edges** tp))

; *Choose EDGE at random*
(**define** e1 (**list-ref** elist 6))

; *Get a list of the VERTEXs associated with that EDGE*
(**define** vlist (**entity:vertices** e1))

; *Choose VERTEX at random*
(**define** v1 (**list-ref** vlist 0))

; *Get a list of EDGEs adjacent to v1*
(**define** adj_edges (**entity:edges** v1))

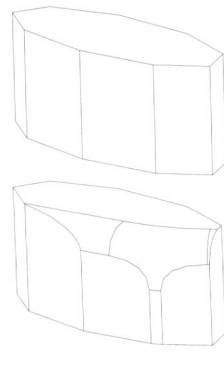

11.7 Entity-Entity Blends

```
; Put constant radius blend attributes on all the adjacent EDGEs
(blend:const-rad-on-edge adj_edges 10)

; Put a VERTEX blend attribute on v1 with setback 20 and bulge 1.5
(blend:on-vertex v1 20 1.5)

; Do the blending
(blend:network (blend:get-network e1))
```

The function **blend:on-vertex** that attaches a blend attribute onto a vertex has two key arguments which specify the shape of the resulting blend:

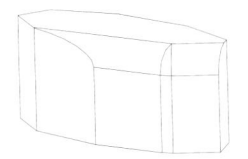

1. **Setback:** the distance at which an EDGE blend is setback from the blend VERTEX. If the setback is specified as **false** then any setback recorded in the EDGE blend attributes are used instead. If an argument of **true** is provided, then an "auto-setback" operation is performed whereby a setback is calculated for each blended EDGE incident with the blended VERTEX on the basis of the local geometry surrounding the vertex and the size of the blend that is applied to the edge. The calculated setback is used in preference to any setbacks associated with the blended EDGE.

Vertex blend produced by "Auto-Setback"

2. **Bulge:** has a value between 0 and 2. This controls the fullness of the vertex blend. If no value is given a default value of 1.0 is used.

11.7 Entity-Entity Blends

EDGEs and VERTEXs are not the only ENTITYs which can be blended. In fact smooth transition surfaces can be generated between any pair of ENTITYs.

The ENTITY-ENTITY blending functionality extends the basic blending facilities by providing API level support for blends which switch between FACE–FACE, EDGE–FACE, FACE–EDGE, EDGE–EDGE and even VERTEX–FACE as they are propagated through a model. For example, Figure 11.9 shows a blend which "rolls" along a path requiring several different types of blend to be created.

The example **ent_ent_bln.scm** demonstrates this facility by creating a blend between two FACEs which have no EDGE in common.

ent_ent_bln.scm

```
; This programs demonstrates the blending of two non-adjacent FACEs
; by creating an ENTITY-ENTITY blend.

; Create part
(define block (solid:block (position -30 -30 -5) (position 30 30 5)))
```

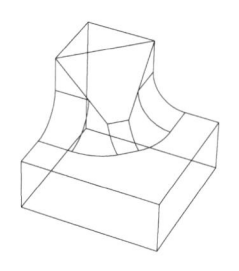

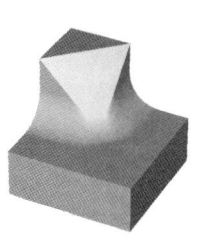

Figure 11.9: Blend rolling round a VERTEX.

```
(define wall (solid:block (position -30 -10 4) (position 30 2 10)))
(define cyl (solid:cylinder (position 0 -15 0) (position 0 50 60) 6))
(solid:unite block cyl wall)

; Pick the two FACEs involved in the blend

; Pick the top FACE of the base block
(define face1 (pick:face (ray (position 0 15 0)  (gvector 0 0 1)) 1))
(entity:set-color face1 4)  ; color it blue

; Pick the cylindrical FACE
(define face2 (pick:face (ray
(position:interpolate (position 0 -15 0) (position 0 50 60)  0.5)
(gvector 0 0 -1)) 1))

(entity:set-color face2 5)  ; color it yellow

; Create a constant radius blend attribute (i.e. vradius object)
(define rad1 (abl:const-rad 5))

; Create an ENTITY-ENTITY blend attribute
; Arguments are the two ENTITYs and the vradius object,
; a position designating a rough starting point for the blend,
; Lastly a convexity flag is given (#t for convex or #f for concave).
(abl:ent-ent-blend face1 face2 rad1 (position 0 15 10) #f)

; Display position of point in dl:view
(dl-item:point (position 0 15 10) )

; Create the blend
(blend:network block)
```

11.8 Healing

The Internet has made the electronic exchange of CAD data between manufacturers using the **same** CAD systems commonplace. However, despite the existence of well established neutral standards for 2D drawings, and 3D models, data exchange between **different** CAD systems is still an error prone task.

The difficulties arise in two ways. Firstly, many of the standard languages for representing 3D CAD data make no attempt to define a solid. IGES for example unambiguously defines the geometry of individual FACEs but does not record which EDGEs are shared by which pairs of FACEs.

Consequently, duplication of ENTITY definitions (i.e. VERTEXs and EDGEs) is inherent in the file format and the first task of any receiving CAD system is the removal of duplicates and the stitching (or re-assembly) of the model.

It is during the re-creation of the adjacency relationships between ENTITYs that the second problem appears. For example, every open EDGE has a start and end VERTEX which lies on its curve. Whether a point lies on a line is not a calculation which has an absolute answer, but rather it depends on the tolerance used to determine if two points are the same. A similar problem can occur when determining if a given EDGE lies on a given SURFACE.

Unfortunately there is a large (around 10^4) variation in the tolerances used by commercial CAD systems. Consequently models which have consistent geometry and topology when exported at one level of precision may lose it when imported into a system that uses a higher precision.

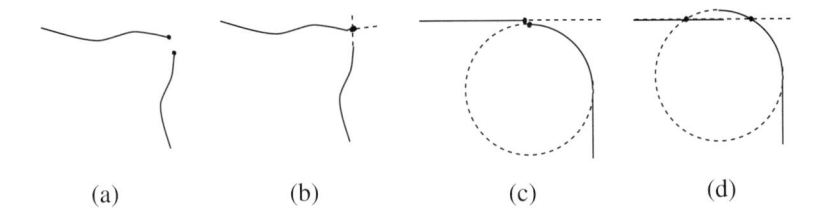

Figure 11.10: The Healing Conundrum.

At first glance the problem looks no harder than tweaking. If VERTEXs do not lie on EDGEs, or EDGEs on FACEs (Figure 11.10(a)), then they could be recalculated by means of the surface/surface intersection operations (Figure 11.10(b)) seen so often in this chapter.

Figure 11.10(c) illustrates a limitation of this approach. Surfaces that meet tangentially at one tolerance may fail to intersect (Figure 11.10(c)) or may intersect in more than one curve (Figure 11.10(d)) when a smaller tolerance is in force. There are two general approaches to dealing with such translation errors:

- Change the data-structure in such a way that the gaps are **healed**.

- Change the modelling algorithms so that they **tolerate** small gaps (i.e. tolerant modeling).

Neither approach is easy, or totally effective. For instance some errors can be fixed by translation of a surface so that it becomes coincident with a blend (see margin). However, movement of geometry may close gaps in one part of the model only to open them elsewhere.

11.8.1 The Healing component

The healing process can be envisaged as a series of stages that progressively improve the consistency of a model's geometry and topology.

1. Initialization: The input model is analysed and attributes containing tolerance values of the various healing processes added to the model.

2. Preprocessing: Initial clean-up by removal of duplicate VERTEXs, zero length EDGEs and zero area FACEs.

3. Geometry Simplification: Conversion of spline surfaces to analytic ones (ie, plane, cone, sphere etc) where possible.

4. Stitching: pairing of adjacent EDGEs and VERTEXs to create topology (ie, FACEs which share bounding EDGEs).

5. Geometry Repair: Adjusts ENTITYs so that geometry and topology (possibly also changed) are consistent. At the end of this phase every:

    - VERTEX lies on the underlying curve of the EDGE it bounds.
    - EDGE lies on two adjoining FACEs.
    - parametric curve lies on its corresponding FACE.
    - pair of EDGEs, if they meet at all, do so at VERTEXs.
    - pair of FACEs, if they meet at all, do so in EDGEs and/or VERTEXs.

6. Postprocessing: Check geometry for consistent orientation and remove redundant FACEs and VERTEXs.

7. Termination: Removal of Healing component attributes.

The last example of this chapter (**heal1.scm**) demonstrates the geometry repair facility (step 5). The program creates a topologically broken BODY by offsetting three FACEs of a blended "wiggle" BODY. Each of the unpicked FACEs creates a separate sheet BODY so the call **hh:combines** forms a single BODY comprising all the shape's constituent parts.

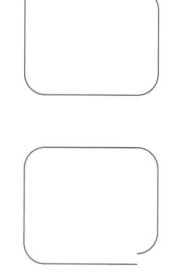

Healing by translation can create new problems elsewhere.

11.8 Healing

heal1.scm

```
; This program uses the healing component to repair blended "wiggle"
; whose FACEs have been unpicked and translated small amounts

;Make a "wiggle"
(define wig (solid:wiggle 50 50 20 1 -2 -2 -1))
(define elist (entity:edges wig))
(solid:blend-edges elist 5)
(solid:massprop wig 0) ;43527
(entity:box wig)

;Unhook and move the spline FACE 0.1 units
(define spline (pick:face (ray (position 0 0 0) (gvector 0 0 1)))))
(define sheet1 (face:unhook spline))
(entity:transform sheet1 (transform:translation (gvector 0 0 0.1)))

;Unhook and move one of the planar side FACE 0.05 units
(define side1 (pick:face (ray (position 0 0 0) (gvector 1 0 0)))))
(define sheet2 (face:unhook side1))
(entity:transform sheet2 (transform:translation (gvector 0.05 0 0)))

;Unhook and move one of the planar side FACE 0.01 units
(define side2 (pick:face (ray (position 0 0 0) (gvector 0 1 0)))))
(define sheet3 (face:unhook side2))
(entity:transform sheet3 (transform:translation (gvector 0 0.01 0)))

; Combines the sheet BODYs into a single BODY with 4 LUMPs
(define ill_body (hh:combine (list wig sheet1 sheet2 sheet3)))

; Attaches healing attributes, must be called before healing
(hh:init-body-for-healing ill_body)

; Cleans the model to be healed by removing thing such as
; zero-length EDGEs, sliver FACEs, and duplicate VERTEXs
(hh:preprocess ill_body)

; Direct all output to "healing_output.dbg"
(debug:file "healing_output")

; Geometry simplification by converting splines surface to analytic
; representations (i.e. planes, cylinders etc) where possible
(hh:simplify ill_body)

; Analyzes the BODY and sets values of required
; options and tolerances for stitching
(hh:analyze-stitch ill_body)

; Stitches FACEs into a single LUMP BODY where possible
```

(**hh:stitch** ill_body)

; *Performs all the geometry related healing operations,*
; *including fixing of EDGE geometries by intersections, snapping*
; *surfaces for fixing tangencies, and refitting spline surfaces*
(**hh:geombuild** ill_body)

; *Last step performs check, and fixs, properties such as*
; *negative FACE areas and duplicate VERTEXs*
(**hh:postprocess** ill_body)

; *Removes healing attributes*
(**hh:terminate-body-for-healing** ill_body)

(**debug:file** "stdout") ; *close file*

; *Check dimensions of the resulting BODY*
(**solid:massprop** ill_body 0);43897
(**entity:box** ill_body)

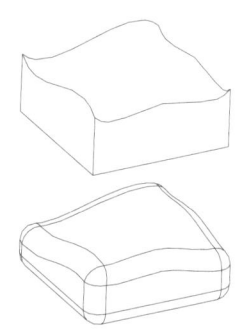

The healing process can be followed by examining the text written to the file "healing_output.dbg".

After initializing the BODY for healing first the report written to this file is generated by the call to **hh:simplify**:

```
SIMPLIFICATION RESULTS :
=========================
      Simplification Tolerance = 0.0001
      no. of initial splines = 5
      no. of final splines = 5

      no. of planes made = 0
      no. of cylinders made = 0
      no. of spheres made = 0
      no. of tori made = 0
      no. of cones made = 0
```

The next output is generated by **hh:analyze-stitch**:

```
STITCH ANALYSIS :
=================
      Input Body Statistics :
        0 Solids
        1 Sheets
        3 Free faces
```

Not to scale

11.8 Healing

```
    Min. Stitch tolerance = 1e-005
    Max. Stitch tolerance = 1
```

After **hh:analyze-stitch** has set the tolerance for all stitching the repair function **hh:stitch** is called and produces the following print-out verifying that the process (in this case) has been successful.

```
STITCH RESULTS :
================
    min_tol = 1e-005
    max_tol = 1
    no. solid lumps made = 1
    no. sheet lumps made = 0
    no. free faces remaining = 0
    no. unshared loops = 0
    no. unshared edges = 0
```

The extension **hh:geombuild** creates an interesting output which first reports the problems, then details the algorithms called to fix them and lastly summarizes the healed data-structure and any remaining problem:

```
GEOMBUILD ANALYSIS :
====================
    geom build tol = 10
    analytic solver tol = 0.1515
    no. of edges = 48
    no. of bad edges = 30
    no. of coedges = 96
    no. of bad coedges = 24
    no. of vertices = 24
    no. of bad vertices = 12
    no. of bad tangent edges = 30
    no. of bad tangent edges analytic = 16
    no. of G1 bad tangent edges analytic = 10
    no. of bad tangent edges uv_uv = 2
    no. of bad tangent edges boundary uv_uv = 2
    no. of bad tangent edges uv_nonuv = 11
    no. of bad tangent edges nonuv_nonuv = 1
    no. of bad tangent edges 3_4_sided = 12
    no. of surfaces = 26
    no. of discontinuous surfaces = 0
    percentage of good geom = 82

 GEOMBUILD CALCULATION RESULTS :
 ===============================

  Analytic Solver :
```

```
        3 degree of snapper graph
        32 analytic tangent junctions resolved
        0 analytic tangent junctions unresolved
        32 analytic intersections resolved
        0 analytic intersections unresolved
        24 vertices resolved (12 intersected, 12 projected)
        12 unstable vertices corrected
        0 vertices unresolved
        0 edges calculated by exact projections
        0 edges calculated by approx projections
        0 coincident snaps resolved

    Isospline Solver :
        3 isospline tangent junctions resolved
        1 isospline tangent junctions unresolved
        3 splines bent to vertices

    Sharp Edge Solver :
        0 sharp edges resolved
          0 intersected
          0 exact projections
          0 approx projections
        0 sharp edges unresolved

        10 vertices resolved
          0 intersected
          10 exact projections
          0 approx projections
        0 vertices unresolved

    Generic Spline Solver :
        4 4-sided patches made
        0 3-sided patches made
        0 failures
        0 unsolvable junctions

    Wrapup Module :
        16 pcurves computed
        0 edges trimmed


GEOMBUILD FIX RESULTS :
========================
Statistics of the healed body after geombuild fix :
        no. of edges = 48
        no. of bad edges = 0
```

```
no. of coedges = 96
no. of bad coedges = 0
no. of vertices = 24
no. of bad vertices = 0
no. of bad tangent edges = 0
no. of bad tangent edges analytic = 0
no. of G1 bad tangent edges analytic = 0
no. of bad tangent edges uv_uv = 0
no. of bad tangent edges boundary uv_uv = 0
no. of bad tangent edges uv_nonuv = 0
no. of bad tangent edges nonuv_nonuv = 0
no. of bad tangent edges 3_4_sided = 0
no. of surfaces = 26
no. of discontinuous surfaces = 0
percentage of good geom = 100
```

The calls to **solid:massprops** verify that the volume of the test object (incorporating three slightly offset FACEs) has increased by a small amount. Note that the healing process carried out in **heal1.scm** can be invoked with a single call to **hh:autoheal**.

11.9 Exercises

1. Create an ENTITY/ENTITY blend between two spheres.

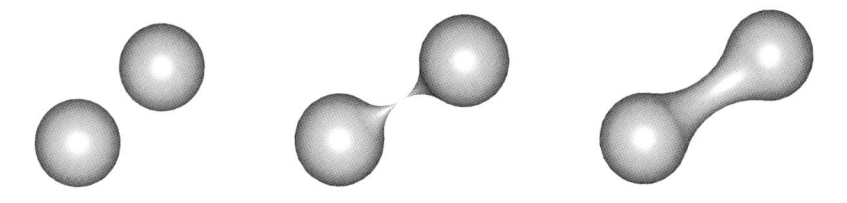

Figure 11.11:   Exercise 1: Blending two spheres.

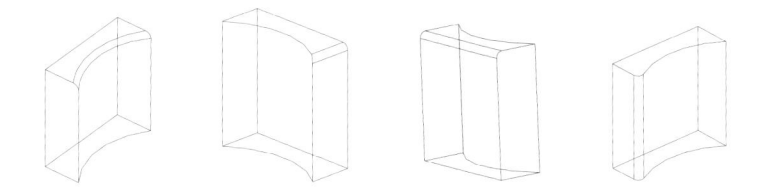

Figure 11.12:   Exercise 2: Blending a curved block.

2. Take a week off and write a program that can blend any EDGE of a curved block shown in Figure 11.12 with a constant radius blend.

Attributes   12

The need for geometric models to hold more in their data-structures than simply the geometry and topology of a shape has long been recognized. Such non-shape, or **attribute**, data ranges from values for properties such as density, color and surface finish (which can be associated with single ENTITYs) through to geometric constraints between two or more ENTITYs. This chapter explores the different ways in which ACIS supports the attachment of user-defined attributes (i.e. classes) to any element of the model data structure. ACIS's method of allowing users to add their own data to ENTITYs exploits C++'s inheritance mechanisms in two ways:

1. Every ACIS ENTITY has a pointer (inherited from the ENTITY base class, see Appendix B) to a list of ATTRIB class instances.
2. Any user-defined class, derived from ATTRIB, can be added to this list.

In other words inheritance is used to guarantee that every ACIS ENTITY has a hook on which data can be hung. Likewise, because all user-defined classes to be placed on this hook appear to the system as **descendants** of the ATTRIB class they are guaranteed to fit and chain together (i.e. form a linked list) with any other type of attribute. Because the ATTRIB class itself is derived from ENTITY[1] they can be debugged, saved, restored and rolled back like any other ENTITY.

For the most part the derivation of a new ATTRIB class follows an identical process to that used to define new ENTITYs (see Chapter 13). Like ENTITYs the user must define an **organization attribute class** as a first level of derivation from the ATTRIB base class. This provides a common label, or name, for all classes used by a given component, application or organization. Examples of this attribute naming scheme are found throughout the ACIS system; for instance all the system defined attributes attached during the boolean and blending processes are derived from the **ATTRIB_SYS** class. Other common labels (or sentinels as they are called) are TSL (for Three-Space Ltd) and ST (for Spatial Corp)[2].

Attribute types:
Simple: holds only alphanumeric data.
Complex: holds pointers to other ENTITYs.
Bridge: holds pointers to external data structures.

---
[1] So it is possible to give attributes attributes!
[2] A registry of attribute identifiers is maintained by Spatial Corp. To ensure no two developers choose the same sentinel writers of new attributes should contact Support@Spatial.com for a unique label.

Unlike other ENTITYs, however, the ATTRIB class supplies several virtual functions which define its behavior when the owning ENTITY is changed during a modeling operation (such as a Boolean unite).

These three virtual methods are:

Split which is invoked when an attribute's owning ENTITY is divided, or partitioned, in some way.

Merge which is invoked when the owning ENTITY combines with another of the same type.

Transform which is invoked when the owning ENTITY is moved, rotated or scaled.

In each case the default action is to do nothing. Rather than discussing the subtleties of these virtual methods the chapter begins by describing two types of "off the shelf" attributes which do not require the user to define any *new* classes. The first example demonstrates the simple labeling of ENTITYs with no defined split/merge/transform function. The second example extends the functionality of these simple attributes by using the "Generic Attribute" component to specify a "change" (i.e. split, merge or transform) behavior. The chapter ends with an example of a full ATTRIB definition involving five separate files.

12.1 Adding a Simple String Attribute

The general nature of the ATTRIB mechanism means that it can be tiresome to use for simple applications where one simply wishes to record, say, a text string, with a particular ENTITY. Because of this the Part Manager component provides a set of generic ATTRIB classes, with default behaviors, which can be used to create a number of simple attributes.

The program **nameattb.cxx** uses the function **add_named_attribute** to add a text string to the FACE of a block.

nameattb.cxx

```
// This program adds a named attribute to the third
// face of a block and then finds it again before saving it.
#include <stdio.h>                              // Input/Output functions.
#include "constrct/kernapi/api/cstrapi.hxx"     // Construction API's
#include "kernel/kernapi/api/kernapi.hxx"       // Declares kernel API's
#include "kernel/kerndata/top/alltop.hxx"       // Topological Classes
#include "ga_husk/pmhusk/nm_attr.hxx"           // Declares add_named_attrib function
#include "kernel/kerndata/lists/lists.hxx"      // Declares the ENTITY LIST class
#include "kernel/kerndata/savres/fileinfo.hxx"  // Declares fileinfo class
void save_ent(char* file_name, ENTITY *ent);    // Defined in save.cxx
```

12.1 Adding a Simple String Attribute

```
void main(){
    api_start_modeller(0);
    api_initialize_constructors();
    BODY* block;
    api_make_cuboid(100,150,200, block);
    FACE *ff = block->lump()->shell()->face();
    int count = 0;
    while(ff != NULL){
        count++;
         if (count == 3)
             add_named_attribute(ff,"the one");
         ff = ff->next();
     }
     ff = block->lump()->shell()->face();
     while(ff != NULL)
       {
         NAMED_ATTRIB *na =
             find_named_attribute(ff, "the one");
         if (na != NULL)
             printf("Found %s \n",na->get_name());
         ff = ff->next();
     } // Save ATTRIB to file
    save_ent("block.sat",block);

    api_terminate_constructors();
    api_stop_modeller();
}
```

Once compiled and run the following text is printed on the screen:

```
Found the one
```

The procedure **find_named_attribute** performs the same task as the more general **find_attrib** function (defined in **kerndata/attrib.hxx**) which searches an ENTITY's linked list of ATTRIB classes looking for a specific type. The following code would also return a pointer to the attribute added by the program:

```
// search if name is not known
    NAMED_STRING_ATTRIB *nna =
        (NAMED_STRING_ATTRIB*)find_attrib(
            ff,
            ATTRIB_ST_TYPE,
            NAMED_STRING_ATTRIB_TYPE);

   if(nna != NULL)
      printf("Found attribute with name %s \n" nna->get_name());
```

Unlike the more specific routines the generic **find_attrib** routine requires both the **organization** (ATTRIB_ST_TYPE) and the **application** specific class name (NAMED_STRING_ATTRIB_TYPE) to be given as arguments.

12.2 Using Generic Attributes

The attributes provided by the Part Management Husk have no specified behavior when their owner is split, merged or transformed (i.e. they do nothing). The Generic Attribute Husk allows users to specify this behavior without the trouble of writing a new ATTRIB class.

The component provides predefined ATTRIB classes for associating a number of simple data types (such as integers, doubles, pointers and strings) with an ENTITY. The action taken when the owning ENTITY is split/merged or transformed is specified by the arguments given to the attribute's constructor. Table 12.1 summarizes the options.

Table 12.1 Generic attribute options.

Action	Description
SplitLose	Lose the attribute
SplitKeep	Keep the attribute on the old ENTITY
SplitCopy	Copy the attribute to the new ENTITY
SplitCustom	Call application supplied routine for name
MergeLose	Lose the attribute
MergeKeepKept	Keep the attribute on the kept ENTITY
MergeKeepLost	Transfer from lost to kept. Lose original on kept
MergeKeepOne	Transfer from lost to kept if none on kept
MergeKeepAll	Transfer from lost to kept. Keep any on kept
MergeCustom	Call application supplied routine for name
TransLose	Lose the attribute
TransIgnore	Leave the attribute unchanged
TransApply	Apply transform (action depends on derived type)
TransCustom	Call application supplied routine for name

The program **genatt.cxx** attaches integer attributes to three blocks and gets the split and merge functions to print out messages when they are invoked.

12.2 Using Generic Attributes

genatt.cxx

```cpp
// This program uses a generic integer attribute to label three orthogonal blocks.
// The blocks are then united and subtracted to show the split/merge behavior.
#include <stdio.h>
#include "kernel/acis.hxx" // Define the precision of modeller data
#include "boolean/kernapi/api/boolapi.hxx" // Declares boolean API's
#include "constrct/kernapi/api/cstrapi.hxx" // Construction API's
#include "kernel/kernapi/api/kernapi.hxx" // Declares kernel API's
#include "ga_husk/attrib/at_int.hxx" // Declares Generic Attrib class
#include "kernel/kerndata/top/alltop.hxx" // Declare topology classes
#include "kernel/kerndata/lists/lists.hxx" // Declare ENTITY_LIST class
#include "ga_husk/api/ga_api.hxx" // Declares GA API's
#include "kernel/kerndata/savres/fileinfo.hxx" // Declares fileinfo class

// Declare Function Prototypes
void save_ent(char*, ENTITY*); //see save.cxx page 72
static void merge_mess(ATTRIB_GEN_NAME*, ENTITY*, logical);
static void split_mess(ATTRIB_GEN_NAME*, ENTITY*);

void main()
{
    // make the test parts
    api_start_modeller(0);
    api_initialize_generic_attributes();
    api_initialize_booleans();

    BODY *xblock, *yblock, *zblock; // Create 3 axis aligned blocks
    api_make_cuboid(100, 14, 14, xblock);
    api_make_cuboid(16, 100, 15, yblock);
    api_make_cuboid(20, 20, 100, zblock);

    // add the integer attributes to each body
    new ATTRIB_GEN_INTEGER(xblock,"sticky",0,SplitCopy, MergeKeepAll);
    new ATTRIB_GEN_INTEGER(xblock,"tellme",1,SplitCustom, MergeCustom);
    new ATTRIB_GEN_INTEGER(yblock,"tellme",2,SplitCustom, MergeCustom);
    new ATTRIB_GEN_INTEGER(zblock,"tellme",3,SplitCustom, MergeCustom);

    // register the custom methods
    set_split_method("tellme",split_mess);
    set_merge_method("tellme", merge_mess);

    // make square hole in yblock
    printf("Start subtract\n");
    api_subtract(xblock,yblock);
    printf("End subtract\n");

    // zblock added to form a cross
    printf("Start unite\n");
```

```
        api_unite(yblock, zblock);
        printf("End unite\n");
        save_ent("yplusz.sat", zblock);

        api_terminate_booleans();
        api_terminate_generic_attributes();
        api_stop_modeller();
}

// Output message when merged
static void merge_mess(ATTRIB_GEN_NAME *attr,
                ENTITY *other_ent, logical delete_owner)
{
        printf("Block %d is merging with Block %d \n",
            ((ATTRIB_GEN_INTEGER*)attr)-> value(),
            ((ATTRIB_GEN_INTEGER*)
             find_named_attrib(other_ent,"tellme"))-> value());
}

// Output message when split
static void split_mess(ATTRIB_GEN_NAME *attr,
                ENTITY *new_ent)
{
        printf("Block %d is being split\n",
            ((ATTRIB_GEN_INTEGER*)attr)-> value());
}
```

When run the program starts by adding generic integer attributes to each BODY. Boolean operations are then used first to create a square hole in block 2 (by subtracting block 1) and then to unite block 3 with block 2 to form a cross. The following output is generated;

```
Start subtract
Block 2 is merging with Block 1
Block 1 is merging with Block 2
End subtract
Start unite
Block 3 is merging with Block 2
Block 2 is merging with Block 3
End unite
```

Notice that merging occurs in both Boolean operations and that *both* of the ENTITYs involved have their merge methods called.

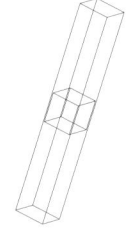

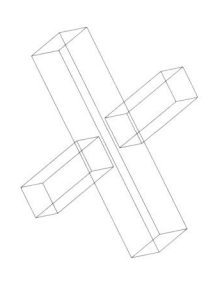

12.3 Adding your own Attributes

During this formation of the cross the "sticky" attribute gets transferred from ENTITY to ENTITY until it is recorded in the SAT file, along with a "tellme" attribute, at the end of the program. These two attributes can be clearly seen in the file **yplusz.sat**:

```
body $1 $2 $-1 $-1 #
integer_attrib-name_attrib-gen-attrib $-1 $3 $-1 $0 \\
                            copy keep_all ig-
nore 6 sticky 0 #
lump $-1 $-1 $4 $0 #
integer_attrib-name_attrib-gen-attrib $-1 $-1 $1 $0 \\
                            custom custom ig-
nore 6 tellme 3 #
shell $-1 $-1 $-1 $5 $-1 $2 #
```

12.3 Adding your own Attributes

The following program, spread across five files, creates a new attribute from the ground up. Four of the files are required to declare and define the new ATTRIB class. The code in **att_org.hxx** and **att_org.cxx** derives an "organization" class (in this instance called **org**) from the ATTRIB class. The actual definition of the new attribute, called ATTRIB_DIM, is contained in **att_dim.hxx** and **att_dim.cxx**. The job of coding all these files is made much easier by the use of several macros. These are:

MASTER_ATTRIB_DECL: Generates nearly all the code required to *declare* a new organization ATTRIB class.

MASTER_ATTRIB_DEFN: Generates nearly all the code required to *define* a new organization ATTRIB class.

ATTRIB_FUNCTIONS: Carries out *declaration* of ATTRIB and ENTITY function.

SAVE_DEF, RESTORE_DEF, ...: A number of standard ENTITY *definition* macros which specify functions used to support ENTITY save, restore and rollback.

The contents of the four attribute definition files can be illustrated as follows:

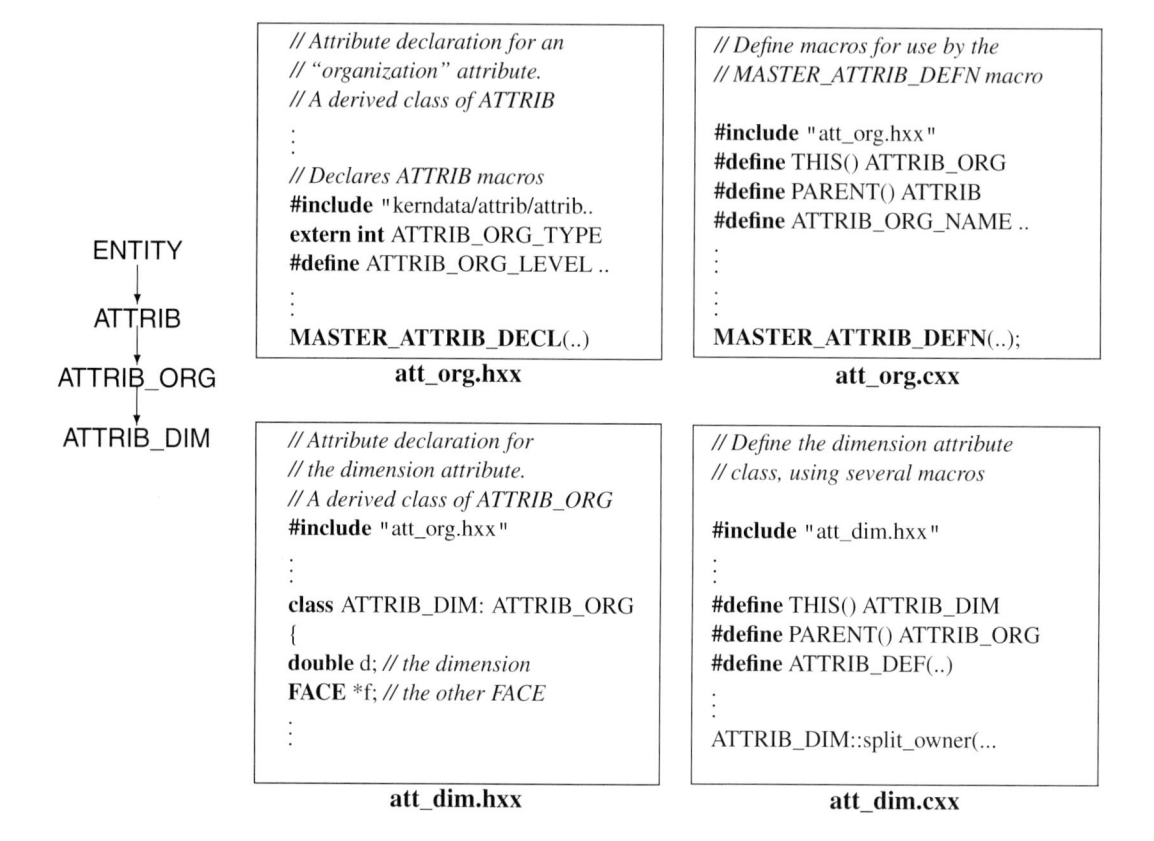

ENTITY
↓
ATTRIB
↓
ATTRIB_ORG
↓
ATTRIB_DIM

```
// Attribute declaration for an
// "organization" attribute.
// A derived class of ATTRIB
:
// Declares ATTRIB macros
#include "kerndata/attrib/attrib..
extern int ATTRIB_ORG_TYPE
#define ATTRIB_ORG_LEVEL ..
:
MASTER_ATTRIB_DECL(..)
```
att_org.hxx

```
// Define macros for use by the
// MASTER_ATTRIB_DEFN macro

#include "att_org.hxx"
#define THIS() ATTRIB_ORG
#define PARENT() ATTRIB
#define ATTRIB_ORG_NAME ..
:
MASTER_ATTRIB_DEFN(..);
```
att_org.cxx

```
// Attribute declaration for
// the dimension attribute.
// A derived class of ATTRIB_ORG
#include "att_org.hxx"
:
class ATTRIB_DIM: ATTRIB_ORG
{
double d; // the dimension
FACE *f; // the other FACE
:
```
att_dim.hxx

```
// Define the dimension attribute
// class, using several macros

#include "att_dim.hxx"
:
#define THIS() ATTRIB_DIM
#define PARENT() ATTRIB_ORG
#define ATTRIB_DEF(..)
:
ATTRIB_DIM::split_owner(...
```
att_dim.cxx

About foreign attributes

> ACIS attributes can happily move from system to system within sat files, though it should be noted that they carry only data, not procedures. In other words it is possible for your CAD system to load a sat file containing FACEs with "wood_grain" attributes and yet display nothing even remotely "knotty" on the screen because it has no procedures for using this data.
>
> In other words the ACIS save and restore mechanism deals elegantly with so-called "foreign" attributes. Although the methods for using the foreign attribute information might not be available, the data remains intact for as long as the ENTITY it is attached to continues to exist.

The fifth file, **att_main.cxx**, contains the program that uses the "dimension" attribute to hold the distance between a pair of parallel, planar FACEs. The program creates a truncated cone and adds an instance of the **ATTRIB_DIM** class to the top FACE which records the distance to the bottom FACE. The **split** behavior of the attribute is invoked when a block is subtracted from the cone forming a slot through the top of the cone.

12.3 Adding your own Attributes 321

att_main.cxx

```cpp
// This program uses a dimension attribute to record the
// distance between the top and bottom faces of a cylinder.
// A cross slot is then added to the top of the cylinder
// causing the dimension attribute to be copied to the new face
#include <stdio.h>
#include "kernel/acis.hxx" // Declare system wide parameters
#include "boolean/kernapi/api/boolapi.hxx" // Declares boolean API's
#include "constrct/kernapi/api/cstrapi.hxx" // Construction API's
#include "kernel/kernapi/api/kernapi.hxx" // Declares kernel API's
#include "baseutil/vector/transf.hxx" //Declare transf class
#include "kernel/kerndata/top/alltop.hxx" // Declare topology classes
#include "kernel/kerndata/geom/plane.hxx" // Declare PLANE class
#include "kernel/kerndata/lists/lists.hxx" // Declare ENTITY_LIST class
#include "att_dim.hxx" // Declare Dimension Attribute
#include "kernel/kerndata/savres/fileinfo.hxx" // Declares fileinfo class

void save_ent( char *filename, ENTITY *ent); //See page 72

void main()
{   // make the test parts
    api_start_modeller(0);
    api_initialize_booleans();
    BODY *cyl, *slot; // Create z-axis aligned cylinder
    api_make_frustum (60, 30, 30, 10, cyl);
    api_make_cuboid(100, 15, 15, slot);
    transf move_up = translate_transf(vector(0,0,25));
    api_apply_transf(slot,move_up);

    // find the top and bottom faces of the cylinder
    FACE *top, *bottom;
    position top_ref, bottom_ref;
    api_find_face(cyl, unit_vector(0,0,1), top);
    top_ref = ((PLANE*)top->geometry())->root_point();
    api_find_face(cyl,unit_vector(0,0,-1),bottom);
    bottom_ref = ((PLANE*)bottom->geometry())->root_point();

    vector v1 = bottom_ref - top_ref;
    double dist = v1.len();
    new ATTRIB_DIM(top,bottom,dist);

    FILE *fp = fopen("attrib1.dbg", "w");
    top->attrib()->debug_ent(fp);
    fclose(fp);

    printf("Create the top slot\n");
    api_subtract(slot,cyl);
```

```
fp = fopen("attrib2.dbg","w");
// go through all the ATTRIBs of all the FACEs
FACE *f = cyl−>lump()−>shell()−>face_list();
while (f)
{ // For each FACE
    ATTRIB *Att = f−>attrib();
    while ( Att )
    { // For each Attrib
        Att−>debug_ent(fp);
        Att = Att−>next();
    }
    f = f−>next_in_list();
}

fclose(fp);

printf("saving the slotted cone\n");

save_ent("dimcyl.sat", cyl);

api_terminate_booleans();
api_stop_modeller();
}
```

Original

First Split

Second Split

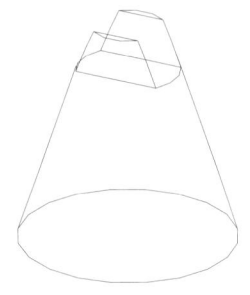

When compiled and run, the program writes the following to the screen:

```
Creating dim
Create the top slot
Creating NULL dim
Splitting
Create dim
Splitting
Create dim
saving the slotted cone
```

At first glance this is not what one would expect. Why has the constructor function of the "dim" class been called four times?

The first instance of the **ATTRIB_DIM** class occurs, as expected, prior to the creation of the slot. The two invocations of the splitting method occur because of the stepwise nature of the Boolean process. During the imprinting process the top FACE is split twice (see margin). When the top FACE is first split the change causes a copy of **ATTRIB_DIM** to be recorded on the current Bulletin Board using the **NULL** constructor (which all ENTITYs provide for such backup).

12.3 Adding your own Attributes

Examination of the **dbg** files shows that the ATTRIB has been successfully copied.

attrib2.dbg

```
dimension **** -746064:
Rollback pointer: -736800
Attribute list   : NULL
Owning entity    : face **** -746112
Previous attrib  : NULL
Next attribute   : NULL
Datum Face       : face **** -746112
Ref Face         : face **** -787352
Dimension        : 60

dimension **** -785232:
Rollback pointer: -741152
Attribute list   : NULL
Owning entity    : face **** -787496
Previous attrib  : NULL
Next attribute   : NULL
Datum Face       : face **** -787496
Ref Face         : face **** -787352
Dimension        : 60
```

The definition of the **ATTRIB_DIM** class requires four files: two for the organization class and two for the attribute itself. The task of writing these is made easier by a number of C++ macros defined in **attrib.hxx**.

The file **att_org.hxx** declares an **organization** class derived from ATTRIB called (for want of something better) **org**.

att_org.hxx

```
// Attribute declaration for a private ''organization'' attribute.
#if !defined( ATTRIB_ORG_CLASS )
#define ATTRIB_ORG_CLASS

#include "kernel/kerndata/attrib/attrib.hxx"

// This attribute type is a derived class of ATTRIB

extern int ATTRIB_ORG_TYPE;
#define ATTRIB_ORG_LEVEL (ATTRIB_LEVEL + 1)
```

```
// Note "NONE" has been used for the library name,
// because this library is not exporting through Spatials
// DECL_ macros to other DLLs.
MASTER_ATTRIB_DECL( ATTRIB_ORG, NONE)

#endif
```

To create a header file for yourself, or your organization, simply replace every occurrence of **ORG** with your own two- or three-letter name (i.e. sentinel). If the code being developed is likely to travel further than your own file store, you should register your sentinel with Spatial Corp's support to ensure it is, and remains, unique.

The actual definition (as opposed to the declaration done in the header file) of the **ATTRIB_ORG** class is done in **att_org.cxx**.

att_org.cxx

```
// This file defines the "standard" organization ATTRIB class
#include "kernel/acis.hxx"
#include "att_org.hxx"
#include "kernel/kerndata/data/datamsc.hxx"

// Define macros for this attribute and its parent,
// for use by the MASTER_ATTRIB_DEFN macro

#define THIS() ATTRIB_ORG
#define THIS_LIB NONE
#define PARENT() ATTRIB
#define PARENT_LIB KERN

// Identifier used to identify a particular entity type.
// The identifier must be unique amongst all attributes
// derived directly from ATTRIB, across all application developers.

#define ATTRIB_ORG_NAME "org"

MASTER_ATTRIB_DEFN( "Dimension Attribute" );
```

This short program simply defines a number of variables (i.e. THIS(), PARENT(), etc.) required by the macro **MASTER_ATTRIB_DEFN**. This, like the header file, is boiler plate code, which is intended to be used as a template by developers.

Things become more interesting with the declaration of the **ATTRIB_DIM** class in **att_dim.hxx**. Here the dimension attribute is given the code **DIM** to distinguish it from other attributes that may be defined by the **ORG** organization.

12.3 Adding your own Attributes

att_dim.hxx

```cpp
// This is the header file for the dimension ATTRIB class
#if !defined( ATTRIB_DIM_CLASS )
#define ATTRIB_DIM_CLASS

#include "att_org.hxx"
#include "kernel/kerndata/top/face.hxx"

extern int ATTRIB_DIM_TYPE;
#define ATTRIB_DIM_LEVEL (ATTRIB_ORG_LEVEL + 1)

class ATTRIB_DIM: public ATTRIB_ORG
{
private:
    double d;
    FACE *f;

public:
    ATTRIB_DIM( ENTITY * = NULL, FACE* = NULL,double = 0.0);
    double dist() const { return d; }
    FACE *face() const { return f; }
    void set_dist( double di) { backup(); d=di; } // backup creates a new bulletin
    void set_face( FACE *fi) { backup(); f=fi; }
    virtual void split_owner( ENTITY * );
    virtual void merge_owner( ENTITY *, logical );
    ATTRIB_FUNCTIONS(ATTRIB_DIM)
};
#endif
```

The **att_dim.cxx** file contains the definitions specified in the header file. A large part of the program consists of defining the value of parameters used by the macro.

att_dim.cxx

```cpp
// This file defines an attribute for recording the distance
// between two planar faces.
#include <stdio.h>
#include <memory.h>
#include "kernel/acis.hxx"      // Declares system wide parameters
#include "att_dim.hxx"   // Declares dimension attribute
#include "kernel/kerndata/data/datamsc.hxx" // Declare attrib macros

// Define macros for this attribute and its parent
#define THIS() ATTRIB_DIM
```

```
#define THIS_LIB NONE
#define PARENT() ATTRIB_ORG
#define PARENT_LIB NONE
#define ATTRIB_DIM_NAME "dimension"

// Implement the standard attribute functions.
ATTCOPY_DEF( "dim attribute" )
LOSE_DEF
    if ( entity() == face() )
        printf("LOSING Reference\n");
DTOR_DEF
DEBUG_DEF
    // Print out a readable description of this entity.
    debug_old_pointer( "Datum Face ", entity(), fp );
    debug_old_pointer( "Ref Face ", face(), fp );
    debug_real( "Dimension", dist(), fp );
SAVE_DEF
    write_ptr( face(), list );
    write_real( dist() );
RESTORE_DEF
    set_face( (FACE*)read_ptr() );
    set_dist( read_real() );
COPY_DEF
    set_face( (FACE*)list.lookup(from->face()) );
    set_dist( from->dist() );
SCAN_DEF
    list.add(face());
FIX_POINTER_DEF
    set_face( (FACE*)read_array(array,(int)face()));
TERMINATE_DEF

ATTRIB_DIM::ATTRIB_DIM(ENTITY* owner, FACE* ref,
                double distance):ATTRIB_ORG(owner)
{
    f=ref;          // Initialize members
    d=distance;
    if (ref == NULL)
      printf("Creating NULL dim\n");
    else
      printf("Creating dim\n");
}

void ATTRIB_DIM::split_owner( ENTITY *new_ent)
{
    backup(); // Record ENTITY for rollback.
    // Duplicate the attribute on the new entity
    printf("Splitting \n");
    new ATTRIB_DIM( new_ent, face(), dist() );
}
```

```
// Virtual function called when two ENTITYs are to be merged.
void ATTRIB_DIM::merge_owner( ENTITY *other_ent, logical delete_owner)
{
    backup(); // Record ENTITY for rollback.
    // If the owner of this attribute is going to be deleted, and
    // there is no dimension attached to the other entity, then
    // the dimension is transferred  to the other entity.
    if (delete_owner)
    {
        ATTRIB *other_att = find_attrib(other_ent, ATTRIB_ORG_TYPE,
                                ATTRIB_DIM_TYPE);
        if (other_att == NULL) // No dimension on other entity, so move it
            move( other_ent );
    }
}
```

Notice that both the **split_owner** and **merge_owner** methods of ATTRIB_DIM start by calling the ENTITY function **backup** to ensure that any subsequent actions can be rolled back. The function **move** (used to transfer ownership of an attribute from one ENTITY to another) is a member of the class ATTRIB. Note that this example has not declared a **trans_owner** function which takes action whenever the owner ENTITY is transformed (see Section 12.5).

12.4 Attribute Pitfalls

Like most powerful tools, attributes require a certain amount of care in their use and implementation. There are three things to note:

1. **Losing referenced ENTITYs:** A complex attribute holds pointers to other ENTITYs. A problem can arise if any of these ENTITYs are deleted during the course of modeling operations. Unless the ATTRIB class contains methods for checking its references, it is all too easy to find pointers directed to free-store. For example, an instance of the **ATTRIB_DIM** class created in **att_main.cxx** could easily find that the datum FACE had been removed by a modeling operation after its creation.

2. **Envisaging the split/merge behaviors:** Anticipating exactly how and when the split/merge functions will be called is not trivial. Both these methods are called during Booleans, with the merge method being invoked near the end of the process (during stitching) and the split method at the beginning of the imprinting process. The meaning of split and merge for ENTITYs other than FACEs and EDGEs can also be hard to envisage.

3. **Analysis invoked by split/merge methods:** There are some limitations on the sort of operations one should try to carry out from within a split/merge method. Obviously calling a Boolean from within a split, or merge, function could lead

to some form of infinite loop. Likewise, because the methods are invoked part way through the Boolean, the state of the model may not necessarily be manifold or closed, which could make certain types of analysis problematic.

12.5 Exercises

1. Write an ATTRIB_DIST class to record the distance between two BODYs (use their centers of gravity as reference points). Define the behavior of the split, merge and translate owner methods as follows:

   Split: If the owning BODY is split, copy the ATTRIB_DIST to both parts (remember to recalculate the centers of gravity).

   Merge: Remove the attribute.

   Translate: Recalculate the distance every time the BODYs are moved.

   Note that this attribute would require both BODYs to be saved and restored at the same time (i.e. in the same ENTITY_LIST given to **api_save_entity_list**).

2. Vary the dimensions of the three blocks created in the program **genatt.cxx** and see if you can predict in what order the split and merge functions will be called.

Customizing ACIS 13

One of the most impressive things about ACIS is the way its designers have used mechanisms available in C++ to allow the basic system to be extended and customized in a manner that is both uniform yet flexible. An even better feature is that much of the coding required to do this is encapsulated inside C++ macros that allow API functions, Scheme extensions and ENTITYs to be added to the system with only a little work.

This chapter starts by demonstrating how a user defined-class can be derived from the ENTITY base class. Why bother doing this? Because you get a lot of functionality for very little effort; any class derived from ENTITY inherits methods for:

1. Save (and restore) to, (and from), SAT files.

2. Debugging using generic ACIS functions.

3. Backing-up on the Bulletin Board system.

4. Recover on error.

The later sections of this chapter cover the creation of a new **API** function and an associated Scheme extension.

13.1 Adding your own ENTITYs

This example creates a new ACIS ENTITY called CSG_NODE which allows binary trees of BODYs to be created. Each GSG_NODE has a **parent**, two **children** and a **type**. The **parent** of the CSG_NODE ENTITY is either NULL (if it's the root) or another CSG_NODE. The two **children** of the node are represented by pointers to ACIS ENTITYs (known as **left_ptr** and **right_ptr**) and it is assumed that these will be either BODYs or other CSG_NODEs. Each node can be either a **PLUS** (i.e.

Boolean UNION) or a **MINUS** (i.e. Boolean SUBTRACT) type allowing the tree to be processed from the root-up by a member function known as **evaluate**.

The example involves five separate files, four (**csg_org.hxx**, **csg_org.cxx**, **csg_node.hxx** and **csg_node.cxx**) to define a new ENTITY class and the fifth, **csg_main.cxx**, which uses it.

The program **csg_main.cxx** not only creates a tree but also evaluates it and then carries out a number of operations to test the generic ENTITY functions.

csg_main.cxx

```
// This program creates a csg type tree by creating instances of
// the CSG_NODE ENTITY class defined in the csg*.cxx and csg*.hxx files.
// The tree once constructed is evaluated twice before being
// saved and restored to and from a SAT file, summarized using a debug
// function. Lastly the tree is deleted and restored by rollback.
#include <stdio.h>                              // Declares file I/O functions
#include "constrct/kernapi/api/cstrapi.hxx"     // Construction API's
#include "kernel/kernapi/api/api.hxx"           // Declares outcome class
#include "kernel/kernapi/api/kernapi.hxx"       // Declares kernel class
#include "kernel/kerndata/top/alltop.hxx"       // Declares topology classes
#include "baseutil/vector/transf.hxx"           // Declares the transform class
#include "baseutil/vector/vector.hxx"           // Declares a vector class
#include "baseutil/vector/unitvec.hxx"          // Declares unit-vector class
#include "kernel/kerndata/lists/lists.hxx"      // Declares ENTITY LIST class for save
#include "kernel/kerndata/bulletin/bulletin.hxx"  // Declares current_bb()
#include "kernel/kerndata/data/debug.hxx"       // Declares Debug Routines
#include "csg_node.hxx"                         // Declares the CSG_NODE ENTITY Class
 #include "kernel/kerndata/savres/fileinfo.hxx" // Declares fileinfo class

void save_ent(char*, ENTITY*);  // see page 72

void main()
{
    api_start_modeller(0);
    api_initialize_constructors();
    BODY *cyl1, *cyl2, *cyl3, *b1;   // Cylinders and a Block
    api_make_frustum(120,20,20,20,cyl1);
    transf rotX = rotate_transf(pi/2,vector(1,0,0));
    api_apply_transf(cyl1,rotX);
    api_make_frustum(120,20,20,20,cyl2);
    api_make_frustum(200,5,5,5,cyl3);
    transf transX = translate_transf(vector(30,0,30));
    api_apply_transf(cyl3,transX);
    api_make_cuboid(100,100,100,b1);

    // Construct a binary tree Using the constructor CSG_NODE(type,parent);
    CSG_NODE *root = new CSG_NODE(PLUS,NULL);
    root->set_right(cyl3);
```

13.1 Adding your own ENTITYs

```
CSG_NODE *sub_node = new CSG_NODE(MINUS,root);
root->set_left(sub_node);
sub_node->set_left(b1);

CSG_NODE *add_node = new CSG_NODE(PLUS,sub_node);
sub_node->set_right(add_node);
add_node->set_left(cyl1);
add_node->set_right(cyl2);

// Test the class --------------------------
BODY *tree_eval;
root->evaluate(tree_eval);
save_ent("tree1.sat",tree_eval);

// Move one of the LEAF BODYs and re-evaluate the tree
transf transY = translate_transf(vector(0,20,0));
api_apply_transf(cyl3,transY);

root->evaluate(tree_eval);
save_ent("tree2.sat",tree_eval);
```

The middle portion of the **csg_main.cxx** program uses the CSG_NODE class's constructor, and **set** functions, to define the tree shown in Figure 13.1.

One important point to note here is that all ENTITYs must be created using the **new** operator. ACIS overloads **new**, allowing a private free list (i.e. supply of dynamically allocated memory) to be used.

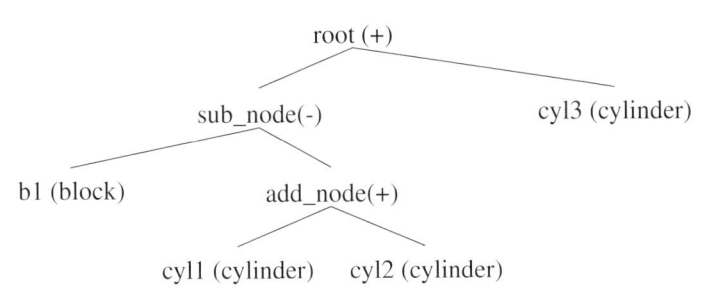

Figure 13.1: The tree defined by the CSG_NODE class.

The margin shows wire frames of the two BODYs generated by evaluating the tree with one of the nodes (the long cylinder) in a slightly different position.

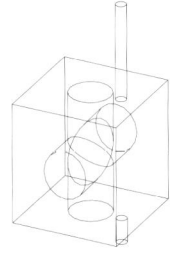

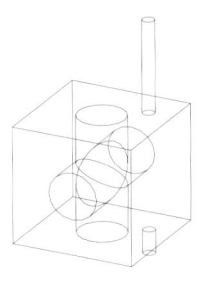

csg_main.cxx (continued)

```cpp
// Test the ENTITY functionality --------------------------

    // 1) Test ENTITY save
    save_ent("root.sat",root);
    // 2) Test ENTITY restore
    ENTITY_LIST input;

    FILE *fp = fopen("root.sat", "r");
    api_restore_entity_list(fp,TRUE,input);
    fclose(fp);

    ENTITY *restored_root = input[0];
    // 3) Test ENTITY debug
    fp = fopen("debug1.dbg", "w");
    debug_size(restored_root,fp);
    fclose(fp);
    // 4) Test ENTITY rollback
    API_NOP_BEGIN  //Macro to roll back changes
      // delete the tree
      api_del_entity((ENTITY*)restored_root);
      // get the last API's bulletin board
      BULLETIN_BOARD *ourbb = current_bb();
      int count = 1;
      // check if "restored_root" is on the board
      if (ourbb != NULL){
        BULLETIN *ourb = ourbb-> start_bulletin();
        while (ourb != ourbb-> end_bulletin())
        {
          count++;
          if (ourb-> old_entity_ptr() == restored_root &&
              ourb-> type() == DELETE_BULLETIN)
                printf("CSG_NODE backup !\n");
          ourb = ourb-> next();
        }
      }
      printf("%d ENTITIES on the BULLETIN_BOARD\n",count);
    API_NOP_END
    printf("api_del_entity rolled-back\n");

    api_terminate_constructors();
    api_stop_modeller();
}
```

13.1 Adding your own ENTITYs

The second part of the **csg_main.cxx** program tests several generic ENTITY behaviors. First the tree is saved in the file **root.sat**. A quick look at the head of this file confirms that the CSG_NODE ENTITY has been correctly added:

root.sat

```
              :
              :
   csg_node-csg $-1 $-1 $1 $2 0 #
   csg_node-csg $-1 $0 $3 $4 1 #
   body $-1 $5 $-1 $6 #
   body $-1 $7 $-1 $-1 #
   csg_node-csg $-1 $1 $8 $9 0 #
   lump $-1 $-1 $10 $2 #
   transform $-1 1 0 0 0 1 0 0 0 1 30 20 30 1 no_rotate .... #
   lump $-1 $-1 $11 $3 #
   body $-1 $12 $-1 $13 #
   body $-1 $14 $-1 $-1 #
```

The **root.sat** is then restored creating a second tree (referred to as **restored_root**) in the process. Next, the **debug** behavior of the new class is tested using the function **debug_size** to summarize the content of the second tree. The output detailing the number and type of ENTITYs in the data structure is written to the file **debug1.dbg** shown below:

debug1.dbg

```
       3 csg_node records,     96 bytes
       4 body records,        128 bytes
       4 lump records,        128 bytes
       2 transform records,   256 bytes
       4 shell records,       160 bytes
      15 face records,        660 bytes
      18 loop records,        576 bytes
      15 surface records,    2616 bytes
      36 coedge records,     1584 bytes
      18 edge records,       1296 bytes
      14 vertex records,      336 bytes
      18 curve records,      2208 bytes
      14 point records,       672 bytes
   Total storage 10716 bytes
```

Once again the new ENTITY is clearly seen playing an equal part in the debugging process which recorded a total of 165 ENTITYs in the tree.

The last test of the new ENTITY is its destruction and resurrection. This is done using the **api_del_entity** function. The immediate effect of this function is to delete the ENTITY, recording its parts on the BULLETIN_BOARD (which is examined to verify that it contains the deleted ENTITY). However, because the operation is enclosed between the macros **API_NOP_BEGIN** and **API_NOP_END** the deletion is undone:

>///Macro to roll back changes
>**API_NOP_BEGIN**
>// delete the tree
>**api_del_entity**(restored_root);
>⋮
>**API_NOP_END**

These macros use the bulletin/rollback mechanism to undo any changes to any ENTITYs made between the **BEGIN** and **END** calls.

When compiled and executed the program writes

```
CSG_NODE backup !
165 ENTITIES on the BULLETIN_BOARD
api_del_entity rolled-back
```

Creating the CSG_NODE Class

The creation of new ENTITY class (CSG_NODE) used in the **csg_main.cxx** program requires four files. Two of these, **csg_org.hxx** and **csg_org.cxx**, define an **organization specific class** which is required before the **application specific class** can be defined in **csg_node.hxx** and **csg_node.cxx**. The job of coding all these files is made much easier by the use of several macros. These are:

MASTER_ENTITY_DECL: Generates nearly all the code required to *declare* a new organization ENTITY class.

MASTER_ENTITY_DEFN: Generates nearly all the code required to *define* a new organization ENTITY class.

ENTITY_FUNCTIONS: Carries out *declaration* of ENTITY function.

SAVE_DEF, RESTORE_DEF, ... : A number of standard ENTITY *definition* macros which specify functions used to support ENTITY save, restore and rollback.

The contents of the four files needed to create this new ENTITY are illustrated as follows:

13.1 Adding your own ENTITYs

```
// Declaration for an
// "organization" ENTITY
// A derived class of ENTITY
  ⋮
// Declares ENTITY macros
#include "kerndata/data/entity..
extern int ENTITY_CSG_TYPE
#define ENTITY_CSG_LEVEL ..
  ⋮
MASTER_ENTITY_DECL(..)
```
csg_org.hxx

```
// Define macros for use by the
// MASTER_ENTITY_DEFN macro

#include "csg_org.hxx"
#define THIS() ENTITY_CSG
#define PARENT() ENTITY
  ⋮
  ⋮
MASTER_ENTITY_DEFN(..
```
csg_org.cxx

```
// Declaration header file for
// the CSG_NODE ENTITY
// A derived class of CSG_ORG
#include "csg_org.hxx"
  ⋮
class CSG_NODE: CSG_ENTITY {
ENTITY *left_ptr;
ENTITY *right_ptr;

CSG_NODE *parent_ptr;
CSG_OP bo_type;
  ⋮
```
csg_node.hxx

```
// Define the methods of the
// CSG_NODE class, using macros

#include "csg_node.hxx"
  ⋮
#define THIS() CSG_NODE
#define PARENT() ENTITY_CSG
#define CSG_NODE_NAME "..
  ⋮
// Main constructor
CSG_NODE::CSG_NODE(....
```
csg_node.cxx

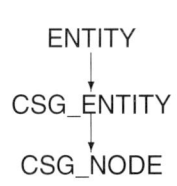

```
ENTITY
  ↓
CSG_ENTITY
  ↓
CSG_NODE
```

In an effort to prevent different developers adding new classes to ACIS with the same names, Spatial insist (not unreasonably) that rather than deriving directly from the ENTITY base class each user and developer first derives what they call an **organization** class.

Each organization class has a unique two- or three-letter name (e.g. STL, TSL, LW) and Spatial Corp's Support Department (who maintain a list of these) are happy to advise on the choice of these **sentinels**.

The program **csg_org.hxx** declares an "organization" class derived from ENTITY.

csg_org.hxx

```
// NOTE:- Standard ACIS header should go here: see STL documentation for details
#if !defined( ENTITY_CSG_CLASS )
#include "baseutil/logical.h"
#include "kernel/kerndata/data/en_macro.hxx"
```

```
#include "kernel/kernel_thread_ctx.hxx"

// Identifier used to find out (via identity() defined below) to what
// an entity pointer refers.
extern int ENTITY_CSG_TYPE;

// Identifier that gives number of levels of derivation of this class
// from ENTITY.
#define ENTITY_CSG_LEVEL (ENTITY_LEVEL + 1)

// The CSG master data structure entity, of which all its private
// specific types are subclasses.
MASTER_ENTITY_DECL( ENTITY_CSG )

#define ENTITY_CSG_CLASS
#endif
```

The program **csg_org.cxx** defines an "organization" class derived from ENTITY.

csg_org.cxx

```
#include <stdio.h>
#include "kernel/acis.hxx"
#include "kernel/dcl_kern.h"
#include "kernel/kerndata/data/datamsc.hxx"   //Declares ENTITY macros
#include "csg_org.hxx"                         //Declares ENTITY parametes

// Identifier used externally to identify a particular entity type.
// This is used within the save/restore system for translating
// to/from external file format.

#define ENTITY_CSG_NAME "csg"

// Macros used to tell the master macro who we are and where we
// stand in the hierarchy

#define THIS() ENTITY_CSG
#define THIS_LIB NONE
#define PARENT() ENTITY
#define PARENT_LIB KERN

// Magic macro to do the rest of the implementation!

MASTER_ENTITY_DEFN( "CSG Node Entities" );
```

13.1 Adding your own ENTITYs   337

Having defined an "organization" class the new CSG_NODE ENTITY class can be derived from it. Once again numerous macros are used, making the process reasonably straightforward. The majority of the member functions required of an ENTITY are needed by the save and restore operation. In the file **csg_node.hxx**, however, all the standard copy, save and restore functions are automatically declared by the **ENTITY_FUNCTIONS** macro.

csg_node.hxx

```
// CSG_NODE.HXX
// NOTE:- Standard ACIS header should go here: see STL documentation for details

#if !defined( CSG_NODE_CLASS )
#define CSG_NODE_CLASS

#include "csg_org.hxx"   // Declares the organization class
#include "kerndata/top/alltop.hxx"

class CSG_NODE;
class ENTITY;

// Define the depth of this class below ENTITY
extern int CSG_NODE_TYPE;
#define CSG_NODE_LEVEL (ENTITY_CSG_LEVEL + 1)

// Define two different types of CSG_NODE
enum CSG_OP {PLUS,MINUS};

class CSG_NODE: public ENTITY_CSG
    {
private:
    // Two branches left and right of this node
    ENTITY *left_ptr;
    ENTITY *right_ptr;
    CSG_NODE *parent_ptr;   // Parent Node
    CSG_OP bo_type;   // PLUS or MINUS

    // Include the standard member functions for all ENTITYs.
    ENTITY_FUNCTIONS( CSG_NODE )
    // Search a private list for this object, used for debugging.
    LOOKUP_FUNCTION

    // Now the CSG_NODE's own member functions
    // Constructor: logged on the Bulletin Board
    // Arguments are a type (i.e. PLUS or MINUS) and parent pointer

    CSG_NODE(); // Default constructor for Bulletin Board rollback
    CSG_NODE(CSG_OP, CSG_NODE*);
```

```
    // Access functions
    CSG_NODE *parent_node() const { return parent_ptr; }
    ENTITY *left_branch() const { return left_ptr; }
    ENTITY *right_branch() const { return right_ptr;}
    CSG_OP op_type() const { return bo_type; }
    // Utility Functions
    void evaluate(BODY*&);
    // Data changing routines.  Each must first backup on the Bulletin Board
    void set_parent( CSG_NODE* );
    void set_type( CSG_OP );
    void set_left(ENTITY*);
    void set_right(ENTITY*);
};
#endif
```

The file **csg_node.cxx** really contains the "beef" of the new ENTITY definition. In this code the save and restore functions are explicitly defined along with the ENTITY's constructors and own particular methods.

The code for the ENTITY member functions is bracketed by macros. For example, the member function for writing the ENTITY to file is defined by the code placed between the macros **SAVE_DEF** and **RESTORE_DEF**.

```
    SAVE_DEF
        // Code for adding member data to the SAT file
        // (pointers to indices) creation list
        write_ptr(parent_node(), list );
        write_ptr(left_branch(), list );
        write_ptr(right_branch(), list );
        write_int( bo_type);

    RESTORE_DEF
```

The order in which the ENTITY definition macros are used is fixed. Also important is the sequence in which ENTITY data is written and read from sat files. It must be the same in both cases!

csg_node.cxx

```
    #include <stdio.h>     // System functions
    #include <logical.h>
    #include <string.h>
    #include <memory.h>
    #include "kernel/acis.hxx"
    #include "csg_node.hxx"
    #include "boolean/kernapi/api/boolapi.hxx"    // Declares boolean API's
    #include "kernel/kernapi/api/api.hxx"         // Declares outcome class
```

13.1 Adding your own ENTITYs

```cpp
#include "kernel/kernapi/api/kernapi.hxx"      // Declares kernel API's
#include "kernel/kerndata/top/alltop.hxx"      // Declare topology classes
#include "kernel/kerndata/data/datamsc.hxx"    // Declares ENTITY macros
#include "kernel/kerndata/data/debug.hxx"      // Declares Debug macros
#include "kernel/kerndata/savres/versions.hxx" // Declares version data

// Include the standard member functions using STL macros
// Declare the class name and its parent.
#define THIS() CSG_NODE
#define THIS_LIB NONE
#define PARENT() ENTITY_CSG
#define PARENT_LIB NONE

// Declare an external name for the entity
#define CSG_NODE_NAME "csg_node"

ENTITY_DEF( CSG_NODE_NAME )
    // Print out a readable description of this entity.
    debug_new_pointer( "Parent Node", parent_node(), fp );
    debug_int( "CSG Node Type", op_type(), fp );
    debug_new_pointer( "Left ENTITY", left_branch(), fp );
    debug_new_pointer( "Right ENTITY", right_branch(), fp );

LOOKUP_DEF
SAVE_DEF
    // Code for adding member data to the SAT file
    // (pointers to indices) creation list
    write_ptr(parent_node(), list );
    write_ptr(left_branch(), list );
    write_ptr(right_branch(), list );
    write_int( bo_type);

RESTORE_DEF
    // Code for reading a line from a save file
    parent_ptr = (CSG_NODE *)read_ptr();
    left_ptr = (ENTITY *)read_ptr();
    right_ptr = (ENTITY *)read_ptr();
    bo_type = (CSG_OP)read_int();

COPY_DEF
    // Code for copying data to new object (using the save list)
    parent_ptr = (CSG_NODE *)list.lookup( from-> parent_node() );
    left_ptr = (ENTITY *)list.lookup( from-> left_branch() );
    right_ptr = (ENTITY *)list.lookup( from-> right_branch() );
    bo_type = (CSG_OP)from-> op_type();

SCAN_DEF
    // Code for adding pointers to the ``pointer-to-indices'' save list
    list.add( parent_node() );
    list.add( left_branch() );
```

```cpp
            list.add( right_branch() );

    FIX_POINTER_DEF
        // Code to convert pointers from array indices to contents
        set_parent( (CSG_NODE *)
                    read_array( array, (int)parent_node() ) );
        set_left( (ENTITY*)read_array( array, (int)left_branch() ) );
        set_right( (ENTITY*)read_array( array, (int)right_branch() ) );

    TERMINATE_DEF

    // Null constructor required for rollback and restore operations
    CSG_NODE::CSG_NODE()
    {
        parent_ptr = NULL;
        left_ptr = NULL;
        right_ptr = NULL;
        bo_type = PLUS;
    }


    // Main constructor
    CSG_NODE::CSG_NODE(CSG_OP type, CSG_NODE *parent)
    {
        set_type(type);
        set_parent(parent);
    }

    // Fix up a copy of this object for a change bulletin, after the
    // object is constructed and copied memberwise.
    void CSG_NODE::fixup_copy( CSG_NODE *rollback ) const
    {
        PARENT()::fixup_copy( rollback );
    }

    // User "destructor" for a CSG_NODE. No type-specific work to do,
    // so all work left to the generic ENTITY destructor.
    void CSG_NODE::lose()
    {
        ENTITY::lose();
    }

    // Final record discard.
    CSG_NODE::~CSG_NODE()
    {
        check_destroy();
    }

    // Member-setting functions. Each ensures that the object has been
    // copied to the BULLETIN_BOARD before making any change.
    void CSG_NODE::set_parent( CSG_NODE *inst)
```

13.1 Adding your own ENTITYs
341

```
{
    backup(); // Before any change to ENTITY data
    parent_ptr = inst;
}

void CSG_NODE::set_type( CSG_OP inst)
{
    backup(); // Before any change to ENTITY data
    bo_type = inst;
}

void CSG_NODE::set_left(ENTITY *left )
{
    backup(); // Before any change to ENTITY data
    left_ptr = left;
}

void CSG_NODE::set_right(ENTITY *right )
{
    backup(); // Before any change to ENTITY data
    right_ptr = right;
}

void CSG_NODE::evaluate(BODY*& result)
{
    // read only so no backup required
    BODY *left, *right;
    outcome check;

    api_initialize_booleans();

    if(left_ptr-> identity() == CSG_NODE_TYPE)
        ((CSG_NODE*)left_ptr)-> evaluate(left);
    else  // it must be a BODY
        api_copy_body((BODY*)left_ptr,left);

    if(right_ptr-> identity() == CSG_NODE_TYPE)
        ((CSG_NODE*)right_ptr)-> evaluate(right);
    else  // it must be a BODY
        api_copy_body((BODY*)right_ptr,right);

    if (bo_type == PLUS)
        check = api_unite(right,left);
    else if (bo_type == MINUS) // subtract right from left
        check = api_subtract(right,left);
    if (check.ok())
        result = left;
    else {
        printf("Error in Tree Evaluation\n");
        result = (BODY*)NULL;
```

```
        }
            api_terminate_booleans();
        }
```

The file ends with **CSG_NODE's** only unique member function called **evaluate**. The method recursively carries out the Boolean operation required for the node type. Notice that the BODYs are copied before being evaluated so that the tree is not changed by the process.

13.2 Creating your own API Function

The creation of one's own personal API function is, like ENTITYs, easily done using macros. The following example creates an API function for calculating the symmetric difference between two BODYs. The program **api_main.cxx** shows how the function is used, while **api_sym.cxx** shows the implementation.

api_main.cxx

```
// This program uses a user-defined API function to
// calculate the symmetric difference between two BODYs
#include <stdio.h>
#include "kernel/acis.hxx"                      // Declares modeller parameters
#include "constrct/kernapi/api/cstrapi.hxx"     // Construction API's
#include "kernel/kernapi/api/api.hxx"           // Declares outcome class
#include "kernel/kernapi/api/kernapi.hxx"       // Declares kernel API's
#include "kernel/kerndata/top/body.hxx"         // Declares BODY class
#include "kernel/kerndata/lists/lists.hxx"      // Declares ENTITY_LIST class
#include "baseutil/vector/transf.hxx"           // Declares the transform class
#include "baseutil/vector/vector.hxx"           // Declares a vector class
#include "api_sym.hxx"                          // Declares new api function
#include "kernel/kerndata/savres/fileinfo.hxx"  // Declares fileinfo class

void save_ent( char *filename, ENTITY *ent); // Save BODYs to file

void main()
{
    api_start_modeller(0);
    api_initialize_constructors();
    // Create two BODYs
    BODY *box, *ball;
    api_make_cuboid(40,40,40,box);
    api_make_sphere(20,ball);
    transf move = translate_transf(vector(12,12,12));
    api_apply_transf((ENTITY*)ball,move);
```

13.2 Creating your own API Function

```
   // Call the API function
   outcome result = api_sym_diff(box,ball);

   if (result.ok())
   {
      save_ent("symbox.sat", box);
      save_ent("symball.sat",ball);
   } else
      printf("Error in api_sym\n");

   api_terminate_constructors();
   api_stop_modeller();
}
```

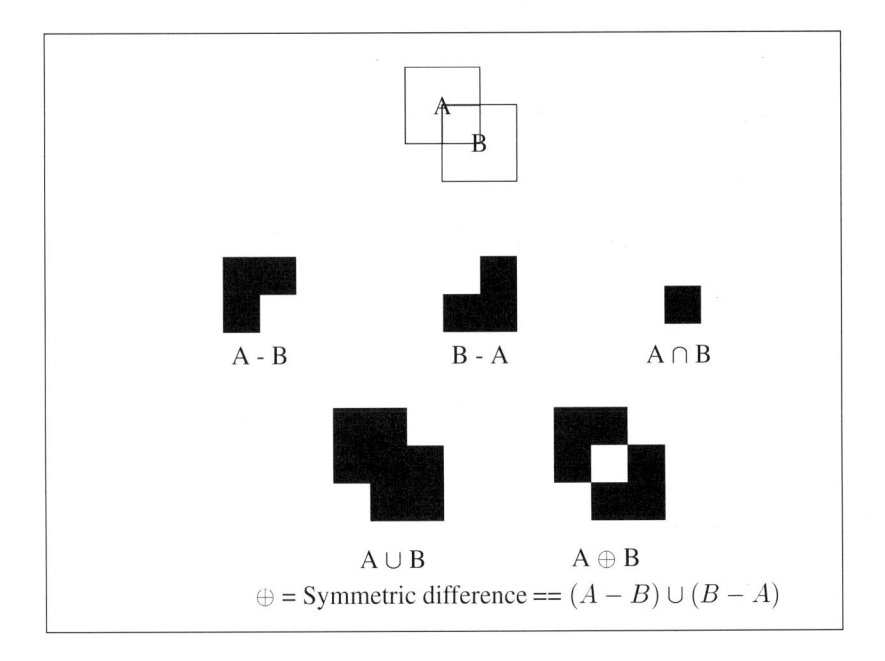

About symmetric differences

A - B   B - A   A ∩ B

A ∪ B   A ⊕ B

⊕ = Symmetric difference == $(A - B) \cup (B - A)$

The file **api_sym.hxx** contains code which specifies the syntax of the API function:

api_sym.hxx

```
// Header file for the Symmetric Difference API
#if !defined( SYM_DIFF )
#define SYM_DIFF
#include "kernel/kernapi/api/api.hxx"
```

```
#include "kernel/kerndata/top/body.hxx"
class outcome;
class BODY;

outcome api_sym_diff( BODY *b1, BODY *b2 );
#endif
```

The actual code for the API is contained in the file **api_sym.cxx**. Most of the API function is placed between the **API_BEGIN** and **API_END** macros which use a combination of error traps and bulletin boards to protect other code from any problems occurring within the procedure.

api_sym.cxx

```
// ****************************************************************
//
// File name: api_sym_diff.cxx (changes model)
//
// Routine name: api_sym_diff
//
// Keywords: Boolean, difference, intersection, subtraction
//
// Action: Computes the symmetric difference of two BODYs
//
// Arguments given:
//        BODY*      b1
//        BODY*      b2
//
// Arguments returned:
//        none
//
// Returns:
//        outcome
//
// Errors:
//        none
//
// Description:
//        Symmetric difference of b1 and b2 is equal to
//      (b1 union b2) minus (b1 intersect b2)
//
//
// ****************************************************************
//
// Include files:
#include <stdio.h>                      // Declares io-functions
#include "kernel/kernapi/api/api.hxx"   // Declare outcome class
```

13.2 Creating your own API Function

```cpp
#include "kernel/kernapi/api/check.hxx"      // Declares argument checking routines
#include "kernel/kerndata/savres/savres.hxx" // Declares non-api copy
#include "boolean/kernbool/boolean/boolean.hxx" // Declares non-api booleans
#include "boolean/kernapi/api/boolapi.hxx"   // Declares booleans API's
#include "kernel/kernapi/api/api.err"        // Declares Error code definitions
#include "api_sym.hxx"                       // Declare this api function

outcome api_sym_diff(BODY* b1, BODY* b2)
{
    BODY *cb1=NULL, *cb2=NULL, *intvol=NULL;
    logical bresult = FALSE;

    api_initialize_booleans();

    API_BEGIN // Macro to set up error handlers and
              // initiate a bulletin board

        cb1 = copy_body_from_body(b1); // Duplicate b1
        cb2 = copy_body_from_body(b2); // Duplicate b2

        bresult = do_boolean(cb1, cb2, INTERSECTION);

    if (bresult)
        intvol = copy_body_from_body(cb2);
    if (intvol != NULL)
        bresult = do_boolean(intvol, b1, SUBTRACTION);
    if (bresult)
        bresult = do_boolean(cb2, b2, SUBTRACTION);

        // result is declared of type outcome in the API_BEGIN macro
        result = outcome( !bresult ? API_FAILED : 0);

    API_END // Macro to terminate error handlers and
            // the current bulletin board

    api_terminate_booleans();

    return result;
}
```

The program **api_sym.cxx** uses several non-API functions for copying BODYs and carrying out Boolean operations. Although these functions do not themselves return **outcomes** they still create bulletins and so support rollback.

13.3 Creating your own Scheme extension

The Scheme AIDE's interpreter offers a means of fast prototyping new applications and painlessly exploring the functionality of the ACIS modeler.

New commands can be added to the Scheme AIDE by writing appropriate C++ functions of your own design. The program **symext.cxx** adds an extension called **solid:symdiff** to the Scheme language which calculates the symmetric difference of the two BODYs given as arguments.

This is done by:

- Creating a C++ program using the extension template shown below.

- Linking the new extension into the Scheme AIDE by recompiling it.

The C++ program is written using several essential macros and procedures which are supplied with the Scheme AIDE.

symext.cxx

```
// This program creates a Scheme extension from
// a user-defined API function to calculate the
// symmetric difference between two BODYs
#include <stdio.h>
#include "kernel/acis3dt.h"      // Declares standard Scheme parameters
#include "scheme/scheme.hxx"     // Declares make_unspecified
#include "kern_scm/ent_typ.hxx"  // Declares get_Scm_Body
#include "pmhusk/ent_utl.hxx"    // Declares Entity modification utils
#include "kernel/geomhusk/trace.hxx"    // Declares ENTER_FUNCTION macro
#include "kernel/geomhusk/ckoutcom.hxx" // Declares check_outcome
#include "kernel/acis.hxx"       // Declares modeler parameters
#include "kernel/kernapi/api/apimsc.hxx" // Declares Kernel API functions
#include "kernel/kerndata/top/body.hxx" // Declares BODY class
#include "api_sym.hxx"           // Declares new API function

ScmObject P_Solid_SymDiff(ScmObject bod1, ScmObject bod2)
{
    // A debugging aid to trace program execution
    ENTER_FUNCTION("P_Solid_SymDiff");

    // Get the two BODYs
    BODY* b1 = get_Scm_Body(bod1);
    BODY* b2 = get_Scm_Body(bod2);

    // Now calculate the difference
    start_entity_modification();
      outcome result = api_sym_diff(b1,b2); // See previous example
      check_outcome(result);
    end_entity_modification(b1, result);
```

```
    // Every Scheme proc must return a result
    // but in this case it is void
    return make_unspecified();
}

SCM_PROC(2,2,"solid:symdiff",P_Solid_SymDiff);
```

Instructions for rebuilding the Scheme AIDE can be found in the "Extending ACIS" section of the documentation. Once the Scheme AIDE is rebuilt one could type:

```
(define tube (solid:cylinder (position 0 0 -50)
                             (position 0 0 50) 10))
(define block (solid:block (position -20 -20 -20)
                           (position 20 20 20)))
; Call new command
(solid:symdiff tube block)
```

A **ScmObject** is the bridge between Scheme and C++. It can contain, or carry, a large number of different *data types* from one environment to the other. The above example shows the **ScmObject** being used to carry a reference to a BODY from the Scheme interpreter to the C++ procedure. The fuction **get_Scm_Body** not only returns a pointer to the BODY but also provides error checking at the Scheme Command level.

The macro **SCM_PROC** is used to associate the name of the Scheme Command, in this case *solid:symdiff*, with the C++ function **P_Solid_SymDiff**. The first two arguments specify the minimum and maximum numbers of arguments.

13.4 Exercise

1. Extend the definition of the CSG_NODE class so that the tree can hold other forms of ENTITY than just BODYs.

Debugging and Error Checking 14

This section looks briefly at mechanisms ACIS provides for debugging ENTITYs and checking the results of API functions. The first example, **error.cxx**, demonstrates how the outcome class returned by all API functions can be interrogated to establish the success or failure of each operation. The next example uses a macro to do the outcome checking. The chapter ends with a demonstration of the **debug_entity** function. In addition to the facilities demonstrated in this chapter's examples ACIS provides several APIs which are useful when bug hunting:

api_checking (Boolean): If TRUE ensures pointer arguments to an API calls are tested to determine whether they are NULL. If they are NULL, a message is printed and the API call returns an error outcome. Checks are also made on certain distances and angles supplied to API calls. Some API calls make more extensive checks internally, but the effect is the same.

api_check_edge (EDGE, list): Checks whether an EDGE defined by a bs3_curve (see Appendix B, Section B.14) is self-intersecting, is twisted, has too much oscillation, has degenerate EDGEs, or is not G^0, G^1 or G^2 continuous. If no errors are found, the NULL list is returned.

api_check_face (FACE, list): Similarly this API checks a spline FACE for various conditions that could cause errors.

api_check_entity (ENTITY, problems, file_ptr): Given an ENTITY, an ENTITY_LIST (called problems) and optionally, a file pointer; this function will traverse the given ENTITY and check for topological, geometric and datastructure problems (see ACIS documentation for a list of the checks).

The exact arguments of these functions can be found in the ACIS documentation.

14.1 Using Outcomes

All API functions return an object of class **outcome** which contains an error number, to indicate the success or failure of the API routine, and a Bulletin-Board pointer.

The program **error.cxx** uses a function called **check_result** to print out the error associated with an API failure.

error.cxx

```
// This program subtracts one torus from a copy of
// itself rotated by the tolerance of the modeler.
#include <stdio.h>                              // Input/Output functions
#include "boolean/kernapi/api/boolapi.hxx"      // Declare boolean API's
#include "constrct/kernapi/api/cstrapi.hxx"     // Construction API's
#include "kernel/kernapi/api/api.hxx"           // Declare outcome class
#include "kernel/kernapi/api/kernapi.hxx"       // Declare kernel API's
#include "kernel/kerndata/top/alltop.hxx"       // Declares topology classes
#include "baseutil/vector/transf.hxx"           // Declares transform class
#include "kernel/kernutil/errorsys/errorsys.hxx" // Declare error functions

void check_result(outcome, char*);

void main()
{
    outcome result = api_start_modeller(0);
    check_result(result,"api_start_modeller");

    result = api_initialize_booleans();
    check_result(result,"api_initialize_booleans");

    result = api_checking(TRUE);
    check_result(result,"api_checking");
    BODY *tor1, *tor2;

    result = api_make_torus(pi*resnor,pi*0.75,tor1);
    check_result(result,"api_make_torus");

    result = api_copy_body(tor1,tor2);
    check_result(result,"api_copy_body");

    transf move = rotate_transf(pi*resnor,vector(1,0,0));
    result = api_apply_transf((ENTITY*)tor1,move);
    check_result(result,"api_apply_transf");

    result = api_boolean(tor1,tor2,SUBTRACTION);
    check_result(result,"api_boolean");

    result = api_terminate_booleans();
    check_result(result,"api_terminate_boolean");

    result = api_stop_modeller();
    check_result(result,"error at the end");
```

14.2 Defining an Error Catching Macro

```
}
void check_result(outcome result, char *mess)
{
    if(!result.ok()){
        err_mess_type err_number=result.error_number();
        printf("Error in %s : %s\n",mess,find_err_mess(err_number));
        }
}
```

Once the program is compiled and run the following text is printed on the screen:

```
Error in api_make_torus : length too close to zero
Error in api_copy_body : segmentation error
Error in api_apply_transf : segmentation error
Error in api_boolean : segmentation error
```

Not surprisingly, after the failure to create the torus things do not go well! The API function, however, *catches* the errors caused by the NULL pointers and allows the program to exit without any problems.

14.2 Defining an Error Catching Macro

Creating a **check_error** function is one way of monitoring the behavior of the API routines. Another method using a macro is shown in the program **api_run.cxx**:

api_run.cxx

```cpp
// This program makes two tori with the same dimensions;
// one is rotated by the resolution of the modeler
// before they are intersected.
// A macro called API_RUN is used to check the results.
#include <stdio.h>                          // Input/Output functions
#include <stdlib.h>                         // Declares exit()
#include "boolean/kernapi/api/boolapi.hxx"  // Declares boolean API's
#include "constrct/kernapi/api/cstrapi.hxx" // Construction API's
#include "kernel/kernapi/api/api.hxx"       // Declares outcome class
#include "kernel/kernapi/api/kernapi.hxx"   // Declares kernel API's
#include "kernel/kerndata/top/alltop.hxx"   // Topological Classes
#include "baseutil/vector/transf.hxx"       // Define transform class

// Error checking macro
#define API_RUN( FUNC_CALL, DO_ON_ERROR)        \
{                                               \
    outcome res = FUNC_CALL;                    \
```

```
        if (!report(res))                                    \
            {printf("Doing error function\n"); DO_ON_ERROR;}  \
    }

    // Function to print out error report
    logical report(outcome res)
    {
        if ( !res.ok())
        {
            err_mess_type err_no = res.error_number();
            printf("API Error: %d in file: %s at line: %d:\n",
                            err_no, __FILE__, __LINE__);
            printf("-------- %s -------\n",find_err_mess( err_no ));
        }
        return res.ok();
    }

    void main()
    {
        API_RUN(api_start_modeller(0), exit(1);)
        API_RUN(api_initialize_booleans(), exit(0);)

        BODY* hoopS;
        API_RUN(api_make_torus(55.0, 25.0, hoopS),exit(0);)

        BODY* hoopB;
        API_RUN(api_make_torus(55.0, 25.0, hoopB),exit(0);)

        // resnor is a very small amount, see page128
        transf rot = rotate_transf(resnor,vector(1,0,0));

        API_RUN(api_apply_transf((ENTITY*)hoopS,rot),exit(0);)

        FACE *faceS = hoopS->lump()->shell()->face_list();
        FACE *faceB = hoopB->lump()->shell()->face_list();

        BODY *int_graph;

        // Do a FACE-FACE intersectioneit
        API_RUN(api_fafa_int(faceB, faceS, int_graph),exit(0);)

        API_RUN(api_terminate_booleans(), exit(0));
        api_stop_modeller();
    }
```

Once this program is compiled and run the following text is printed on the screen:

```
API Error:1400 in file: C:\abe\api_run.cxx at line :24:
```

```
---------------- operation unsuccessful ---------------
-----
Doing error function
```

Since the error function is to **exit**, the program terminates.

> Spatial's documentation is strangely silent on the subject of what the name ACIS means. Here are three possible explanations:
>
> - **A**cis is the name of a character from classical mythology; it is not uncommon for solid modelers to have classical, or historical, names (e.g. Romulus, Medusa, Euclid and SvLis).
>
> - **A**merican **C**ommittee for **I**nteroperable **S**tandards has the correct initials and sounds like the sort of people who should be involved in geometry buses.
>
> - Much of the early code for the **S**ystem was developed by a group whose names are **A**lan, **C**harles and **I**an.
>
> Clearly only one of these explanations can be correct.

About ACIS

14.3 Using the Debug Functions

There are several non-API functions that can assist debugging, the most notable of which are:

debug_entity(ENTITY, file_ptr): Outputs the given ENTITY and all its subsidiary structures to a given file.

debug_size(ENTITY, file_ptr): Outputs the size of store occupied by the given ENTITY and all its subsidiary structures.

debug(file_ptr): Almost every non-ENTITY ACIS class has a debug method that writes its details to a given file (e.g. **position** p1; p1.**debug**(stdout);).

The program **debug.cxx** demonstrates the use of **debug_entity**.

debug.cxx

```
// This program makes a sphere and a torus, and intersects
// their faces to create a wire. It then uses debug routines to
// print out details of their data structures.
    #include <iostream.h>           // Input/Output functions
```

```cpp
#include "boolean/kernapi/api/boolapi.hxx"  // Declares boolean API's
#include "constrct/kernapi/api/cstrapi.hxx"  // Construction API's
#include "kernel/kernapi/api/kernapi.hxx"    // Declares kernel API's
#include "kernel/kerndata/top/alltop.hxx"    // Topological Classes
#include "kernel/kerndata/data/debug.hxx"    // Debug routines
#include "baseutil/vector/transf.hxx"        // Define transform class

void main()
{
    api_start_modeller(0);
    api_initialize_booleans();

    BODY* ball;
    api_make_sphere(50.0, ball);

    BODY* hoop;
    api_make_torus(55.0, 25.0, hoop);

    transf move = translate_transf(vector(25,25,0));

    api_apply_transf((ENTITY*)hoop,move);

    FACE *ballface = ball->lump()->shell()->face_list();
    FACE *hoopface = hoop->lump()->shell()->face_list();

    BODY *int_graph;

    // Face-Face Intersection
    api_fafa_int(ballface, hoopface, int_graph);

    debug_entity((ENTITY*)int_graph,stdout);

    api_terminate_booleans();
    api_stop_modeller();
}
```

Once this program is compiled and run the following text is printed on the screen:

```
================ BODY LIST ================

body 0 32472:
Rollback pointer: 24528
Attribute list   : NULL
Lump list        : NULL
Wire list        : wire 0 102808
Transform        : NULL
Bounding box     : NULL
```

14.3 Using the Debug Functions

```
================ WIRE LIST ================

wire 0 102808:
Rollback pointer: 25008
Attribute list  : NULL
Owning body     : body 0 32472
Next wire       : wire 1 102776
Start coedge    : coedge 0 98960
Bounding box    : NULL

wire 1 102776:
Rollback pointer: 24768
Attribute list  : NULL
Owning body     : body 0 32472
Next wire       : NULL
Start coedge    : coedge 1 98784
Bounding box    : NULL

================ COEDGE LIST ================

coedge 0 98960:
Rollback pointer: 24944
Attribute list   : intcoed 0 101048
Owning entity    : wire 0 102808
Partner          : coedge 2 99004
Previous coedge  : coedge 0 98960
Next coedge      : coedge 0 98960
Edge             : edge 0 94668
Sense            : forward
Parametric form  : NULL

⋮


intcurve 1 92624:
        Rollback pointer: 24592
        Attribute list  : NULL
        Use count       : 1
        Curve type      : interpolated curve:
              : surface-surface intersection
              : periodic B-spline of degree 3
              :      154 control points:
              :(-3.0959405230631, -49.904059476937, 0), 0
              :(-3.0959405230631, -49.904059476937, 0.30793469...
              :(-3.0854576037391, -49.90181466154, 0.630635374...
              :(-3.0396770478105, -49.891964427704, 1.28883866...

⋮


================ POINT LIST ================
```

```
point 0 86544:
Rollback pointer: 24784
Attribute list  : NULL
Use count       : 1
Coordinates     : -49.904059476937, -3.0959405230631, 0

point 1 86496:
Rollback pointer: 24544
Attribute list  : NULL
Use count       : 1
Coordinates     : -3.0959405230631, -49.904059476937, 0

        1 body record,       32 bytes
        2 wire records,      64 bytes
        4 coedge records,   176 bytes
        4 attribute records, 208 bytes
        2 edge records,      88 bytes
        2 vertex records,    48 bytes
        2 curve records,    128 bytes
        2 point records,     96 bytes
Total storage 840 bytes
-----------------------------------------------------------------
-
```

If the line

    debug_size((**ENTITY***)int_graph,**stdout**);

were used in place of the call to **debug_entity**, then only the size summary shown at the end of the above printout would be output: see, for example, the output of **block.cxx** on page 65.

Getting and Compiling the Examples

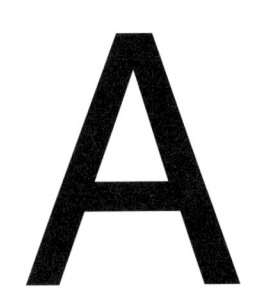

A.1 Getting the Examples

The examples in this book can be downloaded from:

`http://www.hw.ac.uk/mecWWW/research/jrc/research.htm`

The web page contains a separate zip file for each version of ACIS. So examples for ACIS version 6.0 code are located in the file **code60.zip** and for Version 7.0 in **code70.zip** and so on. Generally the only difference between the code in different directories is the names of the included header files or small changes in the syntax of a Scheme extension.

A.2 Compiling the Examples

The examples in this book should work on UNIX, Windows and (with some small modifications, see Page 65) Apple platforms. The exact details of linking and building will vary from user to user so what follows here can only be an outline. The ACIS documentation contains a detailed description of how to link ACIS with the MFC libraries built into Visual C++, which is quite a complex process and beyond the scope of this book. In contrast the following section shows how to create a *console application* (basically a DOS program which executes in a Windows shell) which is all that is required by the programs in this book.

A.2.1 Visual C++ 6.0

These instructions should make a debug version of the first program in the book (block.cxx) that runs using the debug dlls (dynamic link libraries).

1. Preliminaries

The are two methods to make the ACIS debug .dll and .lib files accessible to the application:

(a) Copy the ACIS debug dll library files from the **lib** directory. The default location of this is `Spatial6.0/lib/Nt_dlld/*.dll` to the `windows/system` directory. Or

(b) Extending the system Path environment variable to include the **lib** subdirectory. On Windows NT:

- Right click on the "My Computer Icon"
- Go to "Properties" and select the "Environment" tab
- Enter in Variable: Path
- Enter in Values: %Path%;Spatial6.3/lib/NT/dlld (or what ever)
- Press "Set"

On other Windows OS edit the autoexec.bat by extending the PATH to access the ACIS debug .dll and .lib files (e.g. Path = %Path%;Spatial6.3/lib/NT/dlld).

2. Get the example programs for ACIS6.0 from the web site.

3. Start up Visual C++ 6.0

   (a) Select "New" from the "File" Menu:
   - On the "Project" Tab select "Win32 Console Application"
   - Enter Project Name and Location as prompted.

   (b) Select "Add to Project" from the "Project" Menu:
   - Select "Files" and then Select "block.cxx" from the examples down loaded.

   (c) Select "Set Active Configuration" from the "Build" Menu and select "Win32 Debug"

   (d) Select Settings from the Project Menu:
   - Select the "C++" Tab
     i. Set "Category" to "Preprocessor"
     ii. Add "NT" and "ACIS_DLL" to the "Preprocessor Definitions"
     iii. Add "Additional Include Directories" as required: for block.cxx you need something like:
     `C:/Spatial6.0/cstr6.0,`
     `C:/Spatial6.0/kern6.0,`
     `C:/Spatial6.0/base6.0`
     iv. Set "use runtime library:" to "Debug_Multithreaded_DLL"
   - Select the "Link" Tab
     i. Set "Category" to "Input"

ii. Set "Additional Library Path" to something like `C:/Spatial6.0/lib/Nt_dlld/`
- Select the "Resources" Tab
    i. Add to "Preprocessor definitions" `NT`

(e) Goto the "Build" menu .... compile and run !

4. Other files can be cut from, or inserted into, the project as required.

A.3 Viewing SAT Files with the ACIS AppWizard

Programmers who are using ACIS with Microsoft's Visual C++ can use the ACIS **Application Wizard** (AppWizard) to generate an MFC based application which allows models produced by the example programs to be loaded and viewed in various ways.

When the AppWizard is run from within Visual C++ it asks seven, multiple choice, questions and then automatically generates the source code required to create the type of application specified. It is also possible to extend the default window to include customized buttons and pop-up menus. To use the AppWizard you must first copy the file:

`amfc\aw-i386\AcisAW.aux`

from the ACIS installation directory into,

`Microsoft Visual Studio\Common\MSDev98\Template`

Then consult the ACIS documentation (see under "Interfaces, MFC Interface, Creating a New Application ...") for full instructions. The picture below shows a typical AppWizard interface which was created in less than half an hour.

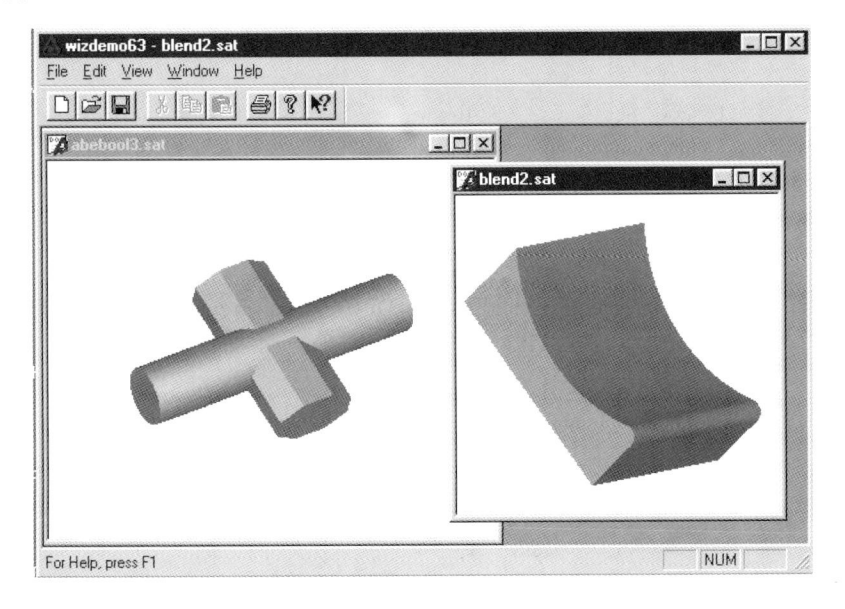

A.4 How to get the ACIS 3D Geometric Modeler

A.4.1 Nonprofit Research and Educational Institutions

Spatial's University Partner Program offers a 3D software component package, which includes the ACIS 3D Geometric Modeler, to nonprofit research and educational institutions. The Program is designed to provide students, faculty, and staff with the opportunity to use the highest quality, state-of-the-art 3D technologies in the classroom and/or for research purposes.

If you are interested in participating in Spatial's University Partner Program, please email productsales@spatial.com for more information. Details on the 3D software component package, offered through the program, are available on Spatial's Web site at http://www.spatial.com.

A.4.2 Commercial Software Developers

Software developers, Original Equipment Manufacturers (OEMs), may receive the ACIS 3D Geometric Modeler and any of Spatial's 3D component products for development and evaluation purposes-with no upfront costs.

Spatial requires that you sign a Development and Evaluation License Agreement before using ACIS and any of their other 3D software component products. This agreement is a contract that is held with Spatial, which enables you to evaluate and develop with their 3D component products, and gain access to related product documentation, free of charge.

What Spatial's Development and Evaluation (D&E) License Agreement entails:

- The D&E Agreement provides you with no-cost evaluation and development until you are ready to ship your product(s). This agreement is renewable yearly and you may terminate it without charge.

- Once you are ready to ship your product(s) or application, Spatial will then convert your D&E Agreement to an OEM Partner Agreement, at which point you become a Spatial OEM Partner and annual and variable partner fees commence. These fees vary based on the Spatial component products used in your application and how you choose to distribute your product(s). In this way, Spatial shares in your success and does not receive revenues until you do, when you begin shipping your application.

- As part of this OEM Partner Program, companies of any size can implement Spatial's quality, high-end 3D software components. Spatial also offers a suite of professional customer services that provide comprehensive technical expertise and development resources to help you successfully implement their components into your product design and development cycle.

To obtain a D&E Agreement from Spatial and more information on their OEM Partner Program, please visit Spatial's Web site at http://www.spatial.com or contact Spatial directly via email at productsales@spatial.com, or via telephone at 303.544.2900 or 800.767.5710.

ACIS Data Structure Class Summary B

This appendix lists in tabular form the member functions of the most commonly used ACIS classes.

Each table is followed by a few brief notes about the class or the working of particular member functions. The layout of the tables reflects the inheritance of functions and data from the base (i.e. **ENTITY**) class.

Note that these tables only provide a summary, rather than an exhaustive definition. The aim is to provide a quick way of jogging the memory. For a definitive description of each class the ACIS documentation should be consulted.

Each table is divided into the following sections:

Data Members: lists the member data referred to by the class's inquiry and change functions.

Constructors: lists the constructors available for each class.

Destructors: lists the destructors available for each class.

Inquiry and Access Functions: lists the access functions for each class. These member functions are nearly all declared to be of type **const**, but to save space this is not shown in the tables.

Change Functions: lists the member functions available for setting the values of the class's data members.

Miscellaneous Functions: lists the functions that don't fit in under any other headings.

Each right-hand table provides names, types and arguments for the **new** or **redefined** member functions in a particular class. Inherited members can be found by reading through the tables listed from *right to left*.

Class:- ENTITY	
Data Members	
	BULLETIN*
	ATTRIB*
Constructors	
	ENTITY
	new
Destructors	
	~ENTITY
	delete
	lose
Inquiry	
	attrib
	check_destroy
	debug_ent
	deletable
	identity
	lookup
	owner
	size
	type_name
	add_method
	apply_transform
	backup
Miscellaneous	
	call_method
	copy_common
	copy_data
	copy_scan
	fix_common
	fix_pointers)
	fixup_copy
	restore_common
	restore_end
	rollback
	save
	save_begin
	save_common
	save_end
	roll_once
Change Functions	
	set_attrib

Class:- CURVE	
Data Members	
	curve*
	int
Constructors	
	CURVE
Destructors	
	lose
	~CURVE
Inquiry Functions	
	deletable
	equation
	identity
	make_box
	trans_curve
	type_name
	use_count
Change Functions	
	add
	operator =
Miscellaneous	
	equation_for_update
	remove

Class:- STRAIGHT		
Data Members		
	straight*	Pointer to a straight object
Constructors		
	public: **STRAIGHT**()	Should be used via the **new** operator
	public: **STRAIGHT**(...)	Makes a STRAIGHT with the given position and direction
	public: **STRAIGHT**(straight const&)	Makes a STRAIGHT entity from a straight object
Destructors		
	public: virtual void **lose**()	Creates a deleted bulletin and posts it on the bulletin_board. Lose methods for attached attributes are also called
	protected: virtual ~STRAIGHT	Invoked by the **lose** method
Inquiry & Access Functions		
	public: unit_vector const& **direction**()	Returns a unit_vector in the direction of the line
	public: curve const& **equation**()	Returns a straight object (derived from curve)
	public: virtual int **identity** (int = 0)	Returns the type identifier STRAIGHT_TYPE if level is unspecified
	public: position const& **root_point**()	Returns a position on the line
	public: curve* **trans_curve**(...)	Transforms the curve's equation (ie, the straight object)
	public: virtual char* **type_name**()	Returns the string "straight"
Change Functions		
	public: void **operator***=(transf const&)	Transforms a STRAIGHT
	public: void **set_direction**(unit_vector)	Sets the STRAIGHTŠs direction to the given unit_vector
	public: void **set_root_point**(position)	Sets the STRAIGHTŠs root point to the given position
Miscellaneous		
	public: curve& **equation_for_update**()	Returns the address of the curve's equation for update operations

For example, if information is required about the *set_attrib* function of the **STRAIGHT** class the following process would find it:

1. The table of the **STRAIGHT** class would be examined.

2. Since no member function of that name is recorded the class summary table to the left (i.e. **CURVE**) is examined.

3. Again since no member function called *set_attrib* is found the summary table to its left (i.e. **ENTITY**) is examined and the function found.

4. The type and arguments of the function can be found by turning to the **ENTITY** class summary.

Methods with arguments shown as (...) have too many arguments for the space available.

B.1 ENTITY

Class:- ENTITY	
Data Members	
BULLETIN*	Pointer to ENTITY's BULLETIN
ATTRIB*	First in a list of ATTRIBs
Constructor	
public: **ENTITY**()	Posts a create BULLETIN on the bulletin board
public: void* operator **new**	Overrides regular **new** operator so a private free list can be used, also calls ENTITY()
Destructor	
protected: virtual ~**ENTITY**()	Should be invoked via **lose** or **delete**
public: void operator **delete**	Calls **protected** destructor (~) and frees memory, but makes no bulletin board record.
public: virtual void **lose**()	Posts a delete BULLETIN on the bulletin board
Inquiry & Access Functions	
public: ATTRIB* **attrib**()	Returns the ATTRIB pointer
protected: void **check_destroy**()	Checks that entity deletion is legitimate
public: virtual void **debug_ent**(FILE*)	Outputs summary of Object
public: virtual logical **deletable**()	TRUE if entity can be safely destroyed by lose( )
public: virtual int **identity**(int)	Returns the type identifier for the instance, its descendants, or its ancestors
public: virtual int **lookup**(logical)	Debug List Utility
public: virtual ENTITY* **owner**()	Returns instance's owner
public: virtual unsigned **size**()	Returns the size, in bytes, of the instance
public: virtual char* **type_name**()	Returns a string containing the class name
public: static MethodFunction **add_method**(...)	Registers an implementation function for a run-time virtual method
public: virtual logical **apply_transform**(...)	Transforms an ENTITY
public: ENTITY* **backup**()	Creates a duplicate object and, if needed, a change BULLETIN
Miscellaneous Functions	
public: virtual logical **call_method**(...)	Executes a virtual method
protected: void **copy_common**(...)	ENTITY copying utility
public: virtual ENTITY* **copy_data**(...)	Creates a new ENTITY and calls copy_common to fill it in
public: virtual void **copy_scan**(...)	Adds dependent entities to the list
protected: void **fix_common**(ENTITY* [~])	Copy Utility, called by fix_pointers
public: virtual void **fix_pointers**(ENTITY* [~])	Used in the copy process
protected: void **fixup_copy**(ENTITY*)	Fixes any pointers in the ENTITY after copying for backup
public: void **restore_common**()	Restores data for an instance from disk to memory
public: void **restore_end**(char*)	Used only by the save/restore system
public: BULLETIN*& **rollback**()	Returns a reference pointer to this instance's bulletin board entry
public: virtual void **save**(ENTITY_LIST&)	Performs a save operation by calling **save_begin**, **save_common**, and **save_end**
protected: void **save_begin**(logical)	Called by the save method
protected: void **save_common**(ENTITY_LIST&)	Called by the save method
protected: void **save_end**(ENTITY_LIST&)	Called by the save method
friend: void **roll_once**(BULLETIN_BOARD*)	Bulletin board system function that calls the roll_notify method
Change Functions	
public: void **set_attrib**(ATTRIB*)	Changes the attribute pointer to point to the new attribute.

Represents: common data and functionality that is mandatory in all classes that are permanent objects in an ACIS model.

Note:

1. The **deletable** function indicates whether an ENTITY is referred to by multiple owners using a **count**. The default TRUE indicates whether this ENTITY is normally destroyed by lose( ) or, if it is shared between multiple owners, gets destroyed when every owner has been lost. Most ENTITYs are destroyed explicitly using lose( ), and so the default returns TRUE.

B.2 BODY

Class:- ENTITY
Data Members
BULLETIN*
ATTRIB*
Constructors
ENTITY
new
Destructors
~ENTITY
delete
lose
Inquiry
attrib
check_destroy
debug_ent
deletable
identity
lookup
owner
size
type_name
add_method
apply_transform
backup
Miscellaneous
call_method
copy_common
copy_data
copy_scan
fix_common
fix_pointers)
fixup_copy
restore_common
restore_end
rollback
save
save_begin
save_common
save_end
roll_once
Change Functions
set_attrib

Class:- BODY	
Data Members	
LUMP*	Pointer to first LUMP in the list
WIRE*	Pointer to first WIRE in the list
TRANSFORM*	Pointer to TRANSFORM between model and world coordinates
Constructors	
public: **BODY**()	Should be used via the **new** operator
public: **BODY**(LUMP*)	Creates a BODY with a LUMP. Sets LUMP back pointers
public: **BODY**(WIRE*)	Creates a BODY with a WIRE. Sets WIRE back pointers
Destructors	
public: virtual void **lose**()	Creates a deleted bulletin and posts it on the bulletin_board
protected: virtual ~BODY()	Invoked by the **lose** method
Inquiry & Access Functions	
public: box* **bound**()	Returns a bounding box if one has been calculated
public: virtual int **identity**(int)	Returns the type identifier BODY_TYPE
public: LUMP* **lump**()	Returns a pointer to the beginning of the list of bounding lumps of a body
public: TRANSFORM* **transform**()	Returns a pointer to the transformation that relates the local to global coordinate system
public: WIRE* **wire**()	Returns a pointer to the start of a list of wires
public: virtual char* **type_name**()	Returns the name "body"
Change Functions	
public: void **set_bound**(box*)	Sets the BODY's bounding box
public: void **set_lump**(LUMP*)	Sets the BODY's LUMP pointer
public: void **set_wire**(WIRE*)	Sets the BODY's WIRE pointer
public: void **set_transform**(TRANSFORM*)	Sets the BODY's TRANSFORM pointer
Miscellaneous	
None	

Represents: a WIRE, SHEET or SOLID BODY.

Notes:

1. It is possible for one BODY to be composed of several disjoint volumes (each represented as a LUMP).

2. A BODY can represent a 2D region (known as a sheet) using double-sided, outside FACEs.

3. A *pure* wire BODY contains only wires, but no lumps, shells, or faces.

4. Wires can represent isolated points, open or closed profiles, and also general wireframe models that have no faces.

5. LUMPs and WIREs are described in a local coordinate system. The local coordinate system is related to the universal coordinate system by a transformation stored within the BODY.

B.3 LUMP

Class:- ENTITY
Data Members
BULLETIN*
ATTRIB*
Constructors
ENTITY
new
Destructors
~ENTITY
delete
lose
Inquiry
attrib
check_destroy
debug_ent
deletable
identity
lookup
owner
size
type_name
add_method
apply_transform
backup
Miscellaneous
call_method
copy_common
copy_data
copy_scan
fix_common
fix_pointers)
fixup_copy
restore_common
restore_end
rollback
save
save_begin
save_common
save_end
roll_once
Change Functions
set_attrib

Class:- LUMP		
Data Members		
	SHELL*	Pointer to the first in a list of SHELLs which bound the LUMP
	LUMP*	Pointer to the next LUMP in a list
Constructors		
	public: **LUMP**()	Should be used via the **new** operator
	public: **LUMP**(SHELL*,LUMP*)	Initializes a new LUMP object with pointers to the start of a SHELL list and the next LUMP in a list
Destructors		
	public: virtual void **lose**()	Creates a deleted bulletin and posts it on the bulletin_board
	protected: virtual ~LOOP	Invoked by the **lose** method
Inquiry & Access Functions		
	public: box* **bound**()	Returns a bounding box if one has been calculated
	public: BODY* **body**()	Returns a pointer to the owning BODY
	public: virtual int **identity**(int)	Returns the type identifier LUMP_TYPE
	public: LUMP* **next**()	Returns a pointer to the next LUMP in a list of LUMPs owned by a BODY
	public: ENTITY* **owner**()	Returns a pointer to the owning BODY
	public: SHELL* **start**()	Returns a pointer to first in a NULL terminated list of SHELLs
	public: virtual char* **type_name**()	Returns the name "lump"
Change Functions		
	public: void **set_bound**(box*)	Sets the LUMP's bounding box
	public: void **set_body**(BODY*)	Sets the LUMP's owning BODY
	public: void **set_next**(LUMP*)	Sets the next LUMP in a list of LUMPs
	public: void **set_shell**(SHELL*)	Sets the first SHELL in the LUMP
Miscellaneous		
	None	

Represents: a connected component of a BODY.
Notes:

1. A LUMP is bounded by one or more SHELLs held in a list. One SHELL must represent the external skin.

2. When an operation (such as Boolean subtraction or intersection) returns a BODY composed of more than one piece, each piece will be represented by a LUMP.

3. A sheet BODY can be composed of a number of LUMPs each containing a number of FACEs.

B.4 SHELL

Class:- ENTITY
Data Members
BULLETIN*
ATTRIB*
Constructors
ENTITY
new
Destructors
~ENTITY
delete
lose
Inquiry
attrib
check_destroy
debug_ent
deletable
identity
lookup
owner
size
type_name
add_method
apply_transform
backup
Miscellaneous
call_method
copy_common
copy_data
copy_scan
fix_common
fix_pointers
fixup_copy
restore_common
restore_end
rollback
save
save_begin
save_common
save_end
roll_once
Change Functions
set_attrib

Class:- SHELL	
Data Members	
FACE*	Pointer to the first in a list of FACEs which form the SHELL
SHELL*	Pointer to the next SHELL in a list
Constructors	
public: **SHELL**()	Should be used via the **new** operator
public: **SHELL**(FACE*,SUBSHELL*,SHELL*)	Initializes a new SHELL object with pointers to the start of SUBSHELL and FACE lists and the next SHELL in a list
Destructors	
public: virtual void **lose**()	Creates a deleted bulletin and posts it on the bulletin_board
protected: virtual ~SHELL	Invoked by the **lose** method
Inquiry & Access Functions	
public: box* **bound**()	Returns a bounding box if one has been calculated
public: FACE* **first_face**()	Returns a pointer to the start of a linked listed of all the FACEs in the SHELL
public: virtual int **identity**(int)	Returns the type identifier SHELL_TYPE
public: SHELL* **next**()	Returns a pointer to the next SHELL in a list of SHELLs owned by a LUMP
public: LUMP* **lump**()	Returns a pointer to the owning LUMP
public: ENTITY* **owner**()	Returns a pointer to the owning LUMP
public: FACE* **face_list**()	Returns a pointer to first in a NULL terminated list of FACEs
public: FACE* **face**()	Returns the first FACE in a complete enumeration of all the FACEs in the SHELL, continued by repeated use of FACE::next_face
public: SUBSHELL* **subshell**()	Returns a pointer to the first SUBSHELL in a list of SUBSHELLs immediately contained within this SHELL
public: WIRE* **wire_list**()	Returns a pointer to the first WIRE of a list of WIREs immediately contained in this SHELL
public: WIRE* **wire**()	Returns the first WIRE in a complete enumeration of all the WIREs in the SHELL, continued by repeated use of WIRE::next()
public: virtual char* **type_name**()	Returns the name "shell"
Change Functions	
public: void **set_wire**(WIRE*)	Sets the SHELLŠs WIRE pointer
public: void **set_bound**(box*)	Sets the SHELL's bounding box
public: void **set_face**(FACE*)	Sets the SHELLŠs FACE pointer to the given FACE
public: void **set_lump**(LUMP*)	Sets the SHELL's owning LUMP
public: void **set_next**(SHELL*)	Sets the next SHELL in a list of SHELLs
public: void **set_subshell**(SUBSHELL*)	Sets the first SUBSHELL in the SHELL
Miscellaneous	
None	

Represents: part of a LUMP's boundary.

Notes:

1. A SHELL is constructed from a connected collection of FACEs, each a bounded or unbounded piece of a single geometric SURFACE.

2. The SHELL is one portion of a LUMP's boundary, and has no internal connection with any other SHELL. If a LUMP has no voids only one SHELL is required.

3. A SHELL composed of single-sided FACEs is said to be **open** and **bounded**, if it has an incomplete boundary (e.g. a missing FACE). It interacts with other BODYs only so far as the defined portions of their SHELLs interact.

B.5 FACE

Class:- FACE		
Data Members		
	SURFACE*	Pointer to the FACE's underlying SURFACE
	SHELL*	Pointer to the FACE's owning SHELL
	FACE*	Pointer to the next FACE in the SHELL's list
	LOOP*	Pointer to the first LOOP in the list
	SIDESBIT	Defines if FACE is SINGLE or DOUBLE_SIDED
	REVBIT	Defines the FACE's sense
	CONTBIT	Defines the containment of double-sided FACEs
Constructors		
	public: **FACE**()	Should be used via the **new** operator
	public: **FACE**(OLD_FACE*,LOOP*,logical)	Creates a FACE, using the given LOOP list, but taking geometry, senses, and shell and subshell owners from the old FACE. If logical is TRUE, updates the SHELL or SUBSHELL FACE list to contain the new FACE
	public: **FACE**(LOOP*,FACE*,SURFACE*,REVBIT)	Creates a FACE initializing its first LOOP, the next FACE on the BODY, the underlying SURFACE geometry, and the sense of the FACE relative to the surface (FORWARD or REVERSED)
Destructors		
	public: virtual void **lose**()	Creates a deleted bulletin and posts it on the bulletin_board
	protected: virtual ~FACE()	Invoked by the **lose** method
Inquiry & Access Functions		
	public: box* **bound**()	Returns a bounding box if one has been calculated
	public: FACE* **next**()	Returns a pointer to the next FACE the list
	public: CONTBIT **cont**()	Returns the containment of double sided FACEs (ie, BOTH_OUTSIDE or BOTH_INSIDE)
	public: SURFACE* **geometry**()	Returns a pointer to the FACE's underlying SURFACE
	public: virtual int **identity**(int)	Returns the type identifier FACE_TYPE
	public: LOOP* **loop**()	Returns a pointer to the first in a list of LOOPs that bounding the FACE
	public: FACE* **next_face**()	Returns the next FACE in a complete enumeration of all the FACEs in the SHELL
	public: FACE* **next_in_list**()	Returns a pointer to the next FACE in the list of FACEs contained directly by a SHELL
	public: ENTITY* **owner**()	Returns a pointer to the owning entity
	public: REVBIT **sense**()	Returns the sense of the FACE (ie, FORWARD or REVERSED) relative to the SURFACE
	public: REVBIT **sense**(REVBIT rev)	Return the sense of the FACE compounded with the sense argument
	public: SHELL* **shell**()	Returns a pointer to the SHELL containing the FACE
	public: SIDESBIT **sides**()	Returns SINGLE_SIDED or DOUBLE_SIDED
	public: SUBSHELL* **subshell**()	Returns a pointer to the SUBSHELL containing the FACE
	public: virtual char* **type_name**()	Returns the name "face"
Change Functions		
	public: void **set_bound**(box*)	Sets the FACE's bounding box to be the given box
	public: void **set_cont**(CONTBIT)	Sets the FACEŠs containment bit to indicate whether the FACE is fully contained within the parent SHELL or not
	public: void **set_geometry**(SURFACE*)	Sets the FACEŠs geometry pointer
	public: void **set_loop**(LOOP*)	Sets the FACEŠs LOOP pointer
	public: void **set_next**(FACE*)	Sets the FACEŠs next FACE pointer
	public: void **set_sense**(REVBIT)	Sets the FACEŠs sense (FORWARD or REVERSED) with respect to the SURFACE
	public: void **set_shell**(SHELL*)	Sets the FACEŠs SHELL pointer
	public: void **set_sides**(SIDESBIT)	Sets SIDESBIT to SINGLE_SIDED or DOUBLE_SIDED
	public: void **set_subshell**(SUBSHELL*)	Sets the FACEŠs SUBSHELL pointer
Miscellaneous		
	None	

Class:- ENTITY
Data Members
BULLETIN*
ATTRIB*
Constructors
ENTITY
new
Destructors
~ENTITY
delete
lose
Inquiry
attrib
check_destroy
debug_ent
deletable
identity
lookup
owner
size
type_name
add_method
apply_transform
backup
Miscellaneous
call_method
copy_common
copy_data
copy_scan
fix_common
fix_pointers)
fixup_copy
restore_common
restore_end
rollback
save
save_begin
save_common
save_end
roll_once
Change Functions
set_attrib

Represents: part of a SURFACE.

B.6 LOOP

Class:- ENTITY
Data Members
BULLETIN*
ATTRIB*
Constructors
ENTITY
new
Destructors
~ENTITY
delete
lose
Inquiry
attrib
check_destroy
debug_ent
deletable
identity
lookup
owner
size
type_name
add_method
apply_transform
backup
Miscellaneous
call_method
copy_common
copy_data
copy_scan
fix_common
fix_pointers)
fixup_copy
restore_common
restore_end
rollback
save
save_begin
save_common
save_end
roll_once
Change Functions
set_attrib

Class:- LOOP	
Data Members	
COEDGE*	Pointer to the first COEDGE in the LOOP
LOOP*	Pointer to the next LOOP in the list
Constructors	
public: **LOOP**()	Should be used via the **new** operator
public: **LOOP**(COEDGE*,LOOP*)	Initializes a new LOOP object with pointers to the start of a COEDGE list and the next LOOP, both of which could be NULL
Destructors	
public: virtual void **lose**()	Creates a deleted bulletin and posts it on the bulletin_board
protected: virtual ~LOOP	Invoked by the **lose** method
Inquiry & Access Functions	
public: box* **bound**()	Returns a bounding box if one has been calculated
public: FACE* **face**()	Returns a pointer to the FACE containing the LOOP
public: virtual int **identity**(int)	Returns the type identifier LOOP_TYPE
public: LOOP* **next**()	Returns a pointer to the next LOOP in a list of LOOPs bounding a FACE
public: ENTITY* **owner**()	Returns a pointer to the owning entity
public: COEDGE* **start**()	Returns a pointer to a COEDGE in the LOOP (the first if the LOOP is not closed)
public: virtual char* **type_name**()	Returns the name "loop"
Change Functions	
public: void **set_bound**(box*)	Sets the LOOPŠs bounding box to be the given box
public: void **set_face**(FACE*)	Sets the LOOP's owning FACE
public: void **set_next**(LOOP*)	Sets the next LOOP in a list of LOOPs
public: void **set_start**(COEDGE*)	Sets the first COEDGE in the LOOP
Miscellaneous	
None	

Represents: an ordered list of connected EDGEs that bounds a FACE.

Notes:

1. A LOOP can always be traversed completely from the start COEDGE by following the COEDGE next pointer until a NULL is returned in an open LOOP, or until the start COEDGE is returned the second time in a closed LOOP.

2. The LOOPs of a FACE are not distinguished as internal or external at the FACE level, because the distinction is not relevant for some surface types (e.g., a cylindrical FACE with two circular LOOPs bounding the ends).

3. FACE LOOPs need not be closed. If not, either open end can be finite or infinite. If either end is infinite, the FACE is infinite. If either end is finite, the FACE is incomplete.

B.7 EDGE

Class:- ENTITY
Data Members
BULLETIN*
ATTRIB*
Constructors
ENTITY
new
Destructors
~ENTITY
delete
lose
Inquiry
attrib
check_destroy
debug_ent
deletable
identity
lookup
owner
size
type_name
add_method
apply_transform
backup
Miscellaneous
call_method
copy_common
copy_data
copy_scan
fix_common
fix_pointers
fixup_copy
restore_common
restore_end
rollback
save
save_begin
save_common
save_end
roll_once
Change Functions
set_attrib

Class:- EDGE		
Data Members		
	VERTEX*	Start of the EDGE
	VERTEX*	End of the EDGE
	COEDGE*	Lying on one the Adjacent Faces
	CURVE*	Geometry of the EDGE
Constructors		
	public: **EDGE**()	Should be used via the **new** operator
	public: **EDGE**(...)	Creates a new EDGE given two VERTICES, a CURVE and a sense
Destructors		
	public: virtual void **lose**()	Creates a deleted bulletin and posts it on the bulletin_board
	protected: virtual ~EDGE	Invoked by the **lose** method
Inquiry & Access Functions		
	public: VERTEX *start()	Returns pointer to start vertex
	public: VERTEX *end()	Returns pointer to end vertex
	public: COEDGE *coedge()	Returns pointer to one of the COEDGE's associated with the EDGE
	public: CURVE *geometry()	Returns the pointer to the underlying CURVE geometry
	public: REVBIT **sense**()	Define whether the EDGE and CURVE share same direction
	public: box* **bound**()	Returns the bounding box if one has been calculated
	public parameter **start_param**()	Returns the parameter value on CURVE at the start vertex (see notes)
	public parameter **end_param**()	Returns the parameter value on CURVE at the end vertex (see notes)
	public: position const &**start_pos**()	Returns the start position of the EDGE
	public: position const &**end_pos**()	Returns the end position of the EDGE
	public: interval **param_range**()	Returns the parameter range (see notes)
	public: virtual char* **type_name**()	Returns the string "edge"
	public: virtual int **identity**(int)	Returns the type identifier EDGE_TYPE
	public: ENTITY* **owner**()	Returns a pointer to the owning entity
Change Functions		
	void **set_start**(VERTEX *)	Sets the EDGE's start VERTEX (calls **backup**)
	void **set_end**(VERTEX *)	Sets the EDGE's end VERTEX (calls **backup**)
	void **set_coedge**(COEDGE *)	Sets the start of a list of COEDGEs corresponding to this EDGE (calls **backup**)
	void **set_sense**(REVBIT)	Sets the EDGE's sense
	void **set_bound**(box *)	Sets the EDGE's bounding box to be the given box (calls **backup**)
	void **set_geometry**(CURVE *)	Sets the geometry of the EDGE to be the given CURVE. Adjusts **use-counts** (calls **backup**)

Represents: part of a FACE's boundary.

Notes:

1. The VERTEX pointer at either or both ends of an EDGE can be NULL, in which case the EDGE is taken to be unbounded in that direction. If the underlying curve is infinite, so is the unbounded EDGE. If the curve is closed, the VERTEX pointers must both be present and represent the same data structure, or both be NULL. In the latter case the EDGE is the whole curve.

2. If the geometry pointer is NULL and both VERTEX pointers refer to the same VERTEX, the EDGE is really an isolated POINT. This occurs at the apex of a CONE.

B.8 COEDGE

Class:- COEDGE		
Data Members		
	COEDGE*	Partner COEDGE
	REVBIT	Relative Direction
	ENTITY*	Instance's Owner
	PCURVE*	Parameterization of the COEDGE's geometry
Constructors		
	public: **COEDGE**()	Should be invoked via the **new** operator
	public: **COEDGE**(...)	Creates and initializes an instance, posting a create BULLETIN on the bulletin board
Destructors		
	public: virtual void **lose**()	Posts delete BULLETIN with backup copy, calls **lose** on every member of the instance's ATTRIB list
	protected: virtual ~COEDGE()	Should be invoked via the **lose** method
Inquiry & Access Functions		
	public: EDGE* **edge**()	Returns pointer to EDGE associated with instance
	public: VERTEX* **end**()	Returns the correct *end* VERTEX from the EDGE relative to the instance's sense
	public: VERTEX* **start**()	Returns the correct *start* VERTEX from the EDGE relative to the instance's sense
	public: VERTEX* **end**(REVBIT)	Returns the correct *end* VERTEX from the EDGE relative to the given sense
	public: VERTEX* **start**(REVBIT)	Returns the correct *start* VERTEX from the EDGE relative to the given sense
	public: PCURVE* **geometry**()	Returns a parameterization of the EDGE's geometry if the COEDGE lies on a parametric surface
	public: virtual int **identity**(int)	Returns the type identifier (COEDGE_TYPE) at given level (COEDGE_LEVEL), or depth, of derviation
	public: LOOP* **loop**()	Returns any owning LOOP of the instance
	public: COEDGE* **next**()	Returns the next COEDGE in a doubly-linked list of COEDGEs
	public: ENTITY* **owner**()	Returns the pointer to the LOOP, WIRE or SHELL that owns the COEDGE
	public: COEDGE* **partner**()	Returns the pointer to a partner COEDGE
	public: COEDGE* **previous**()	Returns the previous COEDGE in a doubly-linked list of COEDGEs
	public: COEDGE* **previous**(REVBIT)	Returns the previous pointer if REVBIT == FORWARD; otherwise returns the next pointer
	public: REVBIT **sense**()	Returns the relationship (FORWARD or REVERSED) between the direction of the COEDGE and that of the underlying EDGE
	public: REVBIT **sense**(REVBIT)	Return the **sense** of the COEDGE compounded with the REVBIT argument
	public: SHELL* **shell**()	Returns the owner of the COEDGE if it is a SHELL; otherwise, it returns NULL
	public: virtual char* **type_name**()	Returns the string "coedge"
	public: WIRE* **wire**()	Returns the owner of the COEDGE if it is a WIRE; otherwise, it returns NULL
Change Functions		
	public: void **set_edge**(EDGE*)	Sets the COEDGE to use the given underlying EDGE and calls **backup**
	public: void **set_geometry**(PCURVE*)	Sets the COEDGE's parameter-space geometry to be the given PCURVE and calls **backup**
	public: void **set_loop**(LOOP*)	Sets the owning ENTITY to be a LOOP
	public: void **set_next**(COEDGE*)	Sets the COEDGE's next COEDGE pointer
	public: void **set_next**(...)	Sets the COEDGE's next COEDGE pointer, taking the sense argument into account
	public: void **set_owner**(ENTITY*)	Sets the COEDGE's owner
	public: void **set_partner**(COEDGE*)	Sets the COEDGE's partner to be the given COEDGE
	public: void **set_previous**(COEDGE*)	Sets COEDGE's previous COEDGE pointer
	public: void **set_previous**(...)	Sets COEDGE's previous COEDGE pointer, taking the sense argument into account
	public: void **set_sense**(REVBIT)	Sets the sense of the COEDGE with respect to the underlying EDGE
	public: void **set_shell**(SHELL*)	Sets the owning entity to be a SHELL
	public: void **set_wire**(WIRE*)	Sets the owning entity to be a WIRE
Miscellaneous		
	public: void **roll_notify**(...)	Flags rollback

Class:- ENTITY
Data Members
BULLETIN*
ATTRIB*
Constructors
ENTITY
new
Destructors
~ENTITY
delete
lose
Inquiry
attrib
check_destroy
debug_ent
deletable
identity
lookup
owner
size
type_name
add_method
apply_transform
backup
Miscellaneous
call_method
copy_common
copy_data
copy_scan
fix_common
fix_pointers)
fixup_copy
restore_common
restore_end
rollback
save
save_begin
save_common
save_end
roll_once
Change Functions
set_attrib

Represents: a boundary segment lying **on** a FACE.

B.9 VERTEX

Class:- ENTITY
Data Members
BULLETIN*
ATTRIB*
Constructors
ENTITY
new
Destructors
~ENTITY
delete
lose
Inquiry
attrib
check_destroy
debug_ent
deletable
identity
lookup
owner
size
type_name
add_method
apply_transform
backup
Miscellaneous
call_method
copy_common
copy_data
copy_scan
fix_common
fix_pointers
fixup_copy
restore_common
restore_end
rollback
save
save_begin
save_common
save_end
roll_once
Change Functions
set_attrib

Class:- VERTEX		
Data Members		
	APOINT*	Location of the VERTEX
	int	Number of EDGEs using the VERTEX
	EDGE*	One of the VERTEX's EDGEs
Constructors		
	public: **VERTEX**()	Should be used via the **new** operator
	public: **VERTEX**(APOINT*)	Creates a new VERTEX give a location
Destructors		
	public: virtual void **lose**()	Creates a deleted bulletin and posts it on the bulletin_board
	protected: virtual ~VERTEX	Invoked by the **lose** method
Inquiry & Access Functions		
	public: int **count_edges**()	Returns the number of EDGES pointed to by the VERTEX
	public: EDGE* **edge**()	Returns NULL if the VERTEX contains more than one EDGE pointer
	public: EDGE* **edge**(int)	Returns EDGE point number, int (eg the first point is accessed by calling edge(0))
	public: logical **edge_linked**(EDGE*)	Returns TRUE if argument uses EDGE
	publice APOINT* **geometry**()	Returns the APOINT
	public: virtual int **identity**(int)	Returns the class type
	public: ENTITY* **owner**()	Returns a pointer to the owning entity
	public: virtual char* **type_name**()	Returns the string "vertex"
Change Functions		
	public: void **add_edge**(EDGE*)	Adds an EDGE to the VERTEX
	public: void **set_edge**(EDGE*)	Replaces all existing pointers with the argument
	public: void **set_geometry**(APOINT*)	Set the APOINT
Miscellaneous		
	public: void **roll_notify**(...)	Inform the Bulletin Board

Represents: the end point of an EDGE.

Notes:

1. The VERTEX may contain pointers to multiple EDGEs to provide access to all the EDGEs at the VERTEX, such as when a body is nonmanifold at a VERTEX, or when an unembedded EDGE dangles from a VERTEX of an otherwise well-formed solid.

2. All EDGEs adjacent to a VERTEX can be found by following pointers through the COEDGEs. This is achieved by following the next, previous, and partner pointers of the COEDGEs of the EDGEs as appropriate for WIREs or embedded EDGEs.

3. If all the EDGEs at the VERTEX are WIREs (i.e. each adjacent to *no* FACEs) or if all are embedded (i.e. each adjacent to two FACEs and in one manifold group), the VERTEX will contain a pointer to a single EDGE.

B.10 CURVE

Class:- ENTITY
Data Members
BULLETIN*
ATTRIB*
Constructors
ENTITY
new
Destructors
~ENTITY
delete
lose
Inquiry
attrib
check_destroy
debug_ent
deletable
identity
lookup
owner
size
type_name
add_method
apply_transform
backup
Miscellaneous
call_method
copy_common
copy_data
copy_scan
fix_common
fix_pointers)
fixup_copy
restore_common
restore_end
rollback
save
save_begin
save_common
save_end
roll_once
Change Functions
set_attrib

Class:- CURVE	
Data Members	
curve*	Pointer to the curve equation (ie, object)
int	Number of EDGEs on the CURVE
Constructors	
public: **CURVE**()	Should be used via the **new** operator
Destructors	
public: virtual void **lose**()	Creates a deleted bulletin and posts it on the bulletin_board
protected: virtual ~CURVE	Invoked by the **lose** method
Inquiry & Access Functions	
public: virtual logical **deletable**()	Indicates whether the CURVE is shared between multiple owners
public: virtual curve const& **equation**	Returns the curve equation for reading only
public: virtual int **identity**(int)	Returns the type identifier CURVE_TYPE
public: virtual box **make_box**(...)	Determines a bounding box for the curve between two points
public: virtual curve* **trans_curve**(...)	Transforms a curve equation by the given transform
public: virtual char* **type_name**()	Returns the curve's name
public: int **use_count**()	Returns the use_count of the CURVE
Change Functions	
public: void **add**()	Increments the CURVE's use_count
public: virtual void **operator*=** (transf const&)	Transforms a CURVE
Miscellaneous	
public: virtual curve& **equation_for_update**()	Returns a pointer to curve's equation for update operations
public: void **remove**(logical lose_if_zero= TRUE)	Decrements the CURVE's use_count. If the use_count becomes 0, the CURVE is deleted

Represents: a variety of curve geometries.

Notes:

1. A use-count is provided so that CURVEs may be multiply referenced. The construction of a new CURVE initializes the use-count to 0.

2. Methods are provided to increment and decrement the use-count and when the use-count reaches 0 the entity is deleted.

3. The **curve** class contains a range of virtual functions which provide access to the specific geometry details.

4. The missing **make_box** arguments are:

   public: virtual box **make_box**(APOINT*,APOINT*, transf*)

B.11 STRAIGHT

Class:- ENTITY
Data Members
BULLETIN*
ATTRIB*
Constructors
ENTITY
new
Destructors
~ENTITY
delete
lose
Inquiry
attrib
check_destroy
debug_ent
deletable
identity
lookup
owner
size
type_name
add_method
apply_transform
backup
Miscellaneous
call_method
copy_common
copy_data
copy_scan
fix_common
fix_pointers)
fixup_copy
restore_common
restore_end
rollback
save
save_begin
save_common
save_end
roll_once
Change Functions
set_attrib

Class:- CURVE
Data Members
curve*
int
Constructors
CURVE
Destructors
lose
~CURVE
Inquiry Functions
deletable
equation
identity
make_box
trans_curve
type_name
use_count
Change Functions
add
operator =
Miscellaneous
equation_for_update
remove

Class:- STRAIGHT	
Data Members	
straight*	Pointer to a straight object
Constructors	
public: STRAIGHT()	Should be used via the **new** operator
public: STRAIGHT(...)	Makes a STRAIGHT with the given position and direction
public: STRAIGHT(straight const&)	Makes a STRAIGHT entity from a straight object
Destructors	
public: virtual void **lose**()	Creates a deleted bulletin and posts it on the bulletin_board. Lose methods for attached attributes are also called
protected: virtual ~STRAIGHT	Invoked by the **lose** method
Inquiry & Access Functions	
public: unit_vector const& **direction**()	Returns a unit_vector in the direction of the line
public: curve const& **equation**()	Returns a straight object (derived from curve)
public: virtual int **identity** (int = 0)	Returns the type identifier STRAIGHT_TYPE if level is unspecified
public: position const& **root_point**()	Returns a position on the line
public: curve* **trans_curve**(...)	Transforms the curve's equation (ie, the straight object)
public: virtual char* **type_name**()	Returns the string "straight"
Change Functions	
public: void **operator*=**(transf const&)	Transforms a STRAIGHT
public: void **set_direction**(unit_vector)	Sets the STRAIGHTŠs direction to the given unit_vector
public: void **set_root_point**(position)	Sets the STRAIGHTŠs root point to the given position
Miscellaneous	
public: curve& **equation_for_update**()	Returns the address of the curve's equation for update operations

Represents: an infinite straight line.
Notes:

1. Full constructor declaration is:
   public: **STRAIGHT**(position const& , unit_vector const&);

2. Both set_* functions have arguments of the type **const&**

B.12 ELLIPSE

Class:- ENTITY
Data Members
BULLETIN*
ATTRIB*
Constructors
ENTITY
new
Destructors
~ENTITY
delete
lose
Inquiry
attrib
check_destroy
debug_ent
deletable
identity
lookup
owner
size
type_name
add_method
apply_transform
backup
Miscellaneous
call_method
copy_common
copy_data
copy_scan
fix_common
fix_pointers
fixup_copy
restore_common
restore_end
rollback
save
save_begin
save_common
save_end
roll_once
Change Functions
set_attrib

Class:- CURVE	
Data Members	
curve*	
	int
Constructors	
CURVE	
Destructors	
	lose
	~CURVE
Inquiry Functions	
	deletable
	equation
	identity
	make_box
	trans_curve
	type_name
	use_count
Change Functions	
	add
	operator *=
Miscellaneous	
	equation_for_update
	remove

Class:- ELLIPSE		
Data Members		
	ellipse	A ellipse object
Constructors		
	public: **ELLIPSE**()	Should be used via the **new** operator
	public: **ELLIPSE**(ellipse const&)	Makes an ELLIPSE entity from an ellipse class
	public: **ELLIPSE**(position, unit_vector,vector,real)	Makes an ELLIPSE from a given center, unit normal, major-axis (vector gives the length and direction), and radius ratio
Destructors		
	public: virtual void **lose**()	Creates a deleted bulletin and posts it on the bulletin_board
	protected: virtual ~ELLIPSE()	Invoked by the **lose** method
Inquiry & Access Functions		
	public: virtual int **identity**(int)	Returns the type identifier ELLIPSE_TYPE
	public: position const& **centre**()	Returns the center of the ELLIPSE
	public: curve const& **equation**()	Returns the curve equation of an ELLIPSE
	public: vector const& **major_axis**()	Returns the major-axis of an ELLIPSE
	public: box **make_box**(...)	Makes a box enclosing a portion of the ELLIPSE between two points lying in the plane of the ELLIPSE
	public: unit_vector const& **normal**()	Returns the normal to the plane of an ELLIPSE
	public: real **radius_ratio**()	Returns the radius ratio of an ELLIPSE
	public: curve* **trans_curve**(...)	Transforms the curve equation of an ELLIPSE
	public: virtual char* **type_name**()	Returns the name "ellipse"
Change Functions		
	public: void **set_major_axis**(vector const&)	Sets the ELLIPSE's major axis to the given vector
	public: void **set_centre**(position const&)	Sets the coordinates of the ELLIPSE
	public: void **set_normal**(unit_vector const&)	Sets the ELLIPSE's planar normal to the given unit_vector
	public: void **set_radius_ratio**(real)	Sets the ELLIPSE's radius ratio to the given value
	public: void **operator*=**(transf const&)	Transforms an ELLIPSE
Miscellaneous		
	public: curve& **equation_for_update**()	Returns address of curve equation for update operations

Represents: an elliptical curve.

Notes:

1. The ELLIPSE class defines an ellipse by its center, a unit normal vector, major-axis vector, and a real giving the eccentricity of the ellipse, i.e., length minor-axis/length major-axis (defaults to 1.0 for a circle).

2. The inherent sense around of the ELLIPSE is given by its normal (using right-hand rule).

B.13 PCURVE

Class:- PCURVE		
Data Members		
	CURVE*	Pointer to an existing CURVE
	int	Type Index
	logical	Negated Flag
	par_vec	parameter space vector
Constructors		
	public: **PCURVE**()	Should be used via the **new** operator
	public: **PCURVE**(pcurve const&)	Makes a PCURVE entity from a pcurve class
	public: **PCURVE**(PCURVE*)	Duplicates a PCURVE
	public: **PCURVE**(...)	Make a PCURVE to point to an existing PCURVE (via a CURVE)
Destructors		
	public: virtual void **lose**()	Creates a deleted bulletin and posts it on the bulletin_board
	protected: virtual ~PCURVE()	Invoked by the **lose** method
Inquiry & Access Functions		
	public: virtual int **identity**(int)	Returns the type identifier PCURVE_TYPE
	public: pcurve const& **def_pcur**()	Returns the definition PCURVE, or NULL if the PCURVE is not private
	public: virtual logical **deletable**()	Indicates whether this entity is normally destroyed by lose() (TRUE), or whether it is shared between multiple owners using a use count
	public: pcurve const& **equation**()	Returns the pcurve object (containing the equation) of a PCURVE
	public: int **index**()	Returns the definition type of the PCURVE
	public: par_vec **offset**()	Returns the par_vec parameter space vector offset
	public: CURVE* **ref_curve**()	Returns the reference CURVE
	public: virtual char* **type_name**()	Returns the string "pcurve"
	public: int **use_count**()	Returns the value of the use-count
Change Functions		
	public: void **add**()	Increments the value of the use-count for the PCURVE
	public: void **negate**()	Negates the PCURVE, either by reversing the PCURVE
	public: void **operator***=(transf const&)	Transforms the PCURVE
	public: void **set_def**(...)	Sets set_def to the nth pcurve of an existing CURVE
	public: void **set_def**(pcurve const&)	Sets def_type to zero, and puts the given pcurve in def
	public: void **shift**(par_vec const&)	Shifts the PCURVE in parameter space by integral multiples of the period on a periodic surface
	public: pcurve* **trans_pcurve**(...)	Constructs a transformed pcurve
Miscellaneous		
	public: void **remove**(logical)	Decrements the value of the use-count for the PCURVE

Class:- ENTITY
Data Members
BULLETIN*
ATTRIB*
Constructors
ENTITY
new
Destructors
~ENTITY
delete
lose
Inquiry
attrib
check_destroy
debug_ent
deletable
identity
lookup
owner
size
type_name
add_method
apply_transform
backup
Miscellaneous
call_method
copy_common
copy_data
copy_scan
fix_common
fix_pointers)
fixup_copy
restore_common
restore_end
rollback
save
save_begin
save_common
save_end
roll_once
Change Functions
set_attrib

Class:- CURVE
Data Members
curve*
int
Constructors
CURVE
Destructors
lose
~CURVE
Inquiry Functions
deletable
equation
identity
make_box
trans_curve
type_name
use_count
Change Functions
add
operator =
Miscellaneous
equation_for_update
remove

Represents: an approximation to a CURVE lying on a SPLINE SURFACE, defined in terms of the surface's u-v parameters.

B.14 INTCURVE

Class:- ENTITY
Data Members
BULLETIN*
ATTRIB*
Constructors
ENTITY
new
Destructors
~ENTITY
delete
lose
Inquiry
attrib
check_destroy
debug_ent
deletable
identity
lookup
owner
size
type_name
add_method
apply_transform
backup
Miscellaneous
call_method
copy_common
copy_data
copy_scan
fix_common
fix_pointers)
fixup_copy
restore_common
restore_end
rollback
save
save_begin
save_common
save_end
roll_once
Change Functions
set_attrib

Class:- CURVE
Data Members
curve*
int
Constructors
CURVE
Destructors
lose
~CURVE
Inquiry Functions
deletable
equation
identity
make_box
trans_curve
type_name
use_count
Change Functions
add
operator =
Miscellaneous
equation_for_update
remove

Class:- INTCURVE	
Data Members	
intcurve*	Pointer to an intcurve object
logical	Reversal Flag
Constructors	
public: **INTCURVE**()	Should be used via the **new** operator
public: **INTCURVE**(intcurve const&)	Makes an INTCURVE entity from an intcurve class
Destructors	
public: virtual void **lose**()	Creates a deleted bulletin and posts it on the bulletin_board
protected: virtual ~INTCURVE()	Invoked by the **lose** method
Inquiry & Access Functions	
public: virtual int **identity**(int)	Returns the type identifier INTCURVE_TYPE
public: curve const& **equation**()	Returns the curve object (containing the equation) of the CURVE
public: box **make_box**(...)	Makes a box enclosing a segment of the INTCURVE between two points, and transforms it
public: curve* **trans_curve**(...)	Transforms the curve's equation
public: virtual char* **type_name**()	Returns the string "intcurve"
Change Functions	
public: void **operator***=(transf const&)	Transforms the INTCURVE
public: void **set_def**(intcurve const&)	Sets the INTCURVE's definition curve
Miscellaneous	
public: curve& **equation_for_update**()	Returns the curve's equation

Represents: an intersection curve as an object in the model.

Notes:

1. An INTCURVE may represent an intersection between two surfaces, the projection of a curve onto a surface, or any other general curve.

2. An INTCURVE records an **interpolated** approximation to a general curve as an **intcurve** (lowercase) and a use-count inherited from the CURVE class.

3. The **intcurve** contains a 3D spline (known as a bs3_curve), a real for a fit tolerance, up to two pointers to the (lowercase) surfaces that describe the curve, and zero, one or two 2D splines (known as bs2_splines) that record interpolated curves in parameter space, depending on whether zero, one or both surfaces are parametric surfaces.

B.15 SURFACE

Class:- ENTITY	
Data Members	
	BULLETIN*
	ATTRIB*
Constructors	
	ENTITY
	new
Destructors	
	~ENTITY
	delete
	lose
Inquiry	
	attrib
	check_destroy
	debug_ent
	deletable
	identity
	lookup
	owner
	size
	type_name
	add_method
	apply_transform
	backup
Miscellaneous	
	call_method
	copy_common
	copy_data
	copy_scan
	fix_common
	fix_pointers
	fixup_copy
	restore_common
	restore_end
	rollback
	save
	save_begin
	save_common
	save_end
	roll_once
Change Functions	
	set_attrib

Class:- SURFACE	
Data Members	
surface*	Pointer to a surface equation (ie, object)
int	Number of FACEs on the SURFACE
Constructors	
public: **SURFACE**()	Should be used via the **new** operator
Destructors	
public: virtual void **lose**()	Creates a deleted bulletin and posts it on the bulletin_board
protected: virtual ~SURFACE	Invoked by the **lose** method
Inquiry & Access Functions	
public: virtual logical **deletable**()	Indicates whether the SURFACE is shared between multiple owners
public: virtual surface const& **equation**	Returns a surface object (which defines the geometry) for reading only
public: virtual int **identity**(int)	Returns the type identifier SURFACE_TYPE
public: virtual box **make_box**(LOOP*)	Determines a bounding box for the **list** of LOOPs given
public: virtual surface* **trans_surface**(...)	Transforms a surface object by the given transform
public: virtual char* **type_name**()	Returns the name "surface"
public: int **use_count**()	Returns the use_count (ie, number of FACEs) of the SURFACE
Change Functions	
public: void **add**()	Increments the SURFACE's use_count
public: virtual void **operator*=** (transf const&)	Transforms a SURFACE
Miscellaneous	
public: virtual surface& **equation_for_update**()	Returns a pointer to a SURFACE's equation (ie, surface object) for update operations
public: void **remove**(logical lose_if_zero= TRUE)	Decrements the SURFACE's use_count. If the use_count becomes 0, the SURFACE is deleted

Represents: a variety of SURFACE geometries.

Notes:

1. SURFACEs are always instantiated as objects of a subclass.

2. One SURFACE can *underlie*, or *support*, a number (known as the use_count) of different FACEs.

3. If the use_count falls to zero then the SURFACE is deleted.

4. The geometry of the SURFACE is defined by the associated **surface** object.

B.16 TORUS

Class:- ENTITY	
Data Members	
BULLETIN*	
ATTRIB*	
Constructors	
ENTITY	
new	
Destructors	
~ENTITY	
delete	
lose	
Inquiry	
attrib	
check_destroy	
debug_ent	
deletable	
identity	
lookup	
owner	
size	
type_name	
add_method	
apply_transform	
backup	
Miscellaneous	
call_method	
copy_common	
copy_data	
copy_scan	
fix_common	
fix_pointers)	
fixup_copy	
restore_common	
restore_end	
rollback	
save	
save_begin	
save_common	
save_end	
roll_once	
Change Functions	
set_attrib	

Class:- SURFACE	
Data Members	
surface*	
int	
Constructors	
SURFACE	
Destructors	
lose	
~SURFACE	
Inquiry & Access Functions	
deletable	
equation	
identity	
make_box	
trans_surface	
type_name	
use_count	
Change Functions	
add	
operator*=	
Miscellaneous	
equation_for_update	
remove	

Class:- TORUS	
Data Members	
torus*	Pointer to a torus object
Constructors	
public: **TORUS**()	Should be used via the **new** operator
public: **TORUS**(...)	Makes a TORUS with the given center, orientation and radii
public: **TORUS**(torus const&)	Makes a TORUS ENTITY from a torus object
Destructors	
public: virtual void **lose**()	Creates a deleted bulletin and posts it on the bulletin_board. Lose methods for attached attributes are also called
protected: virtual ~TORUS	Invoked by the **lose** method
Inquiry & Access Functions	
public: position const& **centre**()	Returns the center of the TORUS
public: surface const& **equation**()	Returns the surface equation of a TORUS for reading only
public: double **major_radius**()	Returns the major radius
public: double **minor_radius**()	Returns the minor radius
public: unit_vector const& **normal**()	Returns the normal (ie, axis)
public: virtual int **identity** (int = 0)	Returns the type identifier TORUS_TYPE if level is unspecified
public: box **make_box**(LOOP*)	Makes a bounding box for the TORUS's entire surface (ignoring the LOOP argument)
public: surface* **trans_surface**(...)	Transforms the surface's equation (ie, the underlying surface object)
public: virtual char* **type_name**()	Returns the string "torus"
Change Functions	
public: void **operator***=(transf)	Transforms a TORUS
public: void **set_center**(position)	Sets the TORUS's position
public: void **set_major_radius**(double)	Sets the TORUS's major radius
public: void **set_minor_radius**(double)	Sets the TORUSs minor radius to the given value
public: void **set_normal**(unit_vector)	Sets the TORUSs normal to the given unit_vector
Miscellaneous	
public: torus& **equation_for_update**()	Returns the address of the torus's equation object for update operations

Represents: a torodial SURFACE.

Notes:

1. Constructor definition is:

    public: **TORUS** (position const&, unit_vector const&, double, double);

    Constructs a TORUS from a center point, normal unit_vector, major radius, and minor radius.

2. All the **set_*** member functions have arguments declared **const&**

B.17 PLANE

Class:- ENTITY
Data Members
BULLETIN*
ATTRIB*
Constructors
ENTITY
new
Destructors
~ENTITY
delete
lose
Inquiry
attrib
check_destroy
debug_ent
deletable
identity
lookup
owner
size
type_name
add_method
apply_transform
backup
Miscellaneous
call_method
copy_common
copy_data
copy_scan
fix_common
fix_pointers
fixup_copy
restore_common
restore_end
rollback
save
save_begin
save_common
save_end
roll_once
Change Functions
set_attrib

Class:- SURFACE	
Data Members	
surface*	
int	
Constructors	
SURFACE	
Destructors	
lose	
~SURFACE	
Inquiry & Access Functions	
deletable	
equation	
identity	
make_box	
trans_surface	
type_name	
use_count	
Change Functions	
add	
operator*=	
Miscellaneous	
equation_for_update	
remove	

Class:- PLANE	
Data Members	
plane	A plane object
Constructors	
public: **PLANE**()	Should be used via the **new** operator
public: **PLANE**(...)	Makes a PLANE from a point and a unit_vector
public: **PLANE**(plane const&)	Makes a PLANE entity from a plane object
Destructors	
public: virtual void **lose**()	Creates a deleted bulletin and posts it on the bulletin_board
protected: virtual ~PLANE()	Invoked by the **lose** method
Inquiry & Access Functions	
public: virtual int **identity**(int)	Returns the type identifier PLANE_TYPE
public: surface const& **equation**()	Returns the surface equation of the PLANE
public: position const& **root_point**()	Returns a point on the PLANE
public: unit_vector const& **normal**()	Returns a unit vector normal to the PLANE
public: surface* **trans_surface**(...)	Transforms the surface equation of the PLANE
public: virtual char* **type_name**()	Returns the name "plane"
Change Functions	
public: void **set_root_point**(position)	Sets the PLANE's root point
public: void **set_normal**(unit_vector)	Sets the PLANE's normal
public: void **operator*=**(transf)	Transforms a PLANE
Miscellaneous	
public: surface& **equation_for_update**()	Returns address of surface equation for update operations

Represents: a planar SURFACE.
Notes:

1. A PLANE is defined by a plane that is in turn given by a point on the plane and a unit normal. The direction of the normal is regarded as the inherent direction of the surface.

2. Missing constructor arguments are:

   public: **PLANE** (position const&,unit_vector const&)

3. Change function arguments are all of type **const&** (this is not shown above).

B.18 SPHERE

Class:- ENTITY
Data Members
BULLETIN*
ATTRIB*
Constructors
ENTITY
new
Destructors
~ENTITY
delete
lose
Inquiry
attrib
check_destroy
debug_ent
deletable
identity
lookup
owner
size
type_name
add_method
apply_transform
backup
Miscellaneous
call_method
copy_common
copy_data
copy_scan
fix_common
fix_pointers
fixup_copy
restore_common
restore_end
rollback
save
save_begin
save_common
save_end
roll_once
Change Functions
set_attrib

Class:- SURFACE
Data Members
surface*
int
Constructors
SURFACE
Destructors
lose
~SURFACE
Inquiry & Access Functions
deletable
equation
identity
make_box
trans_surface
type_name
use_count
Change Functions
add
operator*=
Miscellaneous
equation_for_update
remove

Class:- SPHERE	
Data Members	
sphere	A sphere object
Constructors	
public: **SPHERE**()	Should be used via the **new** operator
public: **SPHERE**(position const&,double)	Makes a SPHERE from a center point and a radius
public: **SPHERE**(sphere const&)	Makes an SPHERE entity from a sphere class
Destructors	
public: virtual void **lose**()	Creates a deleted bulletin and posts it on the bulletin_board
protected: virtual ~SPHERE()	Invoked by the **lose** method
Inquiry & Access Functions	
public: virtual int **identity**(int)	Returns the type identifier SPHERE_TYPE
public: surface const& **equation**()	Returns the surface equation of the SPHERE
public: position const& **centre**()	Returns the center of the SPHERE
public: double **radius**()	Returns the radius of the SPHERE
public: box **make_box**(LOOP*)	Makes a bounding box for this surface (see Notes)
public: surface* **trans_surface**(...)	Transforms the surface equation of the SPHERE
public: virtual char* **type_name**()	Returns the name "sphere"
Change Functions	
public: void **set_centre**(position const&)	Sets the SPHERE's center point
public: void **set_radius**(double)	Sets the SPHERE's radius
public: void **operator*=**(transf const&)	Transforms a SPHERE
Miscellaneous	
public: surface& **equation_for_update**()	Returns address of surface equation for update operations

Represents: a spherical SURFACE.
Note:

1. public: box SPHERE::**make_box**(LOOP*) const; // *list of LOOPs*
   Makes a bounding box for this surface. The box contains the complete underlying surface and ignores the bounding EDGEs, unless the **tight_sphere_box** option is **on**.

B.19 SPLINE

Class:- ENTITY
Data Members
BULLETIN*
ATTRIB*
Constructors
ENTITY
new
Destructors
~ENTITY
delete
lose
Inquiry
attrib
check_destroy
debug_ent
deletable
identity
lookup
owner
size
type_name
add_method
apply_transform
backup
Miscellaneous
call_method
copy_common
copy_data
copy_scan
fix_common
fix_pointers
fixup_copy
restore_common
restore_end
rollback
save
save_begin
save_common
save_end
roll_once
Change Functions
set_attrib

Class:- SURFACE
Data Members
surface*
int
Constructors
SURFACE
Destructors
lose
~SURFACE
Inquiry & Access Functions
deletable
equation
identity
make_box
trans_surface
type_name
use_count
Change Functions
add
operator*=
Miscellaneous
equation_for_update
remove

Class:- SPLINE	
Data Members	
spline	A spline object
Constructors	
public: **SPLINE**()	Should be used via the **new** operator
public: **SPLINE**(spline const&)	Makes a SPLINE entity from a spline class
Destructors	
public: virtual void **lose**()	Creates a deleted bulletin and posts it on the bulletin_board
protected: virtual ~**SPLINE**()	Invoked by the **lose** method
Inquiry & Access Functions	
public: virtual int **identity**(int)	Returns the type identifier SPLINE_TYPE
public: surface const& **equation**()	Returns the surface equation of a SPLINE
public: box **make_box**(LOOP*)	Makes a bounding box for this surface that is surrounded by a loop of EDGEs
public: surface* **trans_surface**(...)	Transforms the surface equation of the SPLINE
public: virtual char* **type_name**()	Returns the name "spline"
Change Functions	
public: void **set_def**(spline const&)	Sets the underlying spline object
public: void **operator*=**(transf const&)	Transforms a SPLINE
Miscellaneous	
public: surface& **equation_for_update**()	Returns address of surface equation for update operations

Represents: a parametric SURFACE.
Notes:

1. A SPLINE records a parametric surface defined by a **spline** object.

2. Each **spline** object holds a pointer to a **spl_sur** and a logical denoting reversal of the sense of the stored surface. A spl_sur contains a use-count (as does the SURFACE) and a pointer to the detailed parametric surface description.

3. All access to the surface data (spline) is through methods for that class.

4. The documentation for **make_box** says :

    public: box SPLINE::**make_box**(LOOP*) const; // list of LOOPs

    Make a bounding box for this surface that is surrounded by a loop of EDGEs. The box contains the complete underlying surface, *and ignores the bounding EDGEs*. If the surface is kept minimal, this is sufficient.

B.20 APOINT

Class:- ENTITY
Data Members
BULLETIN*
ATTRIB*
Constructors
ENTITY
new
Destructors
~ENTITY
delete
lose
Inquiry
attrib
check_destroy
debug_ent
deletable
identity
lookup
owner
size
type_name
add_method
apply_transform
backup
Miscellaneous
call_method
copy_common
copy_data
copy_scan
fix_common
fix_pointers
fixup_copy
restore_common
restore_end
rollback
save
save_begin
save_common
save_end
roll_once
Change Functions
set_attrib

Class:- APOINT	
Data Members	
position	A position object
int	A use-count
Constructors	
public: **APOINT**()	Should be used via the **new** operator
public: **APOINT**(double,double,double)	Creates a new APOINT with x,y and z coordinates
public: **APOINT**(position const&)	Constructs a new APOINT, initializing it using a position object
Destructors	
public: virtual void **lose**()	Creates a deleted bulletin and posts it on the bulletin_board
protected: virtual ~APOINT()	Invoked by the **lose** method
Inquiry & Access Functions	
public: virtual int **identity**(int)	Returns the type identifier APOINT_TYPE
public: position const& **coords**()	Returns the coordinates of the APOINT
public: virtual logical **deletable**()	Indicates whether the APOINT is shared between multiple owners and so can not be deleted
public: virtual char* **type_name**()	Returns the name "apoint"
public: int **use_count**()	Returns the number ENTITIES referencing APOINT
Change Functions	
public: void **add**()	Increments the use_count
public: void **set_coords**(position const&)	Sets the coordinates of the APOINT
public: void **operator***=(transf const&)	Transforms a APOINT
Miscellaneous	
public: void **remove**(logical)	Decrements the use_count. When the use_count reaches zero, the APOINT is deleted

Represents: the position of a VERTEX.

Notes:

1. Cartesian coordinates are assumed, though in principle other coordinate systems might be implemented.

2. A use-count is provided so that APOINTs may be multiply referenced. The construction of a new APOINT initializes the use-count to 0. Methods are provided to increment and decrement the use-count and when the use-count reaches 0 the entity is deleted.

ACIS Hacking Tips

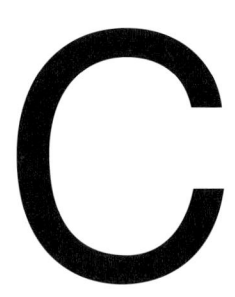

C.1 General Points

- Never use an equality test (i.e. ==) between numbers. Always compare values against a tolerance (i.e. resabs or resnor), see page 130.

- Always program at the highest level of abstraction possible and use API functions in preference to the direct interface.

- If you are not using the BULLETIN_BOARDs, switch logging off. Note **api_logging(FALSE)** will not do this alone! With **api_logging** set to **FALSE** ACIS discards all but the current DELTA_STATE. However, if the program never calls **api_note_state** the current state will become very large indeed. So states should be noted and then removed using **api_delete_ds**.

- If you are using the Bulletin Board be sure not to allow rollback over any initialize statements such as **api_initialise_faceter**.

- Remember that BODYs generally have transforms associated with them.

- Never hesitate to read the actual class and function header files. They frequently contain nuggets of information not found in the documentation.

- Bracket all direct interface calls between **API_BEGIN** and **API_END** macros. These macros enable and terminate ACIS's error handlers. Users should be aware of the following;

    - A variable of type outcome, called **result**, is declared within the API_BEGIN macro. So avoid this variable name for outcomes in your own code.

    - If the code includes multiple API_BEGIN/API_END statements within one function, brackets (i.e. { ... }) should be used to explicitly define the scope of variables (such as "result") defined by the API_BEGIN macro.

– If an error occurs after API_BEGIN has been called ACIS's error handlers will appear to pass execution directly to API_END. Consequently:
  * All clean-up operations should be done after the call to API_END (this will ensure they happen even after errors have occurred).
  * The value of the "result" outcome should always be checked after API_END.
– Since the program's execution must follow reliably between API_BEGIN and API_END no **return** statements are allowed in the code between these macros.

C.2 Finding Example Code

Perhaps the single most useful tip about ACIS programming is this:

"Plagiarize the testbed source code."

This is easily done. Let's say I want to see some examples of the arguments used by **api_ray_test_body**.

1. Go to one of the testbed source code directories, say:

   `blnd\blnd_tb`

2. Use a pattern matching program to search for the function name you are interested in. The exact command will vary with the operating system of the computer being used, for instance:

   UNIX systems users could type
   ```
   fgrep "api_ray_test_body" *.cxx
   ```
   Windows 95/NT users could popup the **Advanced** form of the **Find** utility located on the **Tools** menu of the **Windows Explorer** program.

   Macintosh users can do the same with **CodeWarrior's Find** command.

3. The files listed (blend.cxx, faceprop.cxx, sheet.cxx, uncover.cxx) can be examined to see how the arguments for the API function are declared.

The code for the Scheme extensions (located in `blnd\blnd_scm`) is also a rich source of inspiration.

C.3 The ACIS-Alliance Mailing List

The ACIS-Alliance mailing list copies any e-mail sent to `acis-alliance@cs.columbia.edu` to over 500 ACIS users. The traffic through this list is very variable; sometimes almost nothing for weeks, other times two or three messages a day. This is a good place to ask politely things like "Has anyone out there ever tried to link ACIS to Prolog?". The list is unmoderated.

To get on the list send an e-mail containing the words:

```
subscribe acis-alliance
```
to `majordomo@cs.columbia.edu`. Note: Majordomo is name of the program that manages the mailing list. **Do not** send your *subscribe* message to:

```
acis-alliance@cs.columbia.edu
```

If you do, it will be copied to, and antagonize, 500 people who only a few minutes earlier felt quite neutral about you. Sending an e-mail containing the word *help* to `majordomo@cs.columbia.edu` will return a list of commands majordomo understands.

C.4 support@spatial.com

Spatial's support is generally very efficient and the obvious place to send questions about details and suspected bugs. The shorter and more concise your question, the quicker the response. If the message concerns a suspect bug, then the Scheme AIDE or Test Harness commands needed to reproduce it are appreciated.

All e-mail to `support@spatial.com` should start with a header based on the following:

```
Submitted_by:      George Boole
Company:           MegaCorp
Customer_key:      MEG1
Product/Version:   ACIS 6.2
Email:             gb@megacorp.com
Phone:             314 566 7988
Hardware:          P233
Source/Object:     Object
SystemOS:          Win 95
Compiler:          Visual C++ 6.0

Dear Support,
         Please help me, my program will not run.
```

Customer keys are issued by support on receiving the first e-mail from you.

Bibliography

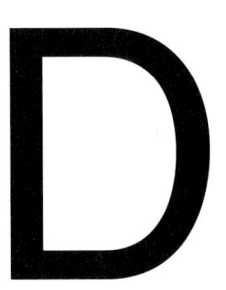

In addition to the ACIS Manuals the following books all give valuable insights into Geometric Modeling:

Computer Graphics Handbook by Michael E. Mortenson, Industrial Press Inc. 1990: Possibly the most useful book on this subject ever written! Each page deals with a different formula or concept; this brevity plus the clear illustrations make it an ideal reference book.

An Introduction to Object-Oriented C++ by Graham Seed, Springer-Verlag, 2001: Everything you wanted to know about & and * but were afraid to ask!

Computer Graphics: Principles and Practice by Foley, van Dam, Feiner and Hughes, Addison Wesley, 1993: A bible for graphics with a very detailed section on splines.

An Introduction to Solid Modeling by Martti Mäntylä, Computer Science Press, 1988: A great introduction to some of the rather deep mathematics underlying B-reps.

Djinn: A Geometric Interface for Solid Modelling by Armstrong et al, 2000 Pub Information Geometers, www.inge.com. This book details a *representation-independent* API for solid modeling and in doing so covers a lot of interesting background theory.

Geometric and Solid Modeling by Christoph Hoffmann, 1989, Pub Morgan Kaufmann, a rich source of background modeling theory including a good discussion on the problems of converting between implicit and parametric forms.

Curves and Surfaces for CAGD: A Practical Guide by Gerald Farin, 4th Edition, Academic Press, 1988: If you only have room for one book on splines then this is the one to choose.

Index Notes

1. Index entries for functions (such as **api_boolean_start**) are created semi-automatically at the <u>end</u> of the example code in the text. Consequently a function may not always appear on the page given. However it will appear in the program that finishes on the stated page.

2. Many geometric items appear several times in the index with slightly different formats (e.g. *CURVE*, *curve class*, *curve:* and *curve*). The convention adopted is:

   - Uppercase names, such as *FACE* and *CURVE*, refer either to ACIS ENTITY classes or C++ Macros.
   - Lowercase names followed by the word "class", such as *curve class*, refer to ACIS's non-ENTITY geometric classes.
   - Lowercase names followed by a colon, such as *curve:*, refer to Scheme extensions.
   - Lowercase names alone, such as *curve*, refer to the generic geometric element of that name.

Index

C^0, C^1, C^2, 196
G^0, G^1, G^2, 195
abl:
    const-rad, 301
    edge-blend, 301
    ell-rad, 301
    pos-rad, 301
ACIS
    alliance listserver, 384
    AppWizard, 359
    attributes, 313
    classes, 37
    components, 42
    file format, 72, 74
    name, 353
    obtaining software, 360
    platforms, ix
    precision, 127
    primitives, 69
    support, 385
    tips, 383
    units, 129
add_named_attribute, 315
advanced blending
    see blending, 301
advanced rendering component, 179
ageing, 268
API
    background, 64
    error checking, 68
    interface, 47
    user defined, 342–345
api_apply_transf, 81, 111
API_BEGIN/API_END, 114, 345, 383
api_body_mass_pr, 85
api_boolean, 96
api_boolean_complete, 156, 158
api_boolean_start, 156
api_change_body_trans, 76
api_change_state, 229
api_check_edge, 349
api_check_entity, 349
api_check_face, 349
api_checking, 349
api_convert_to_spline, 85
api_copy_body, 96
api_cover_circuits, 33
api_cover_wires, 96
api_create_graph_from_faces, 276
api_create_refinement, 171
api_curve_arc, 255
api_curve_line, 255
api_del_entity, 332
api_delete_ds, 383
api_delete_state, 227
api_edfa_int, 134
api_edge_helix, 251, 255
api_edge_law, 260
api_ent_area, 70
api_facet_entity, 171, 175
api_facet_face, 166
api_fafa_int, 136

api_find_cls_ptto_face, 149
api_find_face, 91
api_find_vertex, 93
api_fix_blends, 93
api_get_edges, 171
api_get_face_facets, 166
api_get_faces, 149, 156, 276
api_ihl_clean, 171
api_ihl_compute, 171
api_imprint, 146
api_imprint_stitch, 149
api_initialize_, **46**
api_initialize_booleans, 156, 276
api_initialize_constructors, 65, 66
api_initialize_faceter, 166, 171, 175
api_initialize_sbooleans, 158
api_initialize_sweeping, 251, 255, 260
api_intersect, 85
api_intersect_curves, 286
api_logging, 383
api_make_cuboid, 65, 69
api_make_ewire, 91, 96
api_make_frustum, 69, 81, 224
api_make_planar_disk, 251
api_make_plface, 260
api_make_prism, 69, 73, 91, 156
api_make_pyramid, 69
api_make_sphere, 68, 69, 134, 136, 163, 175
api_make_torus, 69, 136
api_mk_ed_bs3_curve, 132
api_mk_ed_ellipse, 91, 96, 132
api_mk_ed_int_ctrlpts, 202
api_mk_ed_line, 134
API_NOP_BEGIN, 152, 230, 332
API_NOP_END, 152, 230, 332
api_note_state, 227, 229
api_offset_planar_wire, 255
api_point_in_body, 130, 149
api_q_edges_around_vertex, 93
api_ray_test_body, 163
api_remove_face, 149
api_restore_entity_list, 76, 332
api_reverse_face, 96
api_reverse_wire, 91

api_rh_create_background, 175
api_rh_create_light, 175
api_rh_delete_background, 175
api_rh_delete_light, 175
api_rh_render_entities, 175
api_rh_set_background, 175
api_rh_set_background_arg, 175
api_rh_set_light_list, 175
api_rh_set_render_mode, 175
api_rh_terminate, 175
API_RUN Macro, 352
api_save_entity_list, 73, 91, 156
api_save_version, 75
api_selective_boolean_stage1, 158
api_selective_boolean_stage2, 158
api_selectively_intersect, 156
api_set_const_rounds, 93
api_set_default_refinement, 166, 171
api_set_default_vertex_template, 166
api_set_file_info, 73
api_set_int_option, 171
api_set_mesh_manager, 175
api_slice, 145, 152
api_smooth_edge_seq, 93
api_solid_block, 69
api_solid_cylinder_cone, 69, 70, 156
api_solid_sphere, 69
api_solid_torus, 69
api_start_modeller, 64–66, 238
api_stop_modeller, 64, 65, 238
api_str_to_law, 238, 251, 255, 260
api_subtract, 171
api_sw_face_norm, 91
api_sweep_with_options, 91, 251, 255, 260
api_sym_diff, 343
api_terminate_booleans, 156, 276
api_terminate_constructors, 65, 66
api_terminate_faceter, 166, 171, 175
api_terminate_sbooleans, 158
api_terminate_sweeping, 251, 255, 260
api_transform_entity, 260
api_unhook_face, 276
api_unite, 81
api_wiggle, 96, 119

INDEX

api_wire_len, 255
APOINT, 6, 382
    constructor, 114
    coords(), 111, 114
AppWizard, 359
are_parallel, 130
are_perpendicular, 130
ATTRIB, 12, 17, 40, 145, 313
ATTRIB_DIM Example, 322
ATTRIB_FUNCTIONS, 325
ATTRIB_GEN_INTEGER
    constructor, 318
    set_merge_method, 318
    set_split_method, 318
ATTRIB_INTCOED, 145
attributes, 40, 313–328
    complex, 313
    foreign, 320
    generic, 316
    merge method, 314, 318
    simple, 313
    split method, 314, 318
    text string, 314
    transform method, 314
    unknown, 76
    user defined, 319–328

B-rep, 2
    hierarchy, 5
    overview, 3
B-spline, 193–212
    advantages, 9
    blending function, 198
    control points, 197
    knot vector, 198
    NUBS, 9
    NURBS, 9
    surfaces, 207
Bézier
    blending function, 190
    curves, 189
background:
    set-prop, 176, 181, 183
begin (scheme expression), 60
BEGIN_CREATE macro, 101
Bernstein polynomial, 191

blend:
    abl:
        const-rad, 301, 304
        edge-blend, 301
        ell-rad, 301
        ent-ent-blend, 304
        pos-rad, 301
    const-rad-on-edge, 303
    get-network, 301
    network, 301, 303, 304
    on-vertex, 303
    var-rad-on-edge, 301
blending, 12, 91, 93, 297–304
    options, 299
    terminology, 297
BODY, 6, 65, 364
    constructor, 116
    lump(), 104
    transform(), 111
    WIRE, 145
body
    manifold, 16
    multi-dimensional, 16
    nonmanifold, 16
    sheet, 16
    wire, 16
body:
    separate, 97
bool:
    intersect, 55
    nonreg-intersect, 153
    nonreg-union, 153
    nonreg-unite, 278
    subtract, 84
boolean, 81, 84, 127–159
    algorithm, 138–143
    background, 83
    blank BODY, 139
    introduction, 13
    nonregularized, 152
    selective, 154
    sheet-solid results, 96
boundary representation, *see* B-rep
bounded geometry, 3
bounding box, 136
box class, 12, 82

get_*_box, 136
high(), 136, 166
low(), 136, 166
operators, 137
BULLETIN, 17, 221
    DELETE_BULLETIN, 332
    next(), 224
    old_entity_ptr(), 332
    type(), 332
BULLETIN_BOARD, 222
    current, 114
    current_bb(), 332
    disabling, 383
    end_bulletin(), 224, 332
    introduction, 221
    start_bulletin(), 224, 332

cam design, 244
car (scheme expression), 56
Cassini's ovals, 133
cast (C++), 25
cdr (scheme expression), 56
cell:attach, 278
cellular
    adjacency graph, 159, 273
    models, 15
check_result, 351
Chiralities, 287
COEDGE, 6, **7**, 370
    constructor, 116
    edge(), 109
    sense(), 109
    set_previous, 116
color:
    rgb, 177, 181, 183
component, 42–47
    architecture, 42
    dependencies, 46
concave/convex EDGEs, 119
conditional (scheme expression), 54
CONE, 6
cone class, 10
conics, 205
const type (C++), 26
constructors (C++), 35
continuity

    C (parametric), 196
    G (geometric), 195
control points, 188, 197, 202
control polygon, 188
coordinate systems, 97
coordinate_transf, 260
coordinates, *see* api_change_body_trans
copy_body_from_body, 345
create_wire_from_edge_list, 255
cross-section, 94–97, 132
cubic curve design, 193
current_bb(), 224
CURVE, 6, 372
    equation(), 260
    geometry(), 93
    indentity(), 93
curve, 4
    curve intersections, 133
    tracing, 138
curve class, 10
    point_curvature, 260
    point_direction, 260
    test_point, 130
    test_point_tol, 130
    type
        ellipse, 132
        intcurve, 132
curve:
    eval-pos, 61
    from-edge, 61
cvty:
    concave, 121
    convex, 121
    inflect, 121
    knife, 121
    mixed, 121
    tangent, 121
    unknown, 121
C++, 21–48
    classes, 33
        constructors, 35
        friend members, 34
        inheritance, 34
        private members, 33
        protected members, 33

INDEX 393

public members, 33
virtual functions, 34
control
    do-while, 36
    for, 37
    while, 36
data types, 23
enumerated types, 28
exit(), 68
functions, 29
    reference args, 30
general layout, 64
NULL reference argument, 32, 96
overview, 21
type modifier
    casts, 25
    const, 26
    pointers, 24
    references, 26

data-exchange, 72
debug_entity, 354
debug_size, 65, 66, 332, 354
debugging, 349–356
define (scheme expression), 53
deformable surface component, 213
degree of spline curve, 202
degrees_to_radians, 255
DELTA_STATE, 222, 225, 227, 229
    bb(), 227
    next(), 227
direct interface, 103–125
display
    MAC-OS, 173
    MS-Windows, 173
    X-windows, 175
dl-item:point, 304
do (scheme expression), 60
do-while (C++), 36
do_boolean, 345
ds:
    add-spring, 214
    commit, 214
    solve, 214

ed-cvty-info:
    instantiate, 121
EDGE, 4, 6, 7, 369
    constructor, 115
    creation with laws, 257
    end(), 111
    set_geometry, 286
    start(), 111
edge, 4
    radial, 7
    spline, 220
EDGE-FACE intersection, 133, 134
edge:
    bezier, 192
    circular, 246
    circular?, 54
    conic, 206
    ed-cvty-inf, 121
    elliptical?, 54
    law, 295
    linear, 244
    linear?, 54
    spline, 268
    spline?, 54
ELLIPSE, 6, 374
ellipse class, 10
ellipsoids (superquadrics), 263
END_CREATE macro, 101
ENTITY, 17, 37, 363
    attrib(), 40, 318
    bounding boxes, 136
    casting, 76
    copy_scan, 40
    delete, 41
    hierarchy, 22
    identity, 77
    modification by LOP, 288
    save and restore, 39
    set_attrib, 40
    type, 76
    type_name, 40, 145
    use_count(), 11
    user defined, 329–342
ENTITY definition macros
    COPY_DEF, 342
    ENTITY_DEF, 342

FIX_POINTER_DEF, 342
LOOKUP_DEF, 342
RESTORE_DEF, 342
SAVE_DEF, 342
SCAN_DEF, 342
TERMINATE_DEF, 342
entity:
  box, 137, 308
  cells, 278
  coedges, 218
  copy, 100, 153, 206
  debug, 117, 153
  delete-facets, 178
  display, 137
  display-facets, 178
  edges, 56, 61, 93, 109, 308
  faces, 60, 97, 105, 107, 120, 121, 137, 290
  facet, 178
  loops, 120
  set-color, 304
  set-highlight, 185, 278
  set-material, 177, 181
  set-refinement, 178
  set-render-sides, 185
  set-texture-space, 181
  transform, 83, 97, 100, 120, 137, 206, 308
  vertices, 112
  wires, 137
ENTITY_LIST, 73, 76–79
  add(), 109
  array of pointers to, 33
  clear(), 78
  constructor, 77
  count(), 109
  destructor, 77
  functions, 77
  init(), 78
  LIST_ENTRY_DELETED, 78
  next(), 78
  tombstones, 78
enumeration type (C++), 28
env:
  active-part, 101
  set-active-part, 101, 231

set-active-wcs-colour, 99
set-highlight-color, 185, 278
equality testing
  are_parallel, 130
  are_perpendicular, 130
  is_equal, 129
  is_zero, 130
error checking, 68, 349–356
  macros, 352
Euler
  formula, 19
  operators, 140
event:button, 185
Example Programs
  abebool1.cxx, 144
  abebool2.cxx, 146
  abebool3.cxx, 147
  adren.scm, 180
  api_main.cxx, 342
  api_ray.cxx, 163
  api_run.cxx, 351
  api_sym.cxx, 344
  api_sym.hxx, 343
  asweep.cxx, 258
  att_dim.cxx, 325
  att_dim.hxx, 324
  att_main.cxx, 320
  att_org.cxx, 324
  att_org.hxx, 323
  attrib2.dbg, 323
  ball.cxx, 67
  basepos.scm, 61
  bb_size.cxx, 223
  bezier.scm, 192
  blend.cxx, 91
  blend.scm, 93
  block.cxx, 65
  block.scm, 70
  cam-law.scm, 245
  carcdr.scm, 56
  cell_bool.cxx, 157
  cell_graph.scm, 277
  cone.cxx, 69
  coniced.scm, 206
  csg_main.cxx, 330
  csg_main.cxx (continued), 332

INDEX   395

csg_node.cxx, 338
csg_node.hxx, 337
csg_org.cxx, 336
csg_org.hxx, 335
cube.dbg, 66
cutvertex.cxx, 275
debug.cxx, 353
debug1.dbg, 333
diy_tweak.cxx, 283
ds_size.cxx, 226
e_count.cxx, 108
edfaint.cxx, 134
edgeblends.scm, 300
edgetypes.scm, 55
ent_ent_bln.scm, 303
entity_content.cxx, 39
error.cxx, 350
f_count.cxx, 103
f_detail.cxx, 105
f_detail.scm, 60, 106
faceren.scm, 184
facet.cxx, 165
fafaint.cxx, 135
fafaint.scm, 137
feach.scm, 57
fhl.cxx, 170
funnel.scm, 265
genatt.cxx, 317
General layout of C++ code, 64
head.scm, 267
heal1.scm, 307
helical.scm, 248
helix-sw.cxx, 249
hist1.scm, 231
law1.cxx, 236
law_twist.cxx, 253
letvar.scm, 59
letvar.scm (a), 58
letvar.scm (b), 58
lofting.scm, 217
lop_off.scm, 290
lop_offset.scm, 296
mk_sph.scm, 117
move_sep.scm, 289
nameattb.cxx, 314
netsurf.scm, 219

noreg.scm, 153
nurbs.cxx, 201
phl.scm, 172
prism_sw.cxx, 89
psprint.scm, 183
qdint.cxx, 151
rbdblock.scm, xi, 71
render.cxx, 173
Ripple.scm, 214
roll.cxx, 228
rollsave.scm, 231
root.sat, 333
save.cxx, 73
save.sat, 74
save.scm, 79
scherk.scm, 211
section.cxx, 94
section.scm, 97
sel_bool.cxx, 155
slpgrid.scm, 210
Sls_scm.cxx, 51
spin.scm, 181
spsol1.scm, 208
spsol1.scm (continued), 209
suq.scm, 263
surfint.cxx, 131
symext.cxx, 346
taper.scm, 269
tetra.cxx, 113
tetra.cxx (continued), 114–116
tweak_list.scm, 293
tweak_pos.scm, 292
twist.scm, 270
twocones.scm, 84
u-offset.scm, 241
unite.cxx, 80
unite.scm, 83
use_cont.cxx, 122
v_blend.scm, 302
v_coords.cxx, 110
v_coords.scm, 112
vexity.scm, 120
vexity.scm (continued), 120
vol.scm, 87
volume.cxx, 85
wcs.scm, 99

wiggle.cxx, 118
extrusion, *see* sweeping

FACE, 4, 6, 367
    constructor, 116
    creating with laws, 261
    geometry(), 106
    loop(), 106, 109
    next(), 104
    next_in_list(), 109
    picking, 184
    sense(), 124
    type(), 106
face
    double-sided, 11
    external, 11
    internal, 11
    normal, 11
    orientation, 11
    sense, 11
    single-sided, 11
face-edge graph, 5, 273
FACE-FACE intersection, 136
face:
    conical?, 107
    intersect, 137
    law, 264, 265
    planar-disk, 278
    planar?, 107
    plane, 84, 210
    reverse, 210
    sphere, 117
    spherical?, 107
    spline-ctrlpts, 210
    spline-grid, 211, 212
    spline?, 107
    toroidal?, 107
    unhook, 308
faceted hidden line, 169
faceting, 17, 164
    refinement, 167
faceting algorithm, 169
facets
    count, 166
feature recognition, 275
FileInfo, 74

set_product_id, 73
    set_units, 73
filter:
    apply, 97
    type, 97
find_attrib, 171, 315
find_err_mess, 351
find_named_attribute, 315
for (C++), 37
for-each (scheme expression), 57
foreign attributes, 320
free-form surfaces/curves, 187
function prototypes (C++), 29

generic attributes, 316
generic_graph
    components, 276
    cut_vertices, 158, 276
    get_entities_from_vertex, 276
    subtract, 158
geometric continuity, 195
geometry
    intersection, 130
    lowercase, 11
    sharing, 122
    uppercase, 11
George Boole, 139
get_body_box, 96, 136, 166
get_default_stream(), 227
graph theory
    components, 274
    cut vertex, 274, 275
    cycle, 274
    leaf vertex, 274
graph:
    entities, 278
    subgraph_2dcell, 278
    subgraph_3dcell, 278
graphs, 272–278
    cell-adjacency, 273
    face-edge, 273
    vertex-edge, 272
GUI
    picking, 184
gvector:, 71
    from-to, 192

reverse, 97
scale, 97
transform, 182

header files, 45
healing, 305–311
helical offsetting, 249
helix, 251, 255
hh:
    analyze-stitch, 308
    combine, 308
    geombuild, 308
    init-body-for-healing, 308
    postprocess, 308
    preprocess, 308
    simplify, 308
    stitch, 308
hidden line
    faceted, 169
    precise, 172
history, 231
HISTORY_STREAM, 222, 227, 230
    can_roll_back(), 227
    can_roll_forward(), 227
    list_delta_states(), 227
hyperboloids of revolution, 207

IHL_SEGMENT
    visible, 171
illumination models, 164
imprinting, 13, 140, 146
incomplete models, 14
INDEXED_MESH_MANAGER, 175
    SetTransform, 175
inheritance (C++), 22, 34
inquiry
    parametric surfaces, 117
int_surf_surf, 286
INTCURVE, 6, 9, 376
intcurve class, 10
interpolation, *see* spline
    curves, 9
    surface, 210
intersecting shapes, *see* boolean
intersection, 127–159

boxes for approximations, 136
curve-curve, 131
EDGE-FACE, 133
FACE-FACE, 135
graph, 13, 140, 143, 145
surface-curve, 131
surface-surface, 131, 138
test, 150
interval class, 82
is_cylindrical_face, 156
is_equal, 129
is_zero, 130
isometric view, 71

Jacobian matrix, 271

Klein bottle, 278
knot vector, 202
knots, 198

lambda (scheme expression), 59
Lambertian reflection, 164
law class
    binary_law, 238
    constant_law, 260
    domain, 218
    evaluate, 238
    identity_law, 260
    min_rotation_law, 260
    return_dim, 238
    sqrt, 233
    take_dim, 238
    times_law, 260
    twist_path_law, 260
    vector_law, 260
law:
    arguments, 234
    check, 235
    derivative, 235, 242
    face, 264, 265
    nintegrate, 235
    nmax, 242
    nmin, 242
    o, 246
    offset, 242
    piecewise, 244, 246

warp, 268–270
laws, 233–271
    EDGE geometry, 257
    expressions in C++, 235
    FACE geometry, 261
    operator symbols, 240
    relational symbols, 245
    symbols, 262
    syntax, 234
    tree structure, 239
    trigonometric expressions, 236
let (scheme expression), 57
light:
    create-shadows, 181
    set, 177, 181
    set-prop, 177, 181
linear-edge, 153
list-ref, 120, 290
list:
    append, 212
local operations component, 12, 281–297
lofting, 12, 215–218
LOOP, 6, 368
    constructor, 116
    next(), 106, 109
    start(), 109
loop, 4
lop:
    move-faces, 289
    offset-body, 296
    offset-faces, 290
    option
      lop_check_invert, 290
      lop_merge_vertex, 296
      lop_prefer_nearest_sol, 295
      lop_prefer_same_convexity_sol, 293, 295
      lop_repair_self_int, 290, 296
      lop_sort_on_convexity, 295
    tweak-faces, 293, 295
    tweak-faces-init, 295
    tweak-pick-edge-solution, 295
    tweak-query-edge-solutions, 295
lowercase

    classes, 11
    geometry, 10
LUMP, 6, 365
    constructor, 116
    shell(), 104

manifold
    BODY, 16
    properties, 5
marching algorithms, 119
mass properties, 85
    background, 88
MASTER_ATTRIB
    _DECL, 324
    _DEFN, 324
MASTER_ENTITY
    _DECL, 336
    _DEFN, 336
material:
    set-color-prop, 177
    set-color-type, 177, 181
    set-reflection-prop, 181
    set-reflection-type, 181
maths utility classes, 82
matrix class, 82
MESH_MANAGER, 165
MFC, 359
Microsoft Visual Studio, 359
modeling primitives, 12
mouse interaction, 71, 184
multi-dimensional
    BODY, 16
    models, 14
multiple solutions, 287, 292
multiplicity of knots, 198

net surface, 218
nonmanifold
    BODY, 14, 16
    EDGEs, 149
    FACEs, 11
nonregularized booleans, 152
NUBS, 9
number−>string, 246
numerical resolution, 128
NURBS, 9, 201

INDEX

advantages, 9
parameters, 204

object oriented programming, 21
octree modelers, 2
offsetting, 241–249
    BODY, 295
    helical, 247
    planar profiles, 241
option:
    set, 290, 293, 295, 296
        cone_param_lines, 87
            u_param_lines, 87, 206, 210–212
            v_param_lines, 87, 206, 210–212
    sil, 295
orientation of FACEs, 11, 122
outcome class, 68, 350, 351

par-pos, 214
par_xxx classes, 82
parameter
    class, 82
parameter_token, 166
parameterization, 8
parametric
    continuity, 196
    inquiry, 117, 119
    surfaces, 261
part
    history, 230
    manager, 101
    save and restore, 101
part:
    clear, 79, 232
    entities, 185
    load, 79, 232
    new, 101, 231
    save, 79, 232
PCURVE, 6, 375
pcurve class, 10
phl:
    compute, 173
    draw, 173
pi

π, 81, 235
pick:
    face, 185, 214, 289, 290, 293, 295, 308
planar-face:
    normal, 97
PLANE, 6, 379
    constructor, 115
plane class, 10
    constructor, 132
    normal, 124
point, 4
    containment, 130
pointers (C++), 24
poly_vtx
    first, 166
    next, 166
    point, 166
POLYGON, 166
polygon shading, 161
POLYGON_POINT_MESH, 166
POLYGON_VERTEX, 166
position
    class, 70, 82, 85
    constructor, 85
position:
    cylindrical, 98
    interpolate, 304
    spherical, 98
    transform, 182, 212
postscript images, 183
precise hidden line component, 172
precision, 127, 128
primitives
    APIs, 69
    creation, 64–72
    creation via API, 64
    user defined, 113

R-reps modelers, 2
radial edges, 7
rational B-spline, 205
ray
    firing, 161, 162
    tracing, 161
rbd:

clear, 71
line, 71
read-event, 185, 289, 290
recursive (Scheme expression), 55
reference arguments (C++), 30–33
reference type (C++), 26
REFINEMENT
    set_adjust_mode, 166
    set_surf_mode, 166
    set_surface_tol, 166, 171
    set_triang_mode, 166
refinement:
    set-prop, 178
render:
    postscript, 183
    set-background, 176, 181, 183
    set-mode, 181, 183
Render_Color, 175
rendering
    in C++, 173
    in scheme, 175
    in Windows, 173
resabs, 128
resfit, 128
resmch, 128
resnor, 128
REVBIT, 143
RGB Color Model, 181
RH_BACKGROUND, 175
RH_LIGHT, 175
roll:, 231
    delete-all-states, 232
    delete-previous-states, 232
    name-state, 232
rollback, 38, 130, 221
rotate_transf, 81
rubberbanding, 71
ruled surfaces, 267

sab files, 72
sat files, 72
    format, 74
    version, 75
    viewing, 359
saving & restoring
    algorithm, 39

saving and restoring, *see* FileInfo
    algorithm, 75
    models, 72
scaling
    non-uniform, 268
scheme, 49–62
    begin, 60
    conditionals, 54
    define, 53
    do, 61, 182
    expressions, 50
    extensions, 52
    external rep, 52
    for-each, 210
    functions, 53
    lambda, 59
    let*, 58
    lists, 55
        cdr/car, 56, 100
        for-each operator, 57
        length of, 55
        list-ref, 56, 137
    load-path, 59
    print, 55
    recursion, 55
    set, 59, 212
    starting the AIDE, 50
    value of an expression, 60
    variables, 53
        let, 57
scheme extensions
    user defined, 346–347
Scherk's surface, 211
section:, 218
sectioning, 94–97
selective booleans, 154
self-intersection, 290
semi-bounded, 16
sense of a FACE, 122
set! (scheme expression), 59
set-theoretic modelers, 2
shared geometry, 122
sheet BODY, 16
sheet-solid booleans, 96
sheet:
    2d, 97, 278

face, 84, 117, 210, 278
loft-wires, 218
net-wires, 220
planar-wire, 97
SHELL, 6, 366
constructor, 116
face(), 104
face_list(), 109
Shelley,J.H., 216
shelling component, 12, 295, 296
skinning, 215–218
sleep_milliseconds, 175
solid:
blend-edge, 93
blend-edges, 308
block, 55, 69–71, 105, 289
classify-position, 130
cone, 69, 84, 97, 107
cylinder, 69, 83, 153
intersect, 87
massprop, 87, 88, 210, 270, 308
prism, 269, 290
revolve-wire, 192, 206
slice, 97
sphere, 55, 69, 137, 153, 289
subtract, 97, 153, 206
sweep-face, 97
torus, 69, 137
unite, 83, 84, 100, 206
wiggle, 308
solid?, 55
space warp, 267–271
bending, 279
SPHERE, 6, 380
sphere class, 10
sphere:
debug, 117
spine curve, 297
spiric sections, 132
splgrid:
set-point-list, 211, 212
SPLINE, 6, 381
spline, 187–220
blending functions, 188
history, 9
inquiry, 119

surface design, 213
surfaces, 207
spline class, 10
splitting & labelling of FACEs, 141
splsurf:
set-ctrlpt-list, 208
set-u-knot-list, 208
set-u-param, 208
set-v-knot-list, 208
set-v-param, 208
set-weight-list, 208
spring curve, 297
stitching, 12, 141
stitching and separation, 147
STL file format, 186
STRAIGHT, 6, 373
constructor, 114
straight class, 10
strcat, 255
strcpy, 255
string-append, 246
subtracting shapes, *see* boolean
superquadric, 264, 265
SURFACE, 6, 106, 377
equation(), 106
identity(), 106
normal, 11
type_name(), 106
surface
area, 70
B-spline, 207
modelers, 2
of revolution, 267
ruled, 267
surface class, 10
eval_normal, 119
eval_position, 119
int_plane_torus, 132
param, 119
point_normal, 163
surf_surf_int, 132
test_point, 130
test_point_tol, 130
type
plane_type, 124
surface-surface intersection, 138

sweep:law, 295
sweep_options class, 251, 255, 260
    options list, 256
    set_rail_law, 260
    set_twist_angle, 255
sweeping, 12, 88, 249–261
symmetric difference, 343, 345
symtensor
    class, 82
system:sleep, 182

taper
    EDGE(LOP), 288
    plane(LOP), 288
    shadow(LOP), 288
    sweep, 256
    warp, 269
tensor
    class, 82, 85
    constructor, 85
texture-space, 181
tolerance, 127
tolerant modeling, 305
tool BODY, 139
TORUS, 6, 378
torus class, 10
    constructor, 132
transf class, 82, 260
    rotate_transf, 81
    translate_transf, 93
TRANSFORM, 12, 15
    transform(), 111
transform
    * operator, 111
    translate_transf, 111
transform:
    compose, 100, 120
    rotation, 83, 120, 182, 206
    scaling, 212, 268
    translation, 97, 120, 137, 308
Trefoil knot, 295
twisting warp, 270
types (C++), 23

undoing modeler operations, 228
unit_vector
    class, 82, 85
    constructor, 85, 115
uniting shapes, *see* boolean
unknown attributes, 76
uppercase classes, 11
user defined attributes, 319–328

vector
    class, 81, 82, 85, 96
    constructor, 81
    normalise, 114, 260
VERTEX, 4, 6, 371
    constructor, 115
    geometry(), 111
    set_geometry, 286
vertex-edge graphs, 272
vertex:
    position, 112
VERTEX_TEMPLATE, 166
vexity, 119
view class
    set_eye, 175
    set_target, 175
    view_3d_MS, 175
    view_3d_X, 175
view point rotation, 181
view:
    clear, 178
    draw-point, 210
    eye, 182
    gl, 70
    new, 231
    refresh, 182
    set, 71, 98
        eye, 70
        target, 70
        up, 70
        view, 71
    set-eye, 182
    set-size, 183
    set-up, 182
    set-viewport, 183
    target, 182
    up, 182
    viewport, 183
    width, 183

INDEX

volume of models, 85

warping, 267–271
    bend, 279
    effect on volume, 271
    non-uniform scaling, 267
    taper, 269
    twist, 269

wcs:
    model, 98
    set-active, 99
    to-model-transform, 100
    to-wcs-transform, 100
    view, 98

while (C++), 36

WIRE BODY, 16

wire frame modelers, 2

wire-body:, 192, 206, 220, 244, 249, 268
    offset, 242, 244, 246, 249

working coordinate system (wcs), 99

X-windows display, 175

SAXE-COBURG
PUBLICATIONS
mmi